SHELLS ON A DESERT SHORE

The Southwest Center Series

Joseph C. Wilder, Editor

SHELLS ON A DESERT SHORE

Mollusks in the Seri World

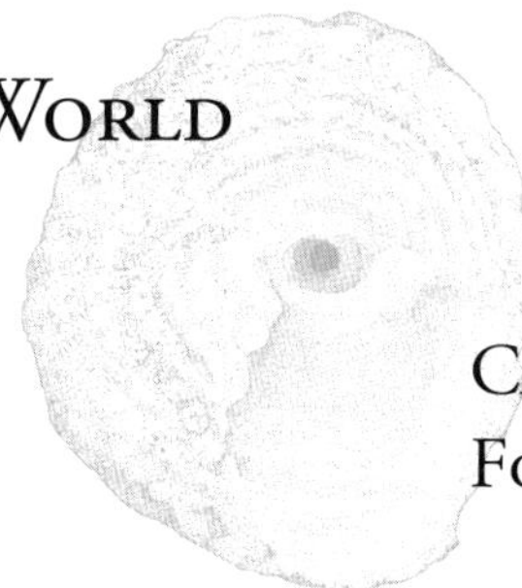

Cathy Moser Marlett

Foreword by Richard S. Felger

The University of Arizona Press
Tucson

The University of Arizona Press

www.uapress.arizona.edu
Library of Congress Cataloging-in-Publication Data
Marlett, Cathy Moser, 1952–
Shells on a desert shore : mollusks in the Seri world / Cathy Moser Marlett, foreword by Richard S. Felger.
pages cm. — (Southwest center series)
Includes bibliographical references and index.
ISBN 978-0-8165-3068-7 (cloth : alk. paper)
1. Seri Indians—Science. 2. Seri Indians—Ethnozoology. 3. Seri Indians—Name. 4. Mollusks—Social aspects—Sonoran Desert. 5. Mollusks—Sonoran Desert—Classification. 6. Seri language—Glossaries, vocabularies, etc. I. Title.
F1221.S43M27 2014
305.897'57—dc23
2013038701

Unless otherwise noted, all artwork and photographs are by the author.

Publication of this book is made possible in part by the Southwest Center of the University of Arizona, and by the proceeds of a permanent endowment created with the assistance of a Challenge Grant from the National Endowment for the Humanities, a federal agency.

Manufactured in the United States of America on acid-free, archival-quality paper containing a minimum of 30% post-consumer waste and processed chlorine free.

19 18 17 16 15 14 6 5 4 3 2 1

To my parents, Ed and Becky Moser
and to the *Comcaac*, who celebrate the sea.

Socaaix Iyat Inoohcö cöicoos

Socaaix iyat cop inoohcö quij xepe iti cöiixi com tcooo maziim áa ya
Socaaix iyat cop inoohcö quij xepe iti cöiixi com tcooo maziim áa ya
Haxölinaail heecto ptiiqui imitaasit com tcooo maziim áa
Haxölinaail heecto ptiiqui imitaasit com tcooo xepe iti cöiixi com iyoometim ita.

Song about Socaaix Bay Point

The tide line of Socaaix Bay Point is beautiful,
The tide line of Socaaix Bay Point is beautiful,
All the different little seashells are really beautiful,
All the different little seashells lie at the tide line.

—composed and sung by Ramón López, 2008

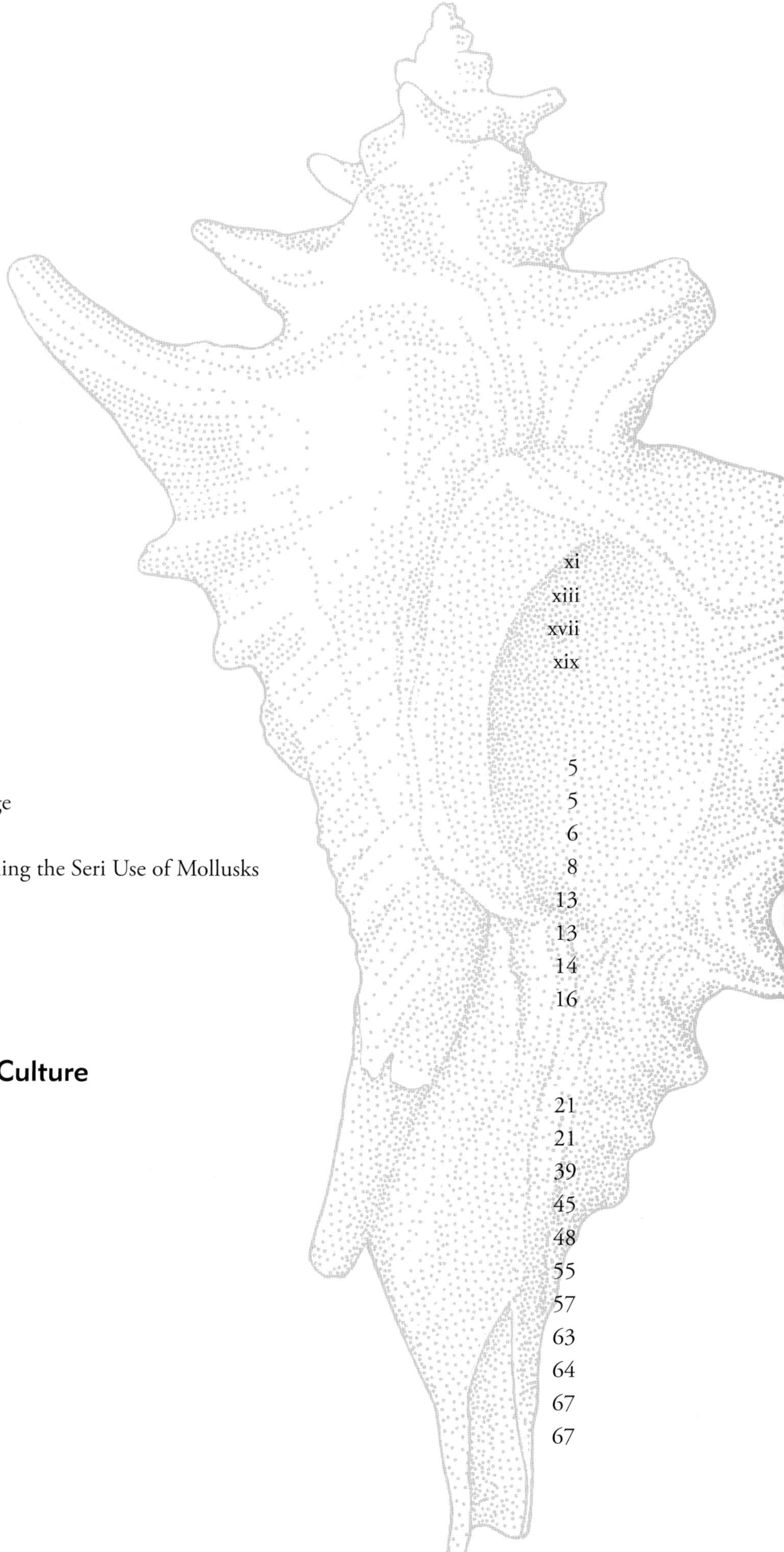

CONTENTS

Part III. Species Accounts

Appendices

TABLES

FOREWORD

Richard Stephen Felger

This book is different from anything you have ever seen. Although most ethnobiological works are authored or co-authored by outsiders, including my own work among the Seris, Cathy Moser Marlett's *Shells on a Desert Shore* results from a lifetime among the *Comcaac*, the Seri people. Cathy grew up in the Seri village of El Desemboque, the daughter of intrepid linguists Mary Beck Moser and Edward Moser. Her childhood language was that of the Seris, as well as English and Spanish. Seri children were among her first friends and the seashore their playground.

How well I remember working with Cathy's parents, being a guest at their Shangri-La small adobe home by the beach in Desemboque. We started collaborating when Cathy was still a child. Becky Moser had a list of some medicinal plants and asked if I knew their scientific names. Twenty-five years later we finalized the Seri ethnobotany book (Felger and Moser 1985). I realized how privileged I was to work with the Mosers and the Seris and to help record a unique slice of knowledge. I want you to know about the fun, laughter, deep interest, and knowledge the Seris have for the natural world. Cathy Marlett's book amply demonstrates these qualities.

Even after the mid-twentieth century Desemboque was a remote village on the Sonora coast. No electricity, no telephone, no running water, and it often took a few weeks for exchange of letters. The outside world was arriving in fits and starts. Significantly, the Seris were able to remain in their homeland, retain much of their traditional knowledge, and deal with the outside world largely on their own terms—not that it has always been easy.

It was an accident of history and isolation that the Seris continued to live off the desert and sea well into the twentieth century. Their homeland, once far more extensive, is too arid for agriculture, but well suited for a livelihood of hunting and gathering. For several centuries their interactions with the outside world were often violent and they suffered unspeakable hardships, yet the people and their culture endured. Traditional food resources included a wealth of sea turtles, fish, and other marine life including mollusks, a wide array of land and even sea plants, and terrestrial animal life. Even today their knowledge of these natural resources is vast, and the sea continues to provide major economic resources.

It was 1951 when Ed and Becky Moser, working with SIL International (the Summer Institute of Linguistics), chose to work among the Seris. At that time the Seri homeland was considered too remote for raising American children, but Ed and Becky believed they were unable to have children. Then Cathy came along about a year later. Along with their Bible translation work, they developed literacy materials and authored or co-authored numerous publications on Seri culture and knowledge. Becky assisted with childbirths and helped care for infirm elderly people. Ed was ever on call to take people for medical care eight hours away over dusty roads.

Cathy spent much of her youth among the Seris and became completely at ease with their ways and the intricacies of the language. She is one of the very few non-Seris fluent in the language, speaking it without an accent. There is a candid photo of Cathy and two young Seri women carrying home firewood in baskets resting on headrings gracefully balanced on their heads (Felger and Moser 1985: figure 9.10). The only way you can tell which one is Cathy is that she is balancing her load with two hands instead of one. Yet she remained an American girl, homeschooled until high school and college in the United States, where she completed a combined degree in biology and art. She married Steve Marlett, an accomplished linguist, who has become an expert in Seri grammar. They raised two sons and have continued to work with the Seris. Steve and Cathy are leaders in helping to keep the language alive and vibrant through a combination of academic and on-the-ground work. In 2005, Becky Moser and Steve Marlett brought to completion a trilingual Seri dictionary, begun by Ed Moser and more than a half-century in the making, with a revised edition appearing in 2010. In print it is a massive book of over nine hundred pages and available in open-access digital format for use by members of the Seri community. Cathy's artwork enhances the dictionary with more than six hundred pen-and-ink illustrations, and her detailed drawings have graced the pages of numerous other publications.

Cathy did not originally set out to write a book about Seri knowledge of mollusks, but it is the natural result of a lifelong interest in shells. Even as a little girl, she was fascinated by the marine life along the then pristine shore of their Gulf of California home. The tide pools and near shore were teeming with sea life and the beach drift heaped with the sea's castoffs, including a potpourri of shells, seaweeds, and early plastitrash. She caught octopuses and dug clams alongside her Seri friends. What child growing up along the shore would not be fascinated by the sea? So it was that Cathy first learned about sea life with her lifelong friends as companions and mentors.

As I am writing this foreword, Cathy reminds me that the Seris were eager to share with her what they knew and that her research on mollusks was fun and rewarding. She would pose a question, in Seri, to a group of women and off they would go. As she told me, "They really loved telling me stuff, maybe because I loved hearing it." Over the years, friends would bring her "new" shells and tell her what they knew about them. The older people, men and women, were especially eager to have their knowledge recorded—they were well aware of the intrinsic value of that knowledge, how fragile it was, and how fast it was disappearing. They recalled many of the wonderful and valuable aspects of the traditional life, but these recollections would sometimes be tempered by memories of the harsh realities of those times. That lesson was brought home to me some years ago on a bright spring day. I was with a group of Seri people who had lived off the desert and sea for the first part of their lives. They were extolling how healthy and strong the people were then. I asked if they would like to return to the "old ways." There was an emphatic, "No. It was hard."

Eventually, Cathy realized that over the decades she had amassed a huge store of traditional knowledge, which she needed to share with the general public as well as the scientific community and preserve for future Seris. She had her own extensive notes and her father's treasure trove of unpublished notes. In addition, she had studied Seri material in ethnological collections and archives in museums and libraries from California to Washington, DC, and Mexico City. Yet conversations with Seris continued to elicit fascinating bits of new information. Although she was aware, as is every researcher, that there is never an end to potential information, in 2008 she knew the time had come to finalize and publish what she had learned. The great diversity and indigenous knowledge of mollusks presented in *Shells on a Desert Shore* brings a voice and life to the silence of shells in the archaeological record. For example, look at the extensive information for the clam, ***haxöl***, *Leukoma grata* (previously named *Protothaca grata*).

Shells on a Desert Shore is a result of Cathy's life with the Seri people and with the shells and even shell-less mollusks of their homeland. It is a work of careful and extensive scholarship, based on decades of her own observations and research and thousands of years of Seri knowledge. This book successfully bridges the gap from oral knowledge to written work.

Every mollusk is described and illustrated with Cathy's original artwork, providing verification and an accurate means of tying indigenous knowledge to scientific biological names. Look at the illustrations—each as fine and meticulous as can be. Look at the Seri names and perhaps also find some of them in the Seri dictionary. For non-Seris the spellings seem exotic and pronunciation hugely challenging. The language is rich and complicated. For example, singular nouns may bear little similarity to the plural form, and verbs typically have over a hundred forms. Cathy Moser Marlett's language skills lend underlying power to *Shells on a Desert Shore.*

This book is a definitive work on local Gulf of California mollusks, let alone Seri knowledge of them. Does this mean that Cathy is done with shells? I suspect she will continue talking to Seris about shells and other marine life and marveling at what she learns long after this book is in our hands. It gives me great pleasure to see this book completed, and I am sure you will share my enthusiasm. There is nothing else like it in the world.

References

Felger, Richard S. and Mary Beck Moser. 1985. *People of the Desert and Sea: Ethnobotany of the Seri Indians.* Tucson: University of Arizona Press.

Moser, Mary B. and Stephen A. Marlett, compilers. 2010. *Comcaac quih yaza quih hant ihiip hac: Diccionario Seri-Español-Inglés.* 2nd edition. Hermosillo: Universidad de Sonora; Mexico City: Plaza y Valdés Editores.

PREFACE

On the eastern shore of the northern Gulf of California the succulent desert touches the sea and sand dunes creep slowly inland, swallowing even the large columnar cactuses until only the exposed tips of their highest arms mark the desert floor where they took root. At a break in a long stretch of such dunes, many bearing evidence of early Gulf peoples, lies the small Seri fishing village of Desemboque.

Desemboque, its Spanish name alluding to its earlier location at the mouth of a nearby usually dry riverbed, is a solitary hamlet enveloped by the smells of the sea and cooking fires—a place where a hard wind can whip the shore sand into shin-stinging bursts, or where on a late spring day a distant mountain lies mirrored on a flat white sea, its reflection interrupted only by pelicans silently gliding just inches above the sea's smooth surface. It is a place where the early morning quiet is broken by the calls of shorebirds and the bubbling thrum of the motors of small boats heading into deeper Gulf waters to bring home their catch. And it is a place made extraordinary by the desert solitude, the dynamic sea, and the special people who live there.

It was only in the mid-twentieth century with more frequent use of the winding dirt tracks—first made through the desert by fish buyers in the 1930s—that the area began its slow emergence from an exquisite isolation, one in which its waters teemed with life, and the shores were rich with sea creatures. In such a place the desert, sea, and dunes held the promise of a wonderful barefoot childhood, and at three months old I was first introduced to my home in the center of this ancient camp the people call *Haxöl Iihom*, 'place of the ***haxöl*** [clams]'.

The sea is the lifeblood of the Seri people—it is a bond unbroken, forged through a long history of dependence on its very rich resources. My own tie was formed early, as our small whitewashed adobe house was only steps away from a constantly changing and limitless seashore world, where waves deposited a remarkable supply of ocean treasures at the high tide line—my favorites were bright orange scallop shells or those of the giant cockles, their deep yellow yet unbleached by the sun.

Many of my childhood memories are closely linked with the world of shells, octopuses, and other interesting creatures. One that is especially vivid is of being taken at the age of four to a beach a few miles away from our home to see something unusual after a storm with heavy seas. Great piles of pearl oysters lined a long section of beach at the high tide line, and a number of Seris were prying open the oysters with knives, in search of pearls. That interesting sight infused by the stink of the dead oysters left an indelible imprint on my young mind, the pearls certainly adding a tantalizing new prize to my list of things to discover.

When the extreme low spring tides exposed hidden tide pools and remote sandbars, there were more uncommon things to find. I still recall my feeling of enchantment when I came upon my first colorful sea slug. Such a beautiful creature moving slowly along a strand of seaweed, its curving, frilled body vividly marked with electric blue, was startling in contrast to the otherwise rather drab rocks and sand. Its common name, Mexican Dancer, is aptly given.

I remember one day coming across vast numbers of very odd-looking transparent creatures with ballooning fringed hoods and wing-like appendages being swept along in a current of the shallow, warm waters bordering Estero Sargento. After capturing one in a jar of seawater with hopes that someone could give it a name, I had no success, as it didn't survive long. But a sketch carefully drawn from memory later proved helpful in providing a name—it was a nudibranch (*Melibe leonina*), a species that I never saw alive again.

There were the hours spent on the rocky intertidal areas hunting for octopuses in their caves to take home to my cat or to cook for myself (ironically, my parents being not particularly fond of seafood). The octopuses were so commonplace to us that the rare sight of a small squid gliding near shore one spring day sent our crowd of kids into a leaping frenzy of excitement. Such an odd creature's unexpected visit helped stretch the boundaries of our little beach world and made me wonder what other interesting creatures were yet to be found.

A swim with my Seri friends often included finding small snails on the rocks beneath our feet and bringing them up to the surface to crack open and consume on the spot. A low tide foray involved eating rock oysters straight from their shells and digging up clams to carry home by the pailful. These simple memories are ones I treasure.

I once rashly insisted on feeding my cat some of the clams despite being warned by Sara Villalobos, a Seri woman I playfully called *himaz* ('my grandmother'), that there was red tide in the area. My mother and I spent much of that night helplessly watching the course of the toxin's awful effects. Somehow through, and probably in spite of, our frenzied and feeble attempts to help him, my cat recovered, leaving me with a much greater respect for the ocean *and* the Seris—and a very well-deserved scolding from Sara the next morning.

My friends learned of my interest in shells and often found unusual ones to show me. It only later occurred to me that the Seri people had unique information about the sea and its creatures, and it just *might* be important to record it. And so what started as a short list of shell names grew to illuminate a fascinating world that was without a doubt quickly disappearing. It soon became obvious that such information from a people with a viable language and vibrant culture could provide a unique glimpse into the lives of those who for millennia inhabited the shores of the Gulf of California and whose stories exist only in the silence of their empty desert camps.

In the years following my father's death in 1976, I explored his detailed Seri lexical files, some of the data accompanied by simple drawings. At times, when a name was strange and archaic, it was like a treasure hunt to find an older Seri who recognized it and could give me a vague description of the shell—or better yet, show me the actual shell. The process evokes the octopus hunts of my childhood, of reaching into a cave, touching something I knew was there, and carefully working it to the surface before it escaped and disappeared forever. Recording such knowledge has been a great delight.

ACKNOWLEDGMENTS

Many Seris shared details with me about their lives and families. Roberto Herrera, Evangelina López, and Sara Villalobos were some of the earliest. Others who followed were: María Luisa Astorga, Victoria Astorga, Ramona Barnett, Rosa Amalia Barnett, Efraín Estrella, Mayra Estrella, Carolina Félix, Juana Herrera, Lorenzo Herrera, Alfredo López, Ramón López, María Luisa Molina, René Montaño, Manuel Monroy, José Ángel Montaño, Cleotilde Morales, Ernestina Morales, Francisco Morales, Josefina Morales, José Juan Moreno, Raquel Moreno, Xavier Moreno, Antonio Robles, Humberto Romero, Angelita Torres, Armando Torres, Chavela Torres, and José Ramón Torres. I am privileged to have played a part in this, a facet of their much larger story.

Others facilitated my journey along various odd paths, helping to bring together some fascinating pieces from the past. I appreciate the help of the collections staff at various institutions close to home: Diane Diddemore, Andrew Higgins, Amy Rule, and Jannelle Weakly at the Arizona State Museum; Erika Castaño and the staff at the University of Arizona Libraries Special Collections; and Katherine Reeve at the Arizona Historical Society. Others farther afield were: Pat Nietfeld, Heather Shannon, Cali Martin, and Bob Hershler at the Smithsonian Institution; Katherine Reagan at the Cornell Library Rare and Manuscript Collections; Chris Travers at the San Diego History Center; Alicja Egbert at the Phoebe A. Hearst Museum of Anthropology; and Alejandro González Villarruel at the Museo Nacional de Antropología in Mexico City, along with Sergio Torres Quintero, Susana Cuevas, and José Carmen Díaz, who helped with a visit to see the Seri items in the collections there.

I thank the following people for their valuable input in such things as reading and commenting on a section, helping with identification or suggesting a resource, and providing a photo or answering a question: Hans Bertsch, Diane Boyer, Richard Brusca, Bob Capen, Fulvio Eccardi, Exequiel Ezcurra, Doug Frasier, Heather Fried, Scott Gable, Lance Gilbertson, Carole Hertz, Jules Hertz, Jim Hills, Linda Humnick, Pat LaFollette, Jorge López Calderón, Natalia Martínez, Jim Mead, Gary Nabhan, Nemer Narchi, Carolyn O'Meara, Rodrigo Rentería, Scott Ryerson, Rafe Sagarín, Jeffrey Seminoff, David Shaul, Enriqueta Velarde, Carolyn White, Ben Wilder, and Teresa Zuberbühler. I am especially grateful to the delightful Astorga-Estrella family, who welcomes us back to Desemboque every time we show up, usually unannounced.

At the Natural History Museum of Los Angeles County, Lindsey Groves first helped me appreciate the vast array of mollusks found in the Gulf of California as represented in the collections there. During a visit to the Santa Barbara Museum of Natural History, Paul Valentich-Scott, Henry Cheney, and Eric Hochberg took time to answer my questions about some odd mollusks and their taxonomy, and have since been generous with their help. Any errors are mine.

The staff at the Prescott College Station in Bahía de Kino made available their library and marine collection; I especially appreciate Abram Fleishman, who let me tag along on a trip to Isla Alcatraz.

Imogene Davis was a tremendous help, giving me access to the family's Edward H. Davis collection. I thank Robert Davis for letting me use an unarchived photograph from his own collection. I am also indebted to Alan

Ferg and Michael Higgins, who provided the opportunity to see the amazing collection of Seri ethnographic material made by William Neil Smith, and allowed me to photograph some of the objects for use in this book.

Xavier Basurto provided insight into the commercial pen shell fishing industry in the Gulf and, most wonderfully, included me in two memorable trips by Seri boat to different shores of Isla Tiburón. Another visit to the island was made with Lynn Johnson during an unforgettable week spent with her while she photographed the Seri people. I greatly appreciate her generous offer to let me include some of her photographs here.

Richard White, more than just a friend, and in the spirit of our grandfather, taught me the fun of *trying* to learn everything; and if not succeeding, at least enjoying the process. Richard cast a critical eye on many of my early paragraphs, never hesitating to give unvarnished comments about my writing as only a family member can.

I am always energized by the contagious enthusiasm of Tom Bowen, a special friend who has walked many Seri paths, both real and metaphorical. His example of clear writing was ever before me.

I would like to express my gratitude to Joe Wilder at the University of Arizona's Southwest Center for his tremendous support during the publication process, and to Allyson Carter, who first handled the manuscript and, together with Amanda Piell and the staff at the University of Arizona Press, worked hard to see it published. And a very special thank you to Anna Mirocha for her help in preparing the manuscript; to Natasha Leullier, who edited it; and Leigh McDonald, who designed the book.

I especially thank Richard Felger, a friend and mentor who early in my tide pool explorations taught me literally to leave no stone turned, for his encouragement—and pushing—to finish the book; and Jean Russell, who touched my life in so many ways. Because of her interest and time, she made a child's world important and sparked my curiosity for that world with the gift of a special metal box for my treasures on which she painted, "Cathy's shells and octopus legs."

Formalizing the linguistic details presented here was made much easier with the help of my husband Steve; certainly, speaking a language and being able to explain it clearly are two very different things. His insightful suggestions and assistance in preparing the manuscript for editing greatly simplified the process, and his enthusiastic input from the very beginning, and willingness to always indulge me in stopping to explore just *one* more beach, added to the fun of research. And thank you Dusty and Gregory, for making our family adventures memorable; your early shell collections played a part in this story, too.

And of course, this journey would have never even begun without the encouragement, inspiration, and love of my parents, Ed and Becky Moser, whose deep personal interest in the *Comcaac* opened paths to better understanding a people and the fascinating world in which they live.

SHELLS ON A DESERT SHORE

PART I

THE SETTING

Figure 1.1. Desemboque, far left background, looking southeast.

INTRODUCTION

In a remote desert camp a seashell lies where it was dropped long ago, perhaps marked by evidence revealing how it was used and who might have left it there—that shell's narrative is repeated throughout the desert and on the extensive dunes that line the Gulf of California and some of its islands. And more than an eloquent reminder of the people who have long walked these shores, the depth and density of the shell deposits clearly speak of the richness of the Gulf's waters and the importance of mollusks in the lives of these people.

The People and Their Language

As one of the last hunter-gatherer groups in North America, the Seri people (*Comcaac*) have long inhabited the Sonoran Desert along the eastern shore of the central Gulf of California in the state of Sonora, Mexico. Estimated to have numbered a few thousand people in past centuries, the group was drastically reduced as a result of sporadic warfare and epidemic diseases, reaching a low of fewer than two hundred individuals by the early twentieth century.[1] Though current population numbers vary, it is probable that there are close to one thousand Seris today, most of whom speak the language in everyday life.[2]

Early history records the Seri territory as stretching from near Guaymas, Sonora, in the south, into an area north of Puerto Libertad and inland perhaps twenty miles or more. The Seris' range, however, was much greater (though less defined), reaching its greatest recorded extent far inland during the seventeenth and eighteenth centuries.[3] Indeed, a number of still-recalled Seri names refer to inland areas that are now sites of Sonoran towns and cities, and even north to the Río Colorado delta. Central to Seri history are the islands of San Esteban and Tiburón, where a great number of place names have been recorded.[4] Other islands of the Midriff Region (the constricted central Gulf) are also named, although these islands were apparently not places of extended residence (figure 1.7, p. 10).[5] And farther away are names on the Gulf's western shore, given by the Seris who visited and perhaps lived there long ago.

With the Spanish expansion into northwestern Mexico in the 1600s, missions, military outposts, and other settlements were established ever closer to the Seri territory. When the Seris took advantage of their proximity and the resources they offered through small-scale raids, the Spanish responded with military force, and thus began more than two centuries of sporadic violent confrontations. As settlements expanded into their territory, and with their numbers and homelands shrinking, the Seris were concentrated on the coastal areas and Isla Tiburón, one of their traditional homes.[6]

While as late as the mid-1950s a group of several families still lived near a permanent water source at Tecomate on the northern shore of Isla Tiburón, most of the people had by then settled in the mainland fishing camp of Desemboque, and later in Punta

Figure 1.2. Juan Barnett Díaz cleaning his catch. Punta Chueca, 2010. Photograph courtesy of Lynn Johnson.

Chueca. By the late 1960s almost all of the Seris resided in these two seaside villages.[7]

Although living a more sedentary lifestyle because of the easy availability of food and manufactured products from outside the community, today the people still rely on desert resources such as cactus fruits and medicinal plants; materials for basketmaking, ironwood (*Olneya tesota*) carving, and jewelry making; construction materials; and firewood. Especially strong is a reliance on the sea, and the highly skilled Seri fishermen provide not only fish and shellfish for consumption within the community, but seafood for the commercial market as well. Mollusks supply both food and shells for stringing into jewelry for sale to tourists, providing a significant source of income since the mid-twentieth century.

The Seris' profound relationship with the natural world is expressed through a rich oral tradition that includes not only simple observations, names, and classification, but folklore, songs, and historical accounts as well. Considering that this group at one time consisted of fewer than two hundred people, no doubt many of them children, one can only be amazed at the wealth of information orally transmitted by these few individuals. And yet, at the same time I realize how much has been lost.

The Seris, who call themselves the *Comcaac* (a plural noun, the singular of which is *Cmiique*),[8] are perhaps best known to the general public today for their fiestas with traditional music and dancing, colorful clothing, and intricate face paintings. Their shell jewelry, baskets, and sculptures in ironwood and stone are a part of many collections as well.[9]

The bonds tying the people to their land, culture, and language are ancient and strong as a result of living in an arid, isolated area with only sporadic outside contact before the early twentieth century. The rather sensational accounts about the Seris in popular literature and newspapers, only some based on fact, effectively helped maintain that isolation. In contrast to most other groups in Mexico, no Seri in at least the last half-century has chosen to leave the country for an extended time. However, several have lived in other parts of Mexico because of marriage to non-Seris, and sponsored travel of musical groups and artisans to the United States and Europe is becoming more commonplace. It is certain, however, that with increased commercial and tourist development, including the construction of highways on the beautiful coast the Seris call home (met with mixed responses by the people), great and rapid change is in the future.

The complex Seri language, known as *Cmiique Iitom*, has long intrigued outsiders.[10] Early twentieth-century studies placed the language in the Yuman family and the larger Hokan family, but more recent scholarship has seriously questioned these relationships, and today the language is considered by some linguists to be an isolate until convincing evidence otherwise can be presented.[11] There is no doubt that the Seris had cultural contact with speakers of Yuman languages, but the lack of more than a handful of clear cognates with them demonstrates the linguistic divide. Unrelated to the neighboring Uto-Aztecan languages of Yaqui (Yoeme), Pima and Papago (O'odham), Ópata, or Eudeve (the latter two no longer spoken), Seri has relatively few loanwords from those languages or Spanish. But the Seris are not newcomers. Evidence such as the abundance of unanalyzable place names suggests that the Seri language has been spoken in this very special part of the world for a very long time.[12]

The Physical Setting

The Gulf of California is a gem lying in a stunning desert setting. Its extensive shoreline, around four thousand kilometers in length,[13] ranges from vast sandy expanses to rocky beaches rich with tide pools, from great sea cliffs to estero mud flats thick with mangroves.[14]

Figure 1.3. Estero Sargento at low tide, with Isla Tiburón in the background. September 2010.

Figure 1.4. Low tide in Desemboque bay, looking south to Cerro Tepopa. January 2007.

Central to the dynamics of the Gulf is the continual supply of nutrient-rich cool water welling up from the deep sea floor, mixing into the warm shallow waters of the Gulf. In addition, the Gulf experiences a remarkable tidal range of more than ten meters at the northern limit and, in the Seri area, of around two meters. Such a range exposes significant intertidal area, giving the people living along the shores easy access to the area's abundant biological resources.

The Gulf can be divided into the northern, central, and southern regions.[15] On its eastern shore, overlapping both the northern and central regions, lies the traditional Seri territory, extending from just north of Guaymas to an area north of El Desemboque del Río San Ignacio,[16] which includes the islands of Tiburón and San Esteban (see figure 1.6, p. 9).

The Gulf as a whole is noted for its remarkable species richness. It is home to about five thousand (named) macroinvertebrates; of these, more than two thousand are mollusks, almost half of which occur in the northern Gulf.[17] In describing the area, Richard Brusca writes: "High nutrient levels, shallow waters, abundant solar radiation, and strong tidal mixing combine to make the northern gulf one of the most productive marine regions in the world."[18] On the biological richness he says:

> Here [past peoples] found a stunning diversity and abundance of shellfish and finfish, easily harvested during the twice-daily tides, and sea turtles that could be captured in shallow waters. . . . To these early collectors of the upper gulf's rich invertebrate biodiversity, the region must have been important beyond imagination.[19]

Indeed, the evidence that is common along the Gulf's shores makes it clear that the area has long been important in providing for the people who live there.

The Seris living in the coastal areas knew well the winds and tides. Before beginning to use wooden boats in the early twentieth century, they navigated the Gulf's waters using balsas (a craft made of three reed bundles lashed together), and according to oral accounts, ventured as far as Baja California using the Midriff Islands, many of which were named, as waypoints (figure 1.7, p. 10).

The people describe an island as *hant xepe imac quiij* 'land in the middle of the sea'. Not surprisingly, the larger Gulf is not explicitly named. The deeper ocean away from the mainland or island shores is referred to as *xepe caacoj* 'large sea', while the well-traveled narrow strait between Isla Tiburón and the mainland, known as the Infiernillo Channel, is known to the Seris as *Xepe Coosot* 'narrow sea' or *Xepe Heeque* 'little sea'.

The word *xepe* refers to either seawater or the sea at large. The tidal extremes are the *xepe caaixaj* 'strong sea' (the spring tide, with the greatest tidal range) and *xepe cmotomn* 'weak sea' (the neap tide, with the smallest tidal range). The high tide is *xepe cpócatoj* 'full sea', and the low tide interval, when the intertidal area is exposed, is the *xepe cöcootij* 'dry sea'. The flood tide is *xepe mquiin* 'sea that returns' and the ebb tide is *xepe quixi* 'sea that finishes'. Two named tidal events are a very low tide at mid-morning, referred to as *hant imizj haxoj cöcootij* 'land that is very dry', and a very high tide during the night, called *xepe ihamoc quih ano toom imquiin* 'sea that returns at night'.

Figure 1.5. A shell midden on Isla Tiburón, 2009.

The area above the high tide line normally not exposed to the tides is the *hant iijöc* 'dry areas of the land'. The high tide line is referred to as *xepe quih iti cöiixi* 'where the sea finishes', or sometimes, *xepe quih iti cöiixi iiqui izooxolcam* 'eighth tide line' (literally, 'the eighth place where the sea finishes').[20] The general beach area below the high tide line is the *xepe iteel* 'edge of the sea'. The *itaaij* is the swash zone, where the advancing wave retreats and the *itaaij iizc* 'next to the *itaaij*' is the exposed wet area. The *xepe án* 'sea area' is the general intertidal zone. The *hast ancoj* 'rocky places' are the rocky intertidal area, while the lowest intertidal and shallow submerged area are together called the *haxoj án*.[21] The deeper submerged area where the sea floor is visible is the *xepe cxatcaj* 'shallow sea', and the deeper ocean is the *xepe cyaail* 'deep sea'.

Significant in the lives of the people were the esteros that provided not only refuge when hiding from their enemies, but were important sources of food as well. Fish, sea turtles, and other creatures could be hunted in the extensive channels, while the great sand flats exposed at low tide were rich with snails and clams that were easily gathered. There are ten esteros in the present-day Seri territory, the largest being Estero Sargento (see figure 1.3, p. 6). Each is named, and includes the word *xtaasi*, a term used not only for esteros, but also for lagoons and large tide pools. The mangrove thickets that characterize the area's esteros are referred to as *pnaacoj án* 'mangrove area'.

Historical References Mentioning the Seri Use of Mollusks

The earliest references to the Gulf's coastal inhabitants and mollusks are found in accounts of forays into the region in quest of purported fabulous treasure: pearls kept by its native peoples. Glimpses into the indigenous cultures are provided by Jesuit missionaries in the seventeenth and eighteenth centuries, although most comments on shell use are limited to observations of personal adornment.[22] One of these Jesuits, Adamo Gilg, included a drawing as an inset to a map in a letter written in 1692. The drawing shows several Seris using what are probably shells as personal adornment (see figure 3.36, p. 58).[23] Another Jesuit, Ignaz Pfefferkorn, writing of his experiences in Sonora in the mid-1700s, noted that shells were used as scrapers to straighten the lower shaft of arrows.[24]

In 1826, Lt. Robert W. H. Hardy, a British Royal Navy veteran who explored the Gulf by sailing ship,

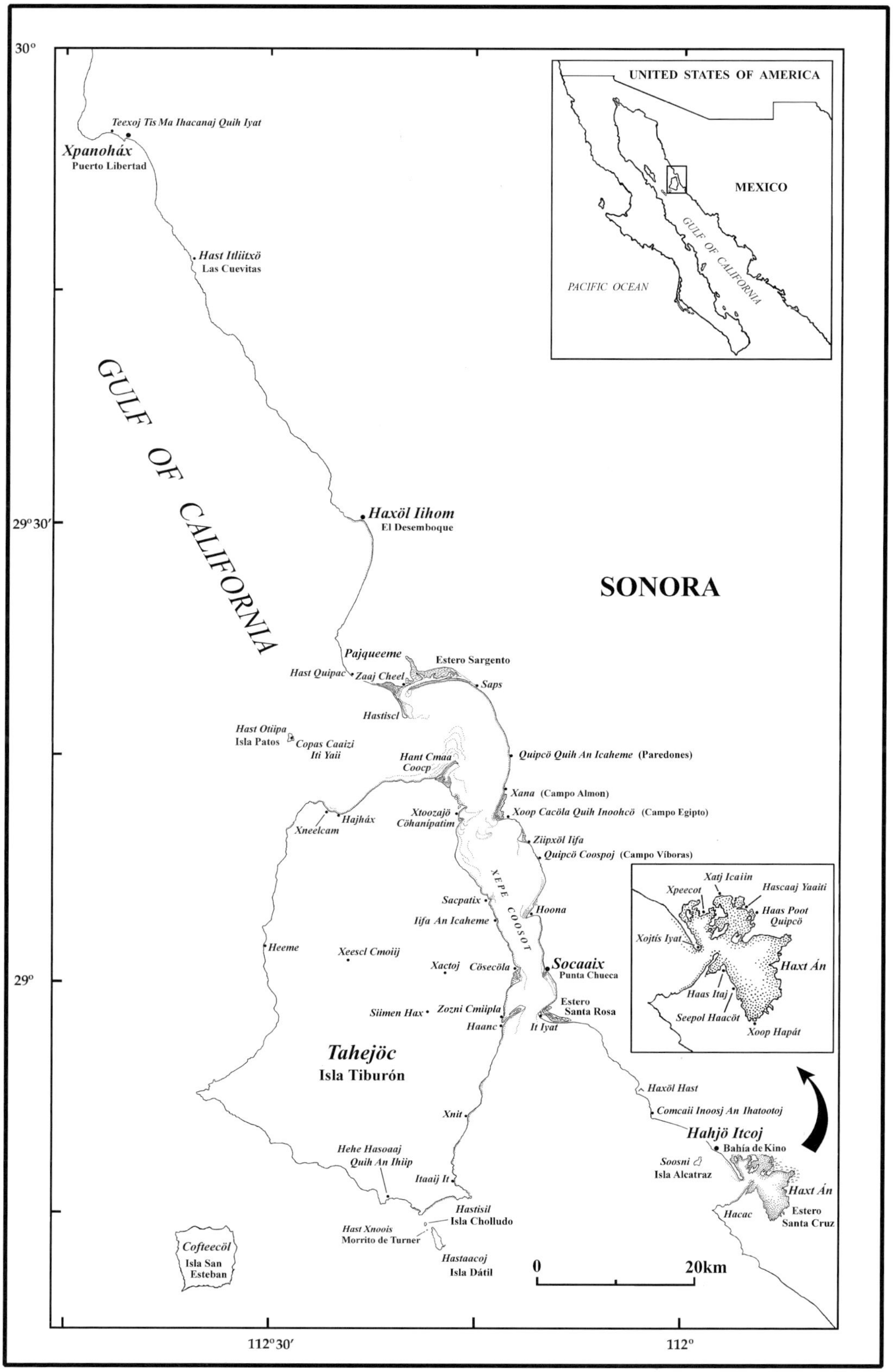

Figure 1.6. The eastern central Gulf coast, including Seri place names cited in the text.

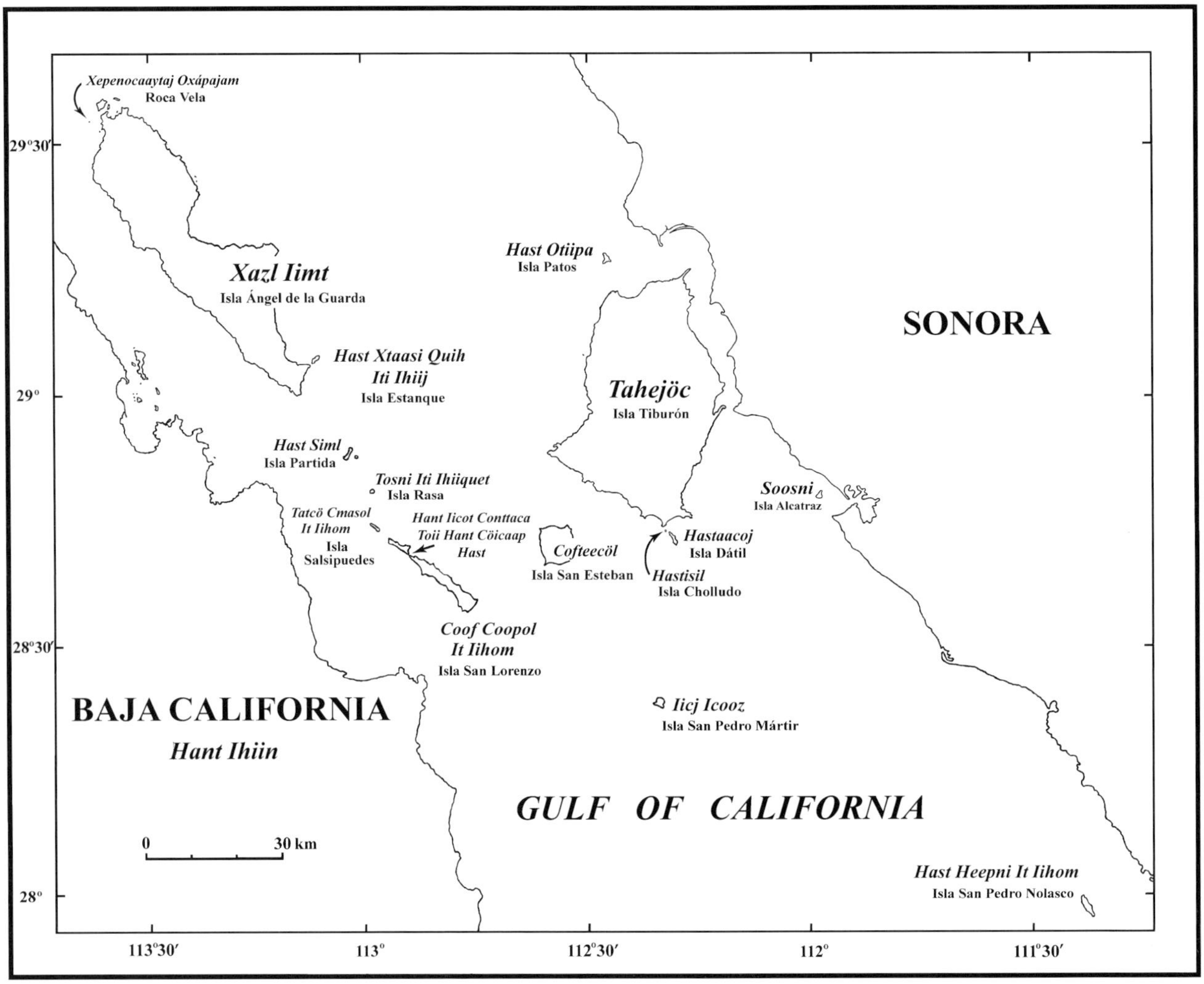

Figure 1.7. The Midriff Islands having recorded Seri names.

visited the Seris on Isla Tiburón. His account mentions shell use, including adornment, as well as a few cursory details of shellfish consumption.[25] Clemens A. Pajeken, a German traveler who briefly visited the area around 1860, included among other deprecating remarks the manner in which the people consumed entirely raw a diet consisting of "oysters, mussels, snakes, with fish and other sea food."[26]

Only a few Seri shell names are recorded in early vocabulary lists. Alphonse Pinart, a French explorer and ethnographer, recorded three in a list made in 1879 (appendix 6).[27]

In his report to the American Bureau of Ethnography W J McGee, a self-trained scientist who made two trips to the Seri area in 1894 and 1895–1896, noted shell use in greater detail, but did not list any Seri shell names. Although McGee's diary is straightforward and detailed, and his published report includes wonderful graphic illustrations and photographs, his report is at times imaginative and marked by racism and dubious details, including purported burial practices involving shells.[28]

Charles Sheldon, an explorer and naturalist, made a weeklong trip into the area, including a visit to Isla Tiburón to hunt deer and bighorn sheep in the winter of 1921–1922. He transcribed a list of ninety-two Seri words, two of which are shell names (appendix 6).[29]

The anthropologist Alfred L. Kroeber made brief mention of the use of shells in his report on a six-day visit he made to the area in 1930, although any shell names he may have recorded were not published.[30] His photographs of the Seris show a few details of shells used in personal adornment.

Edward H. Davis, an ethnographer and collector, made seven visits to the area from 1922 to 1939, writing detailed notes. The notes made in 1936 include transcriptions of Seri names for shells he collected from various locations on the shores of Isla Tiburón and contain some ethnographic information as well (appendix 7).[31]

Notes describing an ethnographic collection made by Gwyneth Harrington in 1941 record the use of shells and include Seri names referencing some of the objects she collected (appendix 6).[32]

William Neil Smith Jr., who studied anthropology at the University of Arizona, did occasional fieldwork with the Seris between 1945 and 1967. His beautiful photographs and work in film, some produced during his early years of living with a group of families on Isla Tiburón, provide an intimate glimpse into the lives of hunter-gatherers. Although he apparently did not record names of shells or details of their use in his notes, some of his work portrays shells being used in everyday life in ways no longer practiced.[33]

In June 1951, the ethnobiologist Alfred Whiting accompanied Smith, spending two weeks in the area, including Isla Tiburón.[34] His diary reveals several instances of shell use, one passage mentioning a child being fed from a shell vessel, with a shell used as a spoon.

For two short periods in 1953 and 1955, Borys Malkin delved more deeply into Seri ethnozoology in doing research for his master's thesis.[35] He identified specimens by scientific and Seri names, although the latter were not published (appendix 6). His study includes mollusks used as food and their related cooking practices, and—apparently a first—Seri ecological and taxonomic information.[36]

William Griffen, an anthropologist, spent a month in Desemboque in 1955 and made a short visit in 1956, also doing research for his master's thesis. He recorded a few details of mollusk use, including consumption as food and use as shell jewelry.[37]

Recent publications note in greater detail the uses and names of some mollusks. Thomas Bowen and Edward Moser describe shell use in pottery making.[38] *People of the Desert and Sea: Ethnobotany of the Seri Indians,*[39] although primarily referencing botany, describes much of Seri culture and the people themselves. The work provides a list of some mollusks and their uses. Finally, the trilingual Seri dictionary, the culmination of over fifty years of documenting the Seri language, contains much cultural information and identifies most of the mollusks presented here.[40]

Figure 2.1. The early days: Edward and Becky Moser transcribing Seri words. Bahía de Kino, 1951. Photographer unknown.

PRELIMINARIES

Background

My parents, Edward and Mary Beck "Becky" Moser, began living in the Seri community in 1951, working with SIL International, a faith-based organization that studies, documents, and assists in developing the world's lesser-known languages.[1] Their linguistic training, personal interest in the people, and inquisitive minds were pivotal in the almost fifty years (Becky) spent there, recording and publishing materials on the language and culture.[2]

As a child, I did not consider our living situation at all unusual, and not surprisingly it was not until leaving for my higher education that I saw it in a new light and began to appreciate the fascinating people in my world, both Seris and others. Our small adobe home in Desemboque was a special place where frequent visitors would enjoy a meal, a bed, or a hot bucket shower behind the kitchen cupboard, since many of them were doing field research. With friends like Ike Russell, an intrepid bush pilot, I was sometimes included in their outings—the best was flying in Ike's plane.[3] It was through these friends that another, more ordered, layer was added to my barefoot explorations of the amazing desert and sea worlds. My interest in recording the Seri knowledge of those worlds later began while doing an independent study of Estero Sargento (see figure 1.3, p. 6) for a college course in biology. In looking through my parents' notes, primarily those made by my father during more than twenty years of working with Roberto Herrera, I was struck by the wealth of information that had been recorded. Although their primary work was in scripture translation, my parents became keenly aware of the importance of cultural documentation and devoted significant effort to recording even minute details of Seri ethnography, both on paper and reel-to-reel tapes. And among the several thousand carefully filed three-by-five-inch paper slips are nearly one hundred relating to mollusks, consisting of the Seri names with an English translation, a brief description, and sometimes a simple sketch of a shell (figure 2.2). Many notes also include additional ethnographic information.

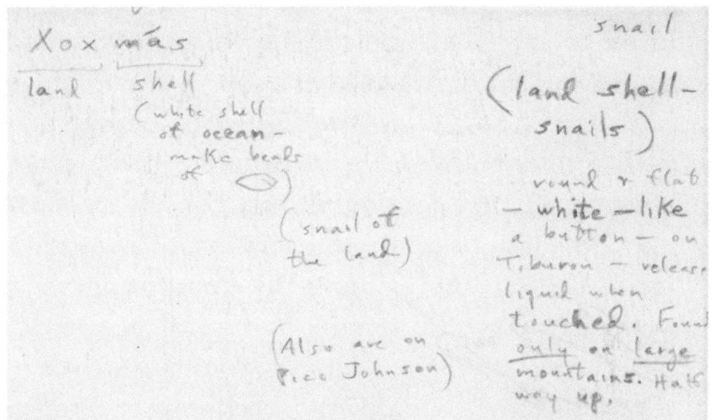

Figure 2.2. File slip made by Edward Moser, detailing information on ***xojmás*** (probably referring to *Sonorella* spp. snails). Date unknown, perhaps between 1960 and 1975.

In the mid-1970s, I began recording the Seri names as well as some cultural information given by an older woman, Sara Villalobos. Later visits to Desemboque provided opportunities to record information both on paper and cassette tape with Eva López and others. The data include folklore, songs, and oral history, as well as names and uses of mollusks. In recent years, my involvement in the process of preparing the Seri dictionary for publication certainly presented many opportunities to clarify data and add new information.[4]

Choosing to focus on the diverse group the scientific world classifies as mollusks was for me simply a matter of interest and convenience, as there is no named Seri category corresponding to this group (see Classification and Naming). My study began with shells; however, a broader focus provided the opportunity to include some other interesting creatures about which the Seris also had information.

My research method primarily involved showing shells to people, as touching something seemed to be the best way to initiate recollections. I rarely used photographs or drawings with the intent of recording yet-uncollected species or Seri names, as in my experience, photographs or line drawings tend to be poor resources for obtaining accurate information. Walking along the shore with a consultant proved to be a wonderful way of gathering information. Nearly all of my research was done in the Seri language, which I have spoken since childhood.

Although most of my consultants were women, a few men, born in the mid-twentieth century when oral narrative was still a viable and commonly used teaching method, were also great sources of information. Younger men are generally not overt in their interest in mollusks (and perhaps more specifically, their shells), and the information obtained from them mainly involved the use of mollusks as bait or details of the commercial harvesting of pen shells (Pinnidae). Women, being more involved in harvesting shellfish for community consumption or simply because of sharing information with another woman, showed more interest in recounting their knowledge in greater detail. In typical Seri style, conversations among a group of women were lively and multilayered, with everyone talking at once—these were times when interesting details about shell use emerged, and they certainly made research highly entertaining.

A recent project to archive texts written by Seri authors has produced some wonderful information.[5] René Montaño wrote about the ***xtiip*** (*Laevicardium elatum*), using as consultants his own parents; this level of intimacy resulted in some unusual and very special data on the shell's use.

Most frustrating, and tantalizing, have been those names surfacing from the past with only vague descriptions, and which are now untraceable. How does one go about finding a "clam with red-colored meat resembling a cross" and named 'cross', recalled only as a description from childhood; a necklace made of shells having the same name as an abandoned seaside camp, mentioned long ago by Chico Romero to his young grandson;[6] a shell called 'woman' surfacing only in an early tape recording; or a shell bearing the archaic name of a camp opposite the north shore of Isla Tiburón? Although the identifications are forever lost, they still provide a glimpse of how the people talked about their ocean world in the past.

The use and knowledge of mollusks in Seri culture is sharply declining. Many of the younger generations can identify only those most commonly eaten or strung into jewelry for the tourist market, and with very little ethnographic detail.

In recent years, I have occasionally seen a few shells in actual nondecorative use, such as a ***xtiip*** shell holding a special clay for use on one's skin or a pigment made from the elephant tree (*Bursera microphylla*) used to decorate fired clay pottery; an ***haxöl icaai*** (*Simomactra dolabriformis*) shell for smoothing clay vessels or clay figures prior to firing; or a section of a ***xtoozajö*** (Vermetidae, worm snail) shell used as a pipe for smoking. Shells are sometimes used on special occasions, such as the traditional fiesta for a leatherback sea turtle (*Dermochelys coriacea*), celebrated on the rare occasion when a leatherback sea turtle is found. During such a fiesta in March 2011, *Laevicardium elatum*, *Nodipecten subnodosus,* and *Dosinia ponderosa* shells contained pigments that were used for painting designs on the turtle (see figure 3.32, p. 56).

The people often keep or display shells that are unusual or attractive, perhaps dredged from deep water by commercial shrimp trawlers or found on distant island shores; even a terrestrial snail from a mountain on Isla Tiburón was someone's small treasure. Sometimes it goes a bit deeper. On a recent trip I met a young woman who chanced in on a conversation I was having about shells. Her face lit up with a big smile, and she very enthusiastically said, "Oh, I love shells too!" This began a conversation about shell names and her showing me a small collection she had made. Because she knew only a few of the shells' better-known Seri names, she was interested to find out that I had been recording their names for a long time and could provide their scientific names. I left with her promising to continue working on her collection and getting information about the shells. This same young woman later went on a short walk in the desert with me. As we walked through an old camp we talked about the names of the shells we saw and how they had been used by her ancestors, perhaps centuries ago. It was a delight to find a young person like her, whose interest might someday bridge traditional Seri knowledge and modern science.[7]

Resources and Data

I relied heavily on the 1971 edition of Myra Keen's book, *Sea Shells of Tropical West America: Marine Mollusks from Baja California to Peru*, for initial identifications. However, considerable revisions in scientific names have taken place since its publication. Carol Skoglund's updates in *The Festivus* (2001,

2002) were most helpful; the majority of gastropod names presented here follow Skoglund (2002). The bivalve names are those given in *Bivalve Seashells of Tropical West America* by Eugene V. Coan and Paul Valentich-Scott (2012). I am including in each species account the earlier name as recorded by Keen (1971), in brackets.

Visits to the malacological collections of the Natural History Museum of Los Angeles County, the Santa Barbara Natural History Museum, the University of Arizona, and the Prescott College Station in Bahía de Kino were very helpful in confirming identifications, as well as for providing a better understanding of mollusk distribution in the Seri region. The ethnographic collections at the National Museum of the American Indian, the National Museum of Natural History, the Arizona State Museum, and the Museo Nacional de Antropología in Mexico City were the source of some wonderful material. The San Diego History Center, the Division of Rare and Manuscript Collections at Cornell University, and the University of Arizona Libraries Special Collections provided an array of unpublished material that substantially enriched this study.

Although this work is clearly not intended as a complete inventory of mollusks found where the Seris live today, I have included the most distinctive species from the mainland coast as well as some found on or around Isla Tiburón, Isla Alcatraz, and Isla Patos. A small number of shells for which I was unable to find Seri names or information are also included for the record.

Some exotic shells have been found in fishing nets, brought up by divers or dredged by commercial shrimp trawlers; many have no commonly used Seri names and there is no information about their use. Although some names have been supplied, a few appear to be descriptive names devised on the spot; such spontaneous naming is unlikely to yield a widely accepted name because the shell is so rarely seen (see Names and Naming, p. 25).

When I include a statement about how common or uncommon a mollusk is, I am referring to the immediate area (between Bahía Kino and Puerto Libertad), rather than to its wider distribution. As with other information, this is based on my own observations or what Seri consultants have shared with me.

Of great help have been both published and unpublished Seri names recorded in the past by ethnographers or other visitors to the region (in one case an interview on cassette tape);[8] some include the scientific identifications (appendix 6). Often such names have supported more recent data or provided clues that helped simplify identification.

Recently digitized portions of films made on Isla Tiburón by William Neil Smith in the early 1950s provide a unique glimpse into the Seris' lives and show several instances of mollusk use.[9] These include mussels being roasted on a beach and shells being used to scrape the interior of a barrel cactus (*Ferocactus tiburonensis* or a related species), to dig a shallow well, and to remove the root bark of a plant that was used to dye basket-making material. Such vignettes from actual past use visually corroborate later descriptions.

In 2011, I had the opportunity to view part of the extensive Seri ethnographic collection acquired by Smith.[10] Most interesting to me were twelve handmade wooden boxes used by the people to store personal items, probably collected on Isla Tiburón. All but one of the boxes included shells among their contents. One box in particular provided something of a puzzle and perhaps a lesson in fieldwork interpretation: in the box were, among other items, an abalone shell (*Haliotis* sp., which does not occur in the Gulf of California), as well as shells of a large clam (*Spisula* sp.) and a knobbed whelk (*Busycon caurica*), both from the Atlantic coast. Also included was the shell of a small, rather rare South African geometric tortoise (*Psammobates geometricus*)—here indeed was an early collector who left us with some unanswered questions.[11]

Seri Traditional Knowledge

In the past, traditional knowledge was only shared through oral narrative—there is no collection, cultural center, lecture series, body of literature, or archive that records and transmits such information. However, a few people have begun recording their traditional knowledge—for instance, about the constellations—in written Seri.

The Seri names of the mollusks discussed here were freely shared and were consistently the same when provided by either sex. In any household when a name was given it was common to have another family member concur or prompt an answer if a name was not recalled.

The people were very transparent with me—if they did not know a shell's name they would say so, and did not invent names to oblige me. In several cases, shells that have recently come into use have been given names in order to be able to refer to them, and one might hear a comment such as, "I call it this because it looks like the . . . shell," with the implication that it is not a well-known shell (for a more detailed discussion, see Names and Naming, p. 25).

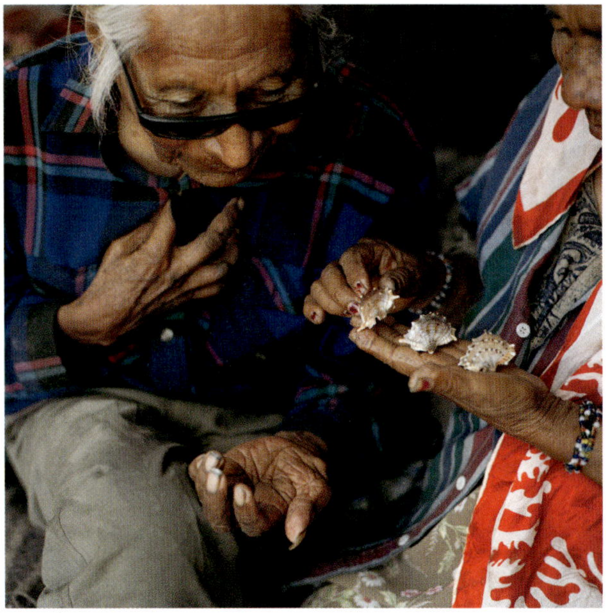

Figure 2.3. Juanita Herrera describing nonnative (*Bufonaria* spp.) shells to her husband, José Ángel Montaño, who had limited eyesight. Punta Chueca, 2010. Photograph courtesy of Lynn Johnson.

The geographical distribution of species also plays a role in the information shared, and familiarity with a shell is certainly influenced by such things as where one's childhood was spent, and what mollusks were eaten or their shells played with. And perhaps relevant are probable changes in mollusk populations and their geographic distribution in past decades, which have occurred naturally or as the result of commercial ventures.[12]

Information about food use is at times less than clear-cut, as the line between what is eaten today and what may have been eaten in the past is often blurred. Records made by early visitors to the area or those made by the Mosers from what they were told by older Seris sometimes reference lesser-known species or even common ones not eaten today. In my own experience, a comment sometimes given is, "We don't eat them, but they were eaten by the Ancestors or the Giants," a statement possibly corroborated by archaeological evidence in old camps or in middens, where the shells referred to are found with shells of mollusks that are commonly eaten today.

The Seri name denotes the whole mollusk, animal and shell. Hence, to be technically accurate, the names of the shell of every shelled mollusk would include the term *inaail* 'its shell'. For example, the giant cockle (*Laevicardium elatum*) is called ***xtiip***, and its shell is called ***xtiip*** *inaail*. However, for the sake of simplicity and reflecting common usage by the Seris, I will refer to each mollusk or its shell by its shorter name, without the word *inaail*.

Throughout the book the glosses (translations) of the Seri names are somewhat modified in order to make more sense and be easier to read than would be a literal translation or a more technical linguistic gloss.

Songs

Songs are the Seri poetry that in simple words evoke strong imagery. They are notoriously difficult to analyze linguistically, as the actual words are separated into syllables and their pronunciations changed to fit the meter. The songs are further complicated by the interpretation given by each singer who, in common practice, explains the song after performing it.

The songs in this book are presented in the following format: the first line contains the actual syllables as they are sung, and the second line gives the Seri words in italics, excluding the vocables (words without translation that complete the meter). The English translation is given separately.[13]

Technical Notes

The terms "mollusk," "shell," and "shellfish" are not synonymous. Although this study deals with the complete animal—body and shell (internal or external)—as well as shell-less mollusks, I most often use the word "shell" since most species presented here are externally shelled. At times the term "mollusk" is most appropriate when referring to the larger group that includes shell-less animals, such as sea slugs and octopuses. When referencing a group of shelled mollusks that are eaten, I occasionally use the term "shellfish."

To make it easier for the general reader, I present the genera and species alphabetically under each family.

As my interest lies in presenting Seri traditional ecological knowledge, I have chosen to not include most common English or Spanish names at the species level, especially since there is considerable variation even among well-known species. Even at the family level, names vary greatly, not to mention how they are written. When I do include a family name in the species accounts, I use a practical approach to data presentation, which may not, however, be consistent from family to family.

For easy reference, throughout the text Seri mollusk names are in boldface and italics; other Seri words are in italics only.

I use the word "Ancestors" to represent the phrase *hantx mocat*, '[people who] come from the base'. *Hantx mocat* refers to people who lived in the recent past, who were not personally known, but who are considered to have been Seris.

The word "Giants" represents the phrase *xiica coosyat* 'singing things'. The phrase refers to beings who existed very long ago, most often characterized by huge stature and whose speech was said to have sounded like singing. Some people consider them to have been Seris as well.[14]

All of the Seri words in this study are written in the community-based writing system.[15] Although the language is not extensively taught in local schools, many people are able to read materials written in their language, and a few are proficient writers. The language is viable, and most households use Seri as their first language; however, it is obvious that increased contact with the world at large and easy availability of such things as television and the Internet clearly put the language at risk.[16]

Photographs and Illustrations

The Seris, their homeland, and material culture have been extensively photographed, and many of the photographs are archived in museum and library collections.[17] It has indeed been rewarding to find and include a few such photographs showing mollusks in actual use in earlier times; I use my own more recent photographs as well. I greatly appreciate having been given permission to include photographs of Seri people in this book. Permission was provided by the subjects themselves or a living family member when the former was not possible.

As I walked through abandoned historic and prehistoric camps in the desert or on ocean dunes, it was interesting to note that what the Seris told me was often confirmed by visible evidence. In this work, I include some photographs taken in such areas to demonstrate the shells in their context of past use.

Unless identified otherwise, the illustrations are my own. Many began as illustrations for the Seri dictionary and most were made from shells in my possession. Those of the mollusks without external shells were created from photographs taken of live animals. The illustrations were detailed in pen and ink, scanned into a computer, and completed in Adobe Photoshop.

The shell lengths are provided in millimeters, measured as shown in figure 2.4. The measurements are of the actual specimens illustrated.

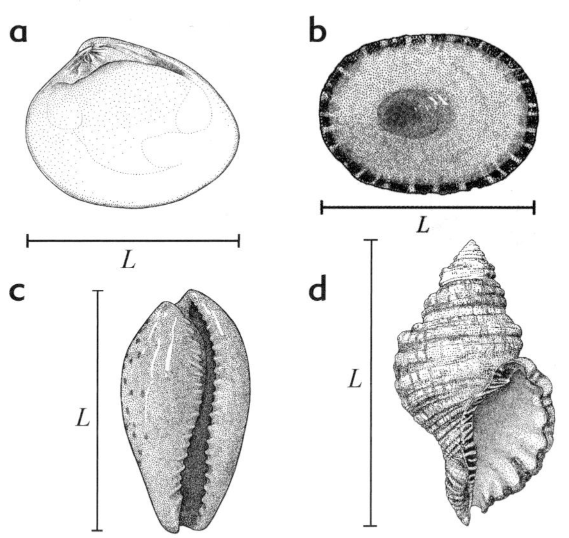

Figure 2.4a–d. Shell measurements: a) bivalves, b) limpets, c) cowries, d) snails.

A Short Pronunciation Guide of Seri

The following explanations of the Seri alphabet are provided to help the reader pronounce Seri words. A concise but more technical description is given in M. Moser and S. Marlett (2010). The bracketed symbols follow the conventions of the International Phonetic Association.

Consonants

c [k]
cö [kʷ] is a labialized voiceless velar stop. Before a vowel it sounds like the *qu* of English *quick*, but before a consonant or at the end of a word one

just rounds one's lips during the pronunciation of the consonant.

f [ɸ] is a voiceless bilabial fricative (similar but not identical to an English *f*).

h [ʔ] is a glottal stop, as used in English between the words of the interjection *Oh-oh.*

j [x] is a voiceless velar fricative, like the *j* in Spanish *jalapeño.*

jö [x^{w}] is a labialized voiceless velar fricative.

l [ɬ] is a voiceless lateral fricative, like the *ll* in *Welsh.*

l̲ [l] is like an English *l*, marked with an underscore because it only occurs in a few words.

m usually [m], but in an unstressed syllable it changes to be more like the following consonant, so it is often pronounced as English *ng* [ŋ] when before a back consonant; the name for the Seri people is *Comcaac* [koŋˈkaak]. It is pronounced as a nasalized *w* when following a back consonant: the word *cmiique* '[Seri] person' is pronounced [kw̃iikɛ].

n [n]

p [p]

qu [k]

r [ɾ] is a voiced alveolar tap, as in the double *t* in the American English pronunciation of *kitty.* Most Seri words with this sound are borrowed from Spanish.

s [s]

t [t]

x [χ] is a voiceless uvular fricative similar to the Madrid dialect for Spanish *j.*

xö [χ^{w}] is a labialized voiceless uvular fricative. The word *haxöl* (*Leukoma grata*) is pronounced [$ʔa\chi^{w}ɬ$].

y [j] is a voiced palatal approximate, just like the *y* in English *yes.*

z [ʃ] is a voiceless postalveolar fricative, similar to English *sh.*

Vowels

a [a] as in *father.*

e is pronounced somewhere between the [ɛ] in *set* and the [æ] in *sat.*

i [i] as in *machine.*

o [o] as in *comb.*

Vowel length is important in Seri. A double vowel indicates that it is pronounced with greater duration. On rare occasions triple vowel length is significant, and three vowels are shown. A sequence of vowels, such as *aai, ai, aii, oe,* is pronounced as one syllable. The *u*, as in *quislitx*, is always a spelling convention that parallels Spanish convention, and is not pronounced.

Stress

One syllable in a word is pronounced with greater prominence than the others; that is, one syllable is stressed. The rules for assigning stress are quite systematic,[18] but they are not particularly relevant to the average reader. However, readers may like to know which syllable is stressed so they may pronounce the words more correctly than not. When an accent is not written, the reader should stress the word according to the following simple rules, to be followed in the order presented (in the examples given, the stressed syllables are underlined):

1. A syllable with a sequence of two or more vowels is stressed. For example, *xiica, ctamoo, xnahooil,* or *xpaleemelc.*
2. Otherwise, if a word ends in a consonant cluster, the final syllable is stressed. For example, *hapafc* or *quislitx.*
3. Otherwise, the penultimate syllable is stressed. For example, *xepe, copas,* or *xtapacaj.*

When the Seri people express their words, as simple as a name given to a shell or as complex as a song about the Earth's creation, it is an eloquent verbalization of their deep feelings for their homeland. The importance of named spaces in their world is obvious, as central to their mythology are beings who gave names to places on their land as well as to the many creatures that inhabit them. Through the complexities and subtleties of the Seri language that provide the means to artfully express their thoughts and emotions, we are given a glimpse of how the people have lived for so long in this beautiful place.

PART II

MOLLUSKS IN THE SERI CULTURE

Figure 3.1. Juanita Herrera gathering *hehe quiinla* (*Senna covesii*), used to make a medicinal tea. Photograph courtesy of Lynn Johnson.

TOPICAL ETHNOGRAPHY

Classification and Naming

An intimate interaction with the biological world surrounding the Seris has given them a deep understanding of it. For example, the people were knowledgable about sea turtles, as they were frequently hunted as a significant source of meat: fourteen named folk taxa were recognized, based on characteristics such as color, body or head size, tail length, and amount of fat the animal contained.[1] The Seri ethnobotany records the people's detailed botanical knowledge, which included over four hundred plant names. A keen sense of observation is evident in the extensive terminology related to the mesquite (*Prosopis glandulosa*), an important plant resource the people relied on for many things. They heavily consumed the pods, naming eight growth stages for the pod itself.[2]

I once watched a Seri friend pick apart the shell of a large limpet and carefully extract a tiny mussel that had bored into its shell, as she was curious about what had made the hole. She also described finding and cutting open a strange intertidal creature long ago, curious as to what it might be. I was intrigued by her inquisitiveness, as such sea creatures, usually too small to attract much interest, are often grouped in the category of "leftovers" or "inconsequential," with species that remain largely unnamed.

More recently, another conversation provided a brief glimpse into how the Seris interact with and perceive their world. In discussing the paper nautilus (*Argonauta* sp.), so uncommon in the area that it is without a known Seri name, a man informed me that he had dubbed it with a name that incorporated two very different creatures whose characteristics it shares, since it did not easily fit with either—a shell and an octopus.

How do the Seris comprehend and give order to what they see? I once naively asked the same man how he would group sea creatures, posing what I thought was a clearly understood question. Hoping for a basic classification, I suggested that perhaps he begin with the largest, most comprehensive group, then move on to succeeding smaller groups. After a rather puzzled look, he obligingly gave a beautiful and lengthy exposition of the members of the different groups, presented in descending order of size—the first group consisting of the largest marine animals, such as whales, dolphins, and sharks, going on to other groups and ending with one composed of the smallest intertidal creatures. I reflected later that it was a rather brilliant example of how not to phrase such a question.

When presenting Seri biological classification in the Seri ethnobotany, Felger and Moser give a basic outline of the larger taxonomic categories and develop botanical classification in greater depth. They suggest the lack of a clear-cut, overt classification system, and point out the obvious problem created by outsiders inventing concepts based on non-Seri perspectives.[3] In my own work a very helpful and well-understood phrase came to light, which could be used to help my understanding of Seri classification.[4] The phrase, *X hant cöquiih* is an idiomatic expression that indicates similarity to category X. When used in a question, it evokes carefully thought-out responses in forming groups—here, of mollusks.[5] The question, *¿Zó haxölinaail hant cöquiihya?* 'What shell[ed mollusk] is it grouped with?' produces a prototype, further reflected in the phrases *iti mota iitax* (pronounced *iti motiitax*) 'beginning' or *yeen oo caap* 'what precedes'. In talking about shells, one consultant explained that there are "typical" ones with which to compare a shell—it could be the largest one, or the most common or well known of a group.

One way that physically similar creatures are linked is through descriptive terms based on form. For example, in the same way as in the scientific world, sharks and rays are not considered *zixcám*, a group including most bony fishes. While the sharks are generically named *hacat* and given additional specific names, the stingrays are grouped as *hacat cmaam* 'female *hacat*', because of their more rounded shape, in contrast to the elongated, streamlined shark form. The rays are then further identified by specific names as well. Such grouping by sexual reference also occurs elsewhere (see Names and Naming, p. 25), including with plants.

Other descriptive names consist of a generic name modified by attributives that name specifics of the group. For example, the generic name for octopus is ***hapaj***. Specific names are then formed by the addition of such attributives as 'large', 'gray', or 'true'.

A similarity might be noted by a comparative description within a name. For example, the common and well-known robustly shelled ***haapx*** snail (*Tegula rugosa*) has a diminutive (unidentified) snail counterpart named ***pohaapx camoz*** 'what thinks it is an ***haapx***'. Such a name form, 'what thinks it is an X', is associated with other animals and plants as well, and reflects a similarity humorously noted as a desire to emulate the "real thing." On the other hand, an attributive such as *áa* 'true' is sometimes used in a name to distinguish a familiar or well-used species from a similar but less common one; or inversely (though not part of a name itself) the term *xahxaii* '[something] like or false' might describe a kind of relationship.

In a few cases, a link is expressed in a compound name consisting of two very different creatures, as in the case of the paper nautilus, mentioned earlier. Another example is the generic name for a bat, *senelcsét*, literally 'butterfly-frigatebird'. As explained to me, the name notes the similar feeding behaviors of bats and frigatebirds—both capture their food on the wing. And interestingly, according to the person explaining the name, the butterfly, and not the bird, is indeed the group to which the bat belongs.

A name might imply a species is toxic and to be avoided. For instance, one barrel cactus (*Ferocactus emoryi*) has a toxic juice that, when drunk as a substitute for water, causes dizziness and stomach upset; the common rock oyster (*Saccostrea palmula*) has a growth form that has adverse effects when eaten. Their Seri names include the attributive *coiicöt* 'that kill' along with their generic names. Other nontoxic species of the same folk genera include the attributive *áa* 'true' and *hapahit* 'which is eaten' (see Ostreidae).

Unknowns might be placed into known categories by the inclusion of an attributive. For example, introduced foods might be grouped with similar-looking native food plants by including *hapéc* 'planted' in their name. Two examples are the name for the potato, *xoját hapéc* 'planted *xoját*' (*Amoreuxia palmatifida*)' and the tomato, *xtoozp hapéc* 'planted *xtoozp*' (*Physalis crassifolia*).[6]

Sometimes a similarity may not be overtly expressed by common terms in names, but by a folkloric account instead. For instance, while we were working on the Seri dictionary and discussing the name of a marine isopod (a scavenger) with the Seri editorial committee, a similar creature, also an isopod, was described as having "run away from the sea," thoroughly disgusted because there wasn't enough sea turtle blood to go around. Its name is *moosni ihaait iisax cöcoaanim* 'who is angry about the sea turtle blood'—the terrestrial isopod we know as the pillbug.

A Classification of Mollusks

The Seris clearly recognize the higher categories of the biological world as: *hehet* (plants); *xiica ccámotam* 'living things' (animals); and *xiica quiistox* 'living things' (humans).[7]

The phrase *xiica ccámotam xepe com ano coom* 'animal sea life' or its shortened form *xepe án icám* 'life of the sea area' distinguishes marine life.[8] Within that group are clearly recognized and named categories such as the bony fishes (*zixcám*), sharks (*hacat*), or sea turtles (*moosni*), among others. Although the group known to the scientific world as marine invertebrates could be loosely described as *xiica ccámotam imítajoj xepe com ano coom* 'living things without bones in the sea', such an expression does not name an overt category and in fact seems somewhat forced. Instead, there are named groups of invertebrates on a par with the bony fishes and others, such as the swimming crabs (*zamt*), shore crabs (*queepl*), jellyfish (*copsiij*), or shelled mollusks (*haxölinaailc*).

The Seris do not name mollusks as a single taxonomic unit—hardly surprising given the array of creatures included in such a group by the scientific world at large. Species within five molluscan classes (themselves unnamed) are recognized by Seri names: the Bivalvia (clams, oysters, mussels, and others); Gastropoda (snails, limpets, and slugs); Polyplacophora (chitons); Scaphopoda (tusk shells); and Cephalopoda (octopuses, squids, and others).

The named category of externally shelled mollusks, and thus including the chitons and tusk shells, is the plural compound *haxölinaailc*, a word that refers to either the live mollusks or their empty shells.[9] The term incorporates the name of the very common

Leukoma grata clam, ***haxöl*** (a primary [unanalyzable] term), and *inaail*, a word meaning the skin of a person or animal, hide, tree bark, rind, a bottle—or simply, a casing. As we see in many places, the term *haxöl* clearly has a broader sense than just the *Leukoma grata* species; sometimes it must simply mean "clam" in a general sense.

As noted earlier, the names for the shell of every externally shelled mollusk would include the term *inaail* 'its shell' (see Preliminaries). Hence in correct usage when referring to just the shell of the *Leukoma grata* clam, one would refer to ***haxöl*** *inaail*. However, in common or colloquial usage when it is obvious that only the shell is referred to, the word *inaail* is sometimes dropped.

The term for shell is the singular compound *haxölinaail*. The word occurs in many descriptive names, for example, ***haxölinaail cooil*** 'blue/green shell' (*Lottia stanfordiana*, *L. dalliana*) and ***haxölinaail cotítzilca*** 'shells that pinch' (*Strombus* spp.).[10] It is interesting to note that with the exception of two species belonging to the Anomiidae family (*Anomia peruviana* and *Placunanomia cumingii*), all mollusks having names incorporating *haxölinaail* are gastropods. In contrast, the term *haxöl* and its possessed form *ihaxöl* (as part of a descriptive name) are used exclusively for bivalves. Possibly important here is what is relevant to the Seris when remarking on the difference between bivalves and gastropods: the internal animal shape of the bivalves is what is noted (see discussion below).

Table 3.1. The use of *haxöl*

haxöl	*Leukoma grata* clam
haxöl *inaail*	*L. grata* shell
haxölinaail	shell (of any shelled mollusk)
haxölinaailc	shelled mollusks (either live or only their shells)
haxölinaail X	(where X is a modifier) a name of a shelled mollusk, in all but two cases referring to gastropods
ihaxöl	X its-*haxöl*; used only for bivalves

Bivalves: The "*Haxöl*" Issue

As there is no clearly named folk taxonomic division distinguishing the bivalves and gastropods, it is important to consider how they are thought about and categorized. Because of my Western point of view I found it easy to unconsciously assume such a division; but how valid are such categories in Seri classification?

While asking questions about how shells are grouped, I found it surprising that no one commented on the number of valves: there was no quick mention of the "clam" creature having two shells that fit together. Although the opposing valves may be loosely described as *pti capxazl* 'what cover each other' or *pte hapxázatol* (pl.) 'what are covered with each other',[11] the phrases can also refer to other objects placed together with their interiors facing each other. In referencing the bivalve shell anatomy, the hinge area is recognized as the place where the shell comes together. The gape between the valves gives the bivalves a "mouth" with two lower lips; however, the shells of snails are not considered to have a mouth, although they do have one (lower) lip (see Anatomy, p. 32).

Perhaps most salient to the concept of the shelled mollusk is that the shell and animal are seen as a unit, rather than being a shell container holding the animal. And although the shells are seen as different in form, there does not seem to be the Western concept of two shells versus one shell.

In working with consultants, I showed them a large collection of the most common shells in the area that included bivalves, gastropods, chitons, and tusk shells. Each consultant separated the shells into piles while commenting on them. The phrase that most closely characterized the bivalves was *ziix* ***haxöl*** *hant cöquiih* 'thing grouped with ***haxöl***', and no gastropod shell was included in that group. However, while one woman grouped all bivalve shells under that phrase, another related it only to those of the unattached bivalves, making a clear distinction between them and the anchored bivalves such as oysters, mussels, and pen shells.

Before it was clear that the number of valves was not a defining characteristic, I asked what made either some or all of the bivalves *ziix* ***haxöl*** *hant cöquiih*, fully expecting a reference to their shells. Interestingly, the concept appears to be based on the shape of the internal animal; the clam's soft anatomy was described, and especially noted was its flat, somewhat elliptical shape. The internal organs were named, but not the two valves.

It is intriguing to consider why ***haxöl*** is central in the terminology relating to shells, and what makes *Leukoma grata* seem to be the prototypical clam. Perhaps what is significant is that the clam is abundant, wide in distribution, and easily harvested, thus an important food. In fact, its shell is ubiquitous in shell middens and desert camps throughout the region.[12]

Among the bivalves there are some commonly recognized generic groups. Some of these groups are the pen shells, the mussels (and mussel-like bivalves), some Venus clams, some oysters, and the pearl oysters. In grouping them, reference was made to the shell's form or sculpture, such as that of the heavy-shelled thorny oyster (*Spondylus limbatus*) or the prominently ribbed shell of the *Nodipecten subnodosus* scallop. However, in grouping together the oysters, one woman noted the different appearance of the internal soft animal, especially its whitish color and mantle edges.

The pen shells are an example of a folk generic and its specific members. ***Seeten*** is the generic name for pen shell, and each member can be described as *ziix* ***seeten*** *hant cöquiih* 'thing grouped with ***seeten***'. The specifics are further identified by descriptive names that include the name ***seeten***, the two most common being ***seeten ctam*** 'male ***seeten***' (*Pinna rugosa*) and ***seeten cmaam*** 'female ***seeten***' (*Atrina tuberculosa*). Others having additional names are ***seeten comihj*** 'smooth ***seeten***' (*Atrina oldroydii*) and ***seeten inaail cquihjö*** 'red-shelled ***seeten***' (*Atrina maura*).

Gastropods: The "***Xpa̱leemelc***" Issue

A Seri name corresponding to the gastropods does not exist, as is true of mollusks in general. This is not surprising given the broad range of creatures included by the scientific world. The externally shelled gastropods are grouped according to the shell form, although it is important to note that reference to "one shell" is not made.

The broadest overt category among the gastropods is *ziix* ***xpa̱leemelc*** *hant cöquiih* 'thing grouped with ***xpa̱leemelc***'. The phrase is given to coil-shelled gastropods, most typically the cone and olive snails, which share a somewhat tubular, pointed shape.[13] The term ***xpa̱leemelc*** is possibly a compound of *xepe* 'sea' and *ípaloj queemelc* 'what move slowly with their tongues', describing the movement of such snails gliding over submerged sand by means of their impressive foot. The name ***iplmateemij***, one of the older names for the *Agaronia testacea* snail, probably shares a similar etymology.

Such a category is somewhat idiosyncratic in that only one person grouped all such snails under the name ***xpa̱leemelc***, while others instead described many as either ***xpa̱leemelc*** *isoj áa* 'really true ***xpa̱leemelc***' or ***xpa̱leemelc*** *xahxaii* '[something] like a ***xpa̱leemelc***'. ***Xpa̱leemelc***, either as a separate name or as part of a descriptive name, is shared by snails of several different families (table 3.2).

Table 3.2. Names that include ***xpa̱leemelc***

mayoocö ixpa̱leemelc	*Agaronia testacea*	'the Yellow-crowned Night Heron's ***xpa̱leemelc***'
oot ixpa̱leemelc	*Agaronia testacea, Crassispira incrassata*	'the coyote's ***xpa̱leemelc***'
xpa̱leemelc	Conidae (*Conus princeps, C. regularis, C. ximenes*)	
xpa̱leemelc caacoj	*Oliva incrassata, O. porphyria, Strombus galeatus*	'large ***xpa̱leemelc***'
xpa̱leemelc cacösxaj	*Fusinus dupetitthouarsi*	'long ***xpa̱leemelc***'
xpa̱leemelc comihj	*Oliva spicata*	'polished ***xpa̱leemelc***'
xpa̱leemelc comihj caacoj	*O. incrassata*	'large ***xpa̱leemelc comihj***'
xpa̱leemelc coospoj	*O. incrassata*	'spotted ***xpa̱leemelc***'
xpa̱leemelc quipcö	*O. incrassata*	'thick ***xpa̱leemelc***'
xpa̱leemelc quicös	*Nassarius brunneostomus*	'spiny ***xpa̱leemelc***'
ziic coopol ixpa̱leemelc	*Crassispira incrassata*	'the Double-crested Cormorant's ***xpa̱leemelc***'

While the broadest category among the gastropods seems to be *ziix* ***xpa̱leemelc*** *hant cöquiih* 'thing grouped with ***xpa̱leemelc***', there are a number of smaller generic taxa whose members include specific names. Examples of such groups are the moon snails, cowries, olive snails, or turret snails (and similar high-spired snails). Members of these groups are further identified by descriptive, specific names.

In addition to having specific names and belonging to different generic groups, small snails such as olive, cerith, and dove snails are categorized as a group by the names ***hast iti cöcooyaj*** 'what travel on rocks', ***hast iti cöquiipax*** 'what climbs up rocks', and ***hast iti cpaainalca*** 'what tumble off rocks'. The shells of such snails, when inhabited by hermit crabs, are found in great numbers on rock surfaces at low tide (figure 3.2). When approached, the hermit crab retreats into the shell, which then often tumbles off its perch.

Figure 3.2. 'What tumble off rocks', small snail shells inhabited by hermit crabs. Desemboque, 2013.

Other Mollusks

The chitons are known generically as ***capoclim***, with the most common species given a specific name.

Not commonly recognized today, the tusk shells are known by one name, ***cozaaij***, a name also given to small worm snails.

The octopuses and squids are categorized as *ziix* ***hapaj*** *hant cöquiih* 'thing grouped with ***hapaj*** (octopus)'. The squids are known only as ***pajaas***, and the octopuses as ***hapaj***; both are primary, generic names. However, being more familiar creatures found in the intertidal zone, the octopuses are further recognized by specific names that include descriptive terms. As noted earlier, the paper nautilus (which no Seri today recalled by name or description) has recently been dubbed a name and included in the generic ***hapaj*** group (see Argonautidae).

Names and Naming

Names make the commonplace personal, perhaps transforming the ordinary into something whimsical to be chuckled over or important enough to be remembered. Such names as 'what the beggar pounds in the sea', 'the firewood my younger sister gathered', 'necklace of the dark place', 'old woman's toenail', and 'what the sea turtle tattooed' give a narrative to shells that one comes upon tossed up on a beach or lying in a remote desert camp. One can almost *hear* them being talked about and used.

When considering names in a society without written records, a basic issue is determining what constitutes a name.[14] A common shell might have a name often used by the community, while on the other hand an uncommon one might be recognized by only one or two people who gave it a name perhaps reflecting a comparison to another, better-known shell. How to judge whether such a phrase is a "true name" versus a spontaneous description? The latter may in time become an equally valid name. One might assume that descriptive names had such inceptions.

One difficulty is that descriptive phrases in Seri have no obvious phonetic distinction from many compounds, such as that heard in the English "white fish" versus "whitefish," where stress provides a clue to their semantic difference—one being a descriptive phrase and the other a name. The issue can usually be resolved by asking questions such as *¿Zó tpai?* 'What is it called?' or *¿Tiix itaasiya?* 'Is that its name?'. Furthermore, names are persistent while descriptive phrases are transitory.

Another related issue is whether one name is more relevant than another. For instance, one name might be extremely common and used by the whole community, while another might be given by only one person, perhaps recalled from a childhood conversation. An example can be seen in the names of two mollusks named in the last century. The well-known name of the slipper snails (*Crepidula* spp.) with their boat-like shell, ***miizj hant quixquim quih iti ihiij*** 'where the navigator sits', is a playful and aptly descriptive name whose creator is often referenced along with the snail's name today. In contrast, another name given by a different person to the *Pholas chiloensis* clam (a much less commonly seen mollusk with a wing-like white shell), ***ziic ccap cooxp ihaxöl*** 'the Tern's ***haxöl***' (where *ziic ccap cooxp* 'white flying bird' is a synonym for *hatj*, *Sterna* sp.) is recalled by only a few family members.

The sociolinguistic situation is important to consider. According to oral accounts, the Seris lived in six geographical bands that reportedly included dialectal variation,[15] and many older people today can identify those from which they descend. One woman informed me that each of her four grandparents was from a different band, explaining that this gave her a broader knowledge base. When someone provides a name, another person might remark, "She's from the —— people; she really knows about this," or perhaps somewhat dismissively, "What does *she* know about that? She's from the desert (inland) people!" Such sociocultural differences have no doubt influenced the information about some species, resulting in multiple names, sometimes evidenced as minor variations in pronunciation as well. For instance, when discussing the names of the *Theodoxus luteofasciatus* snail, María Luisa Astorga, who spent much of her childhood living in the desert areas north of Desemboque, noted that she knew the snail by one name, and it was not until her family settled in Desemboque that she learned another very different name.

One apparent difficulty in collecting names for lesser-known species exists when a name might be given to one mollusk, and no name is supplied for a very similar species. I have been careful to not assume that both species share such a name unless I am told that they do. What appears to be a curious gap may be resolved in the future upon further investigation. An example is provided by two species of the Muricidae. I was given the name *xepe án icös* 'spines of the sea area' for *Pteropurpura centrifuga*, but not to the similar *P. erinacioides*, for which no name was supplied. Another situation concerns the clams of the genus *Tagelus*, where the species illustrated is one of several similar species in the area (see Solecurtidae). The inclusion of only one illustrated and named species does not preclude that others would not share the Seri names.

In working with a Zapotec language of southern Mexico, Eugene Hunn, an anthropologist who has specialized in ethnobiology, uses what he calls the cultural criterion as a method of determining a name; he is less inclined to consider something a name if it is used by a single inventive speaker.[16] While this is helpful, the Seri situation is somewhat different, as the number of people who are knowledgeable about mollusk names has shrunk considerably; consequently, many names are no longer in common use and thus not easily verified. However, because of a long-term relationship with my consultants (and by using the Seri language to collect the information), I feel fairly confident about the names they provided. They did not hesitate to tell me when they did not know a name.

Mollusk names are linguistically either primary or compound. Primary names are unanalyzable and usually well-attested in the community, with the exception of very old names that were recorded from a previous generation and are no longer used. Compound names might be a phrasal compound or a non-phrasal compound noun.

Primary Names

A primary name is not composed of or derived from other words; an example is the English word "clam." It is worth noting here that a primary name is not necessarily the most commonly used name for a particular mollusk.

Primary names such as ***haxt*** (*Crassostrea corteziensis*), ***mas*** (*Olivella dama*) and ***copas*** (*Pinctada mazatlanica*) comprise about twenty-five percent of mollusk names (appendix 2). A good number of these are species having high food or cultural value, something also true of plants, as noted by Felger and M. Moser when discussing Seri plant names: "Large, conspicuous, or culturally important plants tended to have unanalyzable names . . . Approximately 26 percent (110) of the plant names were unanalyzable terms."[17] It is interesting that virtually the same percentage of primary names occurs among mollusks.

As noted by Felger and M. Moser when discussing plant names and referring to Brent Berlin (1973): "Many of the more conspicuous and culturally important ones were known by a name composed of a single word with only one or a few syllables . . . Such terms . . . referred to linguistically as primary names, characteristically suggest considerable cultural antiquity."[18]

Of the five recorded names for *Olivella dama*, a snail represented in the prehistoric shell trade from the Gulf, two are primary (see Olivellidae). On the other extreme, there are a number of small, inconspicuous snails whose names are phrasal compounds that include the term *haxölinaail* 'shell'. These snails did not traditionally have high food or cultural value, but in the last fifty years have become common in shell jewelry. As explained by one consultant, such small creatures were grouped together as *xepe án imiipla* 'the badness (worthless things) of the sea area', and it was not until they were used that they took widely accepted names.

Mollusks with a wide distribution and high cultural value are often individually recognized by Seri names. For example, among the heavily consumed Venus clams, eleven species are named. On the other extreme, the name ***xtoozajö*** covers the entire family of worm snails, which were not eaten. However, one notable exception merited a name: pesky little worm snail shells that, when stepped on, make a perfectly round incision in the sole of the foot. Their name is ***cozaaij*** 'what makes a circular cut', a name also given to the tusk shells, said to produce a similar cut.

Compound Names

A phrasal compound is made up of separate words, as in "flying fish." It is not merely a descriptive phrase, although it may have had its origin as such, but rather is an actual name. A (non-phrasal) compound noun is composed of fused and usually recognizable parts, as in "shellfish" and "cupboard."

Some compound nouns are clearly names because of their forms. For example, ***hapxaacoj*** 'large ***haapx***' from ***haapx*** (*Tegula rugosa*) + *caacoj* 'large', is the name of an unidentified snail similar to, but larger than, the *Tegula rugosa* snail. A spontaneous description would require the phrase ***haapx*** *caacoj* 'large ***haapx***'.

Some names are potentially misunderstood, especially when composed of separate words, since they could be confused with noun phrases that have merely a descriptive function. For example, the phrase *ziix ccam* is a descriptive phrase that can be translated as 'thing that is alive' or, with a slight phonetic difference, a phrasal compound name meaning 'animal'. A related name, a compound noun etymologically formed from the same words, is *zixcám*, meaning 'fish'.

A phrasal compound name, then, is the most ambiguous, since its structure fits either an actual name or a spontaneous description.

Compound names, usually descriptive, often highlight the physical characteristics of the shell. Such names contain terms that refer to its shape: ***cascanoohcö*** (deeply rounded) (*Euvola vogdesi*); color: ***haxölinaail cheel*** (red) (*Trivia solandri*); size: ***xpa̲leemelc caacoj*** (large) (*Strombus galeatus* and others); texture: ***haxölinaail cnoosc*** (rough) (*Nassarius* spp.); or similarity to something: ***mosniil inoosj*** (a claw) (*Diodora inaequalis*).

A few names describe a characteristic sound or movement, as in ***taijitiquiixaz*** 'what tinkle in the surf (swash zone)' (*Heterodonax pacificus*); or ***xepe yamaasij*** 'what is rolled by the sea' (*Strombus galeatus* and others).

Some mollusks are named for their cultural function, such as ***haxöl icaai*** 'clamshell for making pottery' (*Simomactra dolabriformis*) or for a physiological effect, as in ***hapos cahaaxat*** 'what irritates the throat' (*Chama buddiana*).

Curiously, some very distinctive snails, such as the striking little murex *Vokesimurex tricoronis*, are not recognized today by name by many Seris, and when named, the names vary widely. On the other hand, the singular and uncommon Gulf sundial (*Architectonica nobilis*) has only one, well-known name. The difference is perhaps due to the latter having a very special use.

Animal-Related Names

A name might be based on association with an animal, perhaps because of the shell's color or shape. For example, one name of the attractively marked *Agaronia testacea* snail, ***mayoocö ixpa̲leemelc*** 'the Yellow-crowned Night Heron's ***xpa̲leemelc***', suggests a similarity in the shell's color and markings to the bird's plumage, while the ***mooxon iixz*** 'the scorpionfish's pet' (*Cardites laticostatus*) shell is strikingly similar to the shape and markings of the fish's fin.

Other animal-related names describe some physical attribute or action of the animal. The vividly marked *Columbella strombiformis* snail has two such names, both referring to a variety of black sea turtle (*Chelonia mydas*): ***mosniil itaajc*** 'the *mosniil* sea turtle's saliva' and ***mosniil oozt*** 'what the *mosniil* sea turtle tattooed'. A white clam, *Psammotreta cognata*, is called ***sleecoj cooxp ihaxöl*** 'the Snowy Egret's ***haxöl***'; another of its names, ***haxöl icaai sleecoj iti yazám*** '***haxöl icaai*** (*Simomactra dolabriformis*, a similar white shell) stepped on by a heron' notes a posterior depression or fold on its shell that gives the appearance of its having been stepped on.

A name might reflect the mollusk as being eaten by an animal, hence it "belongs" to that animal. The ***hatj ihaxöl*** 'the Tern's ***haxöl***' (*Tellina regia*) is said to be eaten by the Tern.

Table 3.3. Sex-based names

Many shells have multiple names; the names included below may be one of several given to that shell.

Shells that include *ctam* 'male' in a name or are otherwise said to have a "male" shape or texture	**Shells that include *cmaam* 'female' in a name or are otherwise said to have a "female" shape or texture**
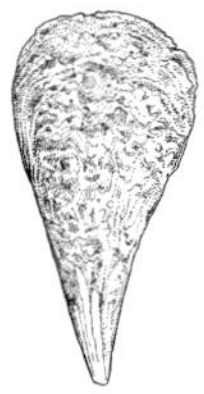 *Pinna rugosa* ***seeten ctam***	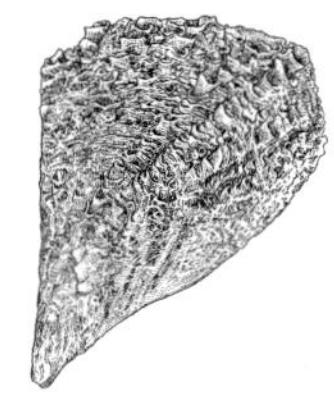 *Atrina tuberculosa* ***seeten** cmaam*
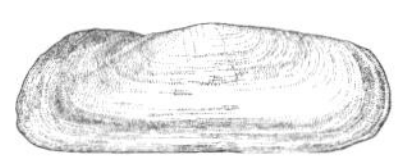 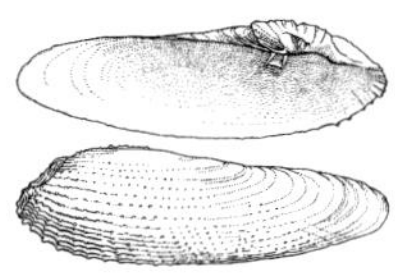 *Tagelus peruvianus* *Pholas chiloensis* ***haxöl icaai ctam***	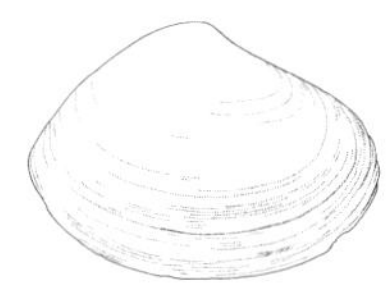 *Simomactra dolabriformis* ***haxöl icaai***
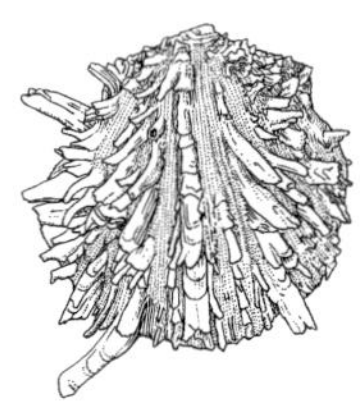 *Spondylus leucacanthus* ***teexoj ctam***	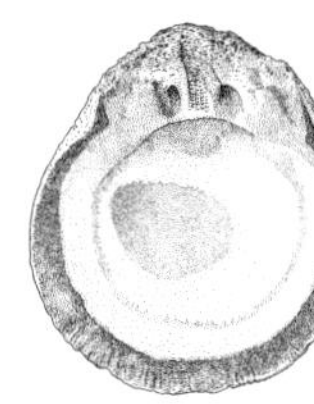 *Spondylus limbatus* ***teexoj***
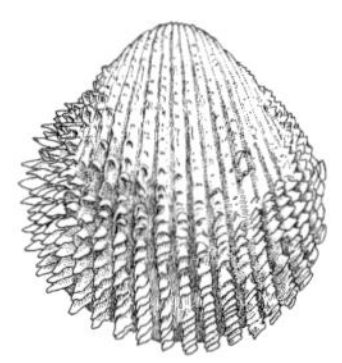 *Trachycardium consors* ***zacz ctam***	 *Trachycardium procerum* ***zacz***
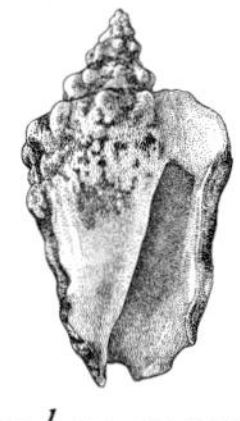 *Strombus granulatus* ***haxölinaail cotitla ctam***	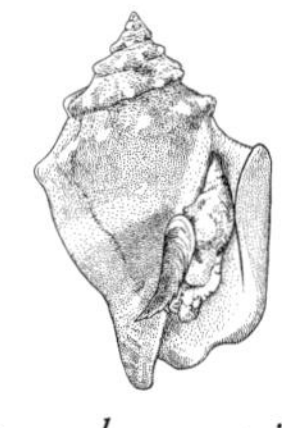 *Strombus gracilior* ***haxölinaail cotitla cmaam***

Shells that include *ctam* 'male' in a name or are otherwise said to have a "male" shape or texture	Shells that include *cmaam* 'female' in a name or are otherwise said to have a "female" shape or texture

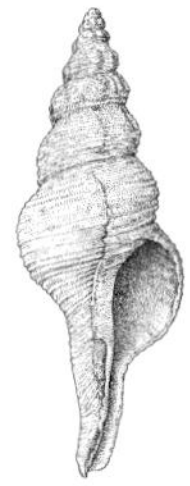

Fusinus dupetitthouarsi
xasecö ctam

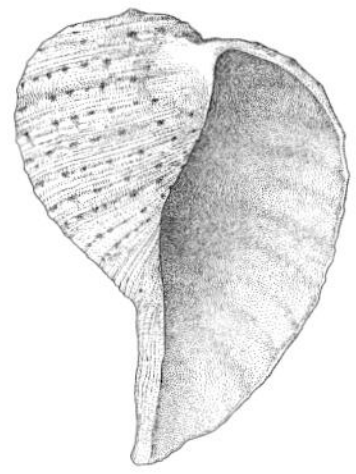

Ficus ventricosa
xasecö

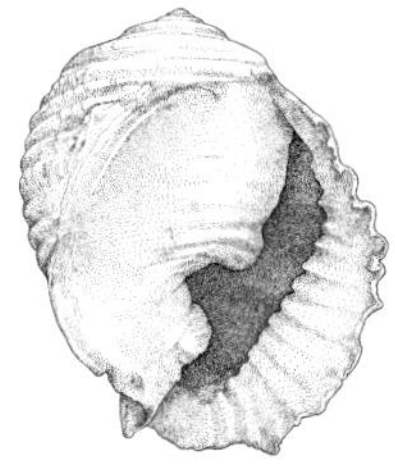

Malea ringens
xasecö cmaam

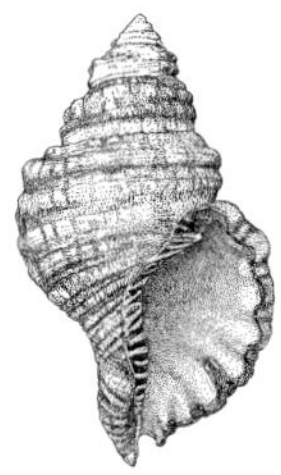

Cymatium keenae
xepe yamaasij ctam

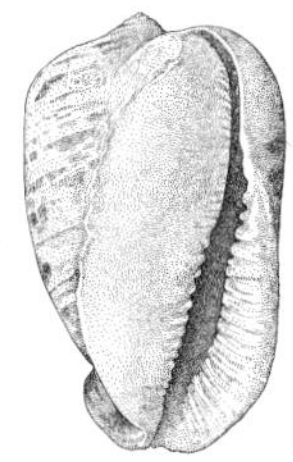

Cypraecassis coarctata
xepe yamaasij

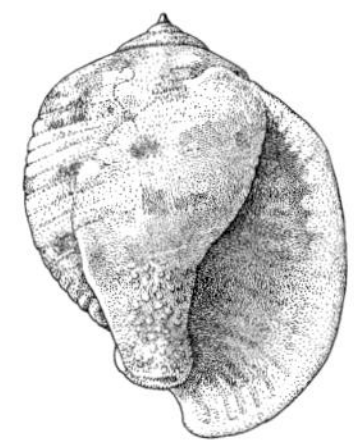

Semicassis centiquadrata

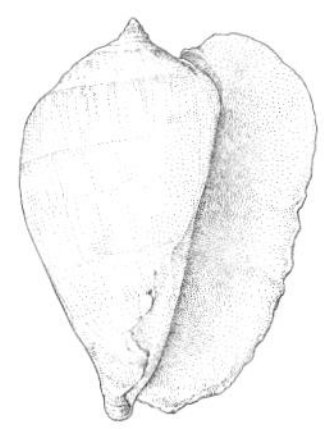

Strombus galeatus

Malea ringens

Naming Through Folklore

Accounts from Seri folklore relate how some mollusks were named. Some are linked to comments made while they were still people and being named at the world's creation (see Mythology and Folklore). A few names are based in accounts relating conversations between the shells themselves. Such names might reflect the mollusk's appearance or movement. For example, the globular bubble snail *Bulla gouldiana* is named ***cacaapxom*** 'what fattens (something)'. The name was given by the Name Giver when the person who was to become that snail pointed out that he had fattened "those others" present who were also receiving names.

The ***hant iiha cooml*** 'what investigate the land' snail (*Conus princeps*) was sent out soon after the earth's formation to test the hardness of the newly formed land. Such a task is graphically evoked by the snail's slow, delicate gliding over the sand's surface.

Names Referencing Sex

Names referencing sex are sometimes used to describe mollusks; this is based on the shell's shape or surface sculpture (table 3.3). When distinguished by sexual reference it is usually the smaller shell, perhaps more squat or round that suggests a "female" shape, while the elongate, thinner shape is characteristically "male." The surface sculpture might provide an important distinction as well. María Luisa Astorga, referring to the human form, explained that females are smoother, softer, and prettier; males are rough-skinned and whiskered. Sex-based naming also occurs in other domains, such as plants and other animals.[19]

An example of sex-based naming distinguishing shells within a family can be seen in the Pinnidae (pen shells), where one species is named ***seeten ctam*** 'male ***seeten***' (*Pinna rugosa*), while another is ***seeten cmaam*** 'female ***seeten***' (*Atrina tuberculosa*). Sometimes clams from different families are distinguished in the same way. For example, a similarity is noted between the ***haxöl icaai*** 'clamshell for making pottery' (Mactridae, *Simomactra dolabriformis*) and two other shells sharing the name ***haxöl icaai ctam*** 'male ***haxöl icaai***': *Pholas chiloensis* (Pholadidae) and *Tagelus* sp. (Solecurtidae). The latter two are distinguished from the former by their "male" shape.

Names Referencing Size

A number of species are named with reference to size and incorporate the terms for "small" or "large" in their names. For example, two limpets have shells with blue interiors: *Lottia dalliana* and *L. stanfordiana*. Both have names referring to the color, but to distinguish between them, the larger *L. dalliana* also has a well-used name that incorporates *caacoj* 'large'.

Sometimes modifiers referring to size or sex are used in names given to a shell that is little known when comparing it to one that is better known. Such names require additional investigation to determine whether they are actual names in common use or simply spontaneous descriptions.

Personally Given Names

A shell might be dubbed a certain name even though a more recognized name is already known. One family gave the name ***zojiime*** to *Costoanachis coronata* snails gathered at a beach camp named *Zoj Iime*, where they were abundant. Although the family knew the shell's more common name, ***haxölinaail quimaaxat*** 'gray-brown shell', they enjoyed referring to it by their own name.

When Juanita Herrera was left with one unpurchased necklace of strung ***mayoocö ixpa̱leemelc*** (*Agaronia testacea*) shells, she spontaneously dubbed the shell *captasa* 'unlucky', even though she knew its commonly used name. In another instance, María Luisa Astorga gave that same name to *Mitra tristis* shells she found in beach drift because she thought they weren't attractive. She indicated that she did not know a common Seri name for them.

The *Pholas chiloensis* clam, uncommon in the present-day Seri region, has as one of its names ***ziic ccap cooxp ihaxöl*** 'the Tern's ***haxöl***'. According to Cleotilde Morales, her mother, María Antonia Colosio, named the clam long ago.

There is no evidence that the names in such examples have come into widespread use. However, there is one clear instance where a name did become widely used and its author still recognized by name: the slipper limpets were named by Loreta Marcos, who called them ***miizj hant quixquim quih iti ihiij*** 'where the navigator sits' because the shell resembles a boat with a seat.[20]

Names Referencing the Coyote

In Seri folkore the coyote, *oot*, is a gullible, greedy creature. Coyote ownership is manifested in the name of something that varies from the norm; as one person explained, the Name Giver named the "real" thing, but the coyote owns a poor copy. As further explained by Raquel Moreno, the coyote thinks something is attractive or better and grabs it for himself. As is often the case, such things are perhaps smaller or abnormal in shape or color, and are humorously noted as indeed being what would attract a coyote. For instance, the hard wood formed around a knothole of the elephant tree and left behind when the softer surrounding wood decays often resembles a tiny boat. The object, called *oot icanoaa* 'the coyote's boat', sometimes becomes a child's toy.[21]

A name might be based on a characteristic of the coyote, such as its gullibility. For instance, a mirage is called *oot iix* 'the coyote's water'. It is said that the coyote sees the water on a desert dry lake bed and runs toward it. Upon arriving, it does not find water. It turns around, then sees water again in the distance. Again, the coyote runs toward it and so on, back and forth.

The coyote's name is associated with marine invertebrates as well. For example, the large edible blue swimming crab, *zamt* (*Callinectes bellicosus*) is mirrored by the *oot izamt* 'the coyote's swimming crab' (*Cronius ruber*, and perhaps other similar species), a much smaller, feisty crab that is not eaten. Two other crabs, found beneath intertidal rocks, are the small *queepl*, a grapsoid crab, and the *oot iqueepl* 'the coyote's *queepl*', a rock slider (*Petrolisthes armatus*) that scuttles about, its curious black, flat body pressed closely against a rock.

Table 3.4. Seri mollusk names referencing the coyote

Name	Species "belonging" to the coyote	Species or objects bearing the original name
cmatjij oot quih yaa	'***cmatjij*** belonging to the coyote', *Hipponix antiquatus*	***cmatjij***, *Lottia dalliana*
cotitzil oot quih yaa	'***cotitzil*** belonging to the coyote', *Strombus granulatus*	***haxölinaail cotitzil***, *Strombus gracilior*
oot icamazjij	'the coyote's ***camazjij***', *Tivela byronensis*	***camazjij***, *Megapitaria squalida*
oot icopas	'the coyote's ***copas***', *Isognomon janus, I. recognitus*	***copas***, *Pinctada mazatlanica*
oot ihaxöl	'the coyote's ***haxöl***', *Chione tumens*	***haxöl***, *Leukoma grata*
oot iquiit	'the coyote's ***quiit***', *Carditamera affinis*	***quiit***, *Carditamera affinis*
oot isatoj	'the coyote's ***satoj***', *Mytilus californianus, Modiolus tumbezensis, Entodesma pictum*	***satoj***, *Modiolus capax*
oot iscám	'the coyote's balsa', *Crepidula* spp.	(a balsa)
oot iseeten	'the coyote's ***seeten***', *Atrina maura, A. oldroydii*	***seeten***, Pinnidae
oot iteexoj	'the coyote's ***teexoj***', *Spondylus leucacanthus*	***teexoj***, *Spondylus limbatus*
oot ixpa̱leemelc	'the coyote's ***xpa̱leemelc***', *Agaronia testacea, Crassispira incrassata*	*various snail species with the name* ***xpa̱leemelc***
oot izacz	'the coyote's ***zacz***', *Trachycardium consors*	***zacz***, *Trachycardium procerum*
oot yácmolca	'the coyote's fetish', *Morum tuberculosum*	(a fetish)

Mollusks said to belong to the coyote are shown in table 3.4. The names suggest a similarity to a well-known mollusk with an established, usually primary, name. Two names denote a resemblance to an object.

Although expressions and names involving the coyote can appear to be spontaneous and somewhat humorously given, the names presented here are apparently well established.

Changes in Names

It has been interesting to observe or be told about historic changes in names, as such information can provide a glimpse into how names may have changed in the past.

One recent change has happened as the result of avoiding references to explicit sexual anatomical terms in a way now considered vulgar; as a Seri consultant explained, such references were often made in the past. The common name for the *Trivia solandri* snail was ***hasit camós*** 'puckered vulva'. Today the snail's acceptable and widely used name, ***haxölinaail cheel*** 'red shell', references the shell's color.

Some names have been changed because of name taboo. When a person had recently died, as a mark of respect his name was not spoken.[22] If the deceased had been named after something, then following his death that thing was renamed using a descriptive phrase containing the words *ziix* 'thing' or *xiica* 'things' in order to avoid saying the deceased's name. For example, one of the old names of the common olive snail (*Olivella dama*) was ***mas***. When a man named *Mas* died,[23] the snail was renamed ***xiica cooxp*** 'white things', the name in common use today.

As noted earlier, primary names often represent mollusks having high cultural or food value and have been retained through time. When the Seris were spread over a large territory, well-used names that changed in one area due to name taboo may not have completely disappeared from use in other areas.

Naming Newly Encountered Species

Puerto Libertad is a non-Seri fishing town near the northern edge of the traditional Seri territory. As late as the 1950s, Seris periodically lived there and continued to visit the area in later times as tourist numbers increased. Only recently, however, have Seri people been living there for greater lengths of time. Since around 2004, a few young people have attended the *preparatoria* (high school) in town, and adult Seris have lived with them. While there, the women began gathering unusual shells not common to the beaches to the south where most Seris live today. As the shells have been shared and used in shell jewelry, they have acquired names suggesting their similarity to more familiar shells (see Psammobiidae, Terebridae, and Ovulidae).

Sometimes a large or unusual shell is not recognized as being from the immediate area or is from deeper waters and seldom seen. Such shells are often described as being from the *xepe caacoj* 'large (open) sea' or *xepe cyaail* 'deep sea'. In one case a very unusual shell was dubbed a name in reference to *Hant Ihiin* (Baja California), a faraway place of strange things and where the Giants were said to have lived (see Anomiidae).

Plural Names

Some shells are called by their plural form even when a single specimen is referred to, as in ***xiica cooxp*** 'white things' (*Olivella dama*) and ***haxölinaail coopl*** 'black shells' (*Mitrella ocellata*). In the latter, the singular form *haxölinaail* is followed by the plural form *coopl* 'black', rather than the singular *coopol*. Such small shells share the characteristic of being seen in greater quantity, as when gathered for use in necklaces. Similar plural names are given to food items occurring in quantity: *xiica coopol* 'black things' (coffee [beans]), *xiica coosot* 'narrow things' (rice), and *xiica is cquihjö* 'things with red seeds' ([shelled, dried] beans).

Anatomy

Seri terms for the molluscan anatomy are for the most part those used with other animals; see the terms in table 3.5 and figures 3.3–3.6.[24] For example, the internal parts of a mollusk have such names as *itaamalca* 'its horns or antlers' (tentacles) or *ipl* 'its tongue' (foot). In one case, a compound noun, *itojipl*, literally 'its eye's tongue', refers to the ctenidia (gills).

In discussing molluscan anatomical terms used in native languages of North America, the early American ethnologist John P. Harrington noted that "nowhere is there any term which is applied to mollusk body parts which is not applied to similar body parts of animals which are not mollusks."[25] Interestingly, there are two such Seri terms specific to mollusks: *yahipxat*, the ink or ink sac of the octopus and squid, and *itaqueecöj*, the bivalve adductor muscle.[26] Although *itaqueecöj* refers to the adductor muscle of all bivalves, it is most often used for those of the pen shells or other larger bivalves; the muscles of smaller bivalves are usually just referred to as *ipxasi* (its flesh or meat).

The term *ipxasi* is given to the internal animal of the externally shelled mollusk, as well as the octopus and squid. Although some people include the sea hare (*Aplysia californica*) as being fleshy, others do not. As one person explained, the animal is "fatty" or "watery," not solid.[27]

As previously mentioned, the external shell of a mollusk is *inaail*, a term that can refer to the skin of other mollusks, such as that of the octopus or sea hare. The shelled mollusk's interior cavity is the *ipxasi an ihiih* 'where its flesh is'. The muscle scars are the *itaqueecöj iti yáxalca* 'where the *itaqueecöj* sit'.

Because bivalves have shells that can gape open, they are considered to have a mouth, *iteen*, and two lips, *itamoolc* 'its lower lips' (the term *itacops* 'its upper lip' is not used in mollusk anatomy). Even though the shells of gastropods are not considered to have a mouth, their outer lip is referred to as *itamooj* 'its lower lip'.

Table 3.5. Anatomical terms

Parts of the shell	**Seri name**	**Literal translation**
anterior and posterior edge (bivalve); anterior and posterior ends of aperture (gastropod)	*itaseej ihiyáx*	'corner of its mouth'
exterior surface	*hapx iicp*	'its outside'
spicules (as in shell of *Pinna rugosa*)	*zatx*	'its glochids'
interior cavity	*ipxasi an ihiih*	'where its flesh is'
interior surface	*an iicp*	'its inside'
outer lip (gastropod); ventral edges (bivalve)	*itamooj*, *itamoolc* (pl.)	'its lower lip(s)'
spines	*icös*	'its spine(s)'
Bivalve shell		
adductor muscle scar	*itaqueecöj iti ihiij*	'where the adductor muscle sits'
anterior and posterior margins	*cöipatj iicp*	'its sides'
auricles	*isloj*	'its ears'
dorsal (angled) margin on either side of the beak (not used for a round bivalve shell)	*islaaaj*	'its shoulder'
hinge area	*iti pte hapxazl*	'where [the shells] come together'
hinged valves	*pti capxazl*	'with what they cover each other'
ventral margin	*ihita*	'its sharp edge'
Univalve shell		
apex	*itís*	'its tip'
body whorl and/or siphonal canal	*itx*	'its buttock'
operculum	*iti yahimoz*	'[a hard thing] with which it covers itself'

Parts of the shell	**Seri name**	**Literal translation**
operculum (long, horny, such as that of *Strombus gracilior*)	*inoosj*	'its claw'
umbilicus	*islitx*	'its ear canal'

Internal anatomy of the shelled mollusks

adductor muscle	*itaqueecöj*	——
byssal fibers (pen shell or mussel)	*ilít*	'its hair'
	imas	'its pubic hair'
	itaa	archaic, probably 'its hair'
ctenidia (gills)	*itojipl*	'its gills'
eyes	*itoj*	'its eyes'
foot	*ipl*	'its tongue'
foot muscle with byssal gland (of pen shell)	*iti cöihiti*	'what it connects with'
mantle edges (bivalve)	*iteeloj*	'its edges'
mucus or secretion	*itaajc*	'its saliva'
periostracum (fibrous)	*iteems*	'its whiskers'
radular dart	*yacotni*	'what it stabs with'
siphon	*xepe iisi*	'with which it drinks seawater'
stomach	*ano yahit*	'in which it eats'
tentacles or eye stalks	*itaamalca*	'its horns'
visceral mass	*itoozj*	'its intestines'

Cephalopod anatomy

arms	*inloj*	'its arms'
beak	*itaast*	'its teeth'
eggs	*iipx*	'its eggs'
eyes	*itoj*	'its eyes'
fins (squid)	*íseloj*	'its pectoral fins'
gills	*iit iixac*	'its lice's nits'
ink, ink sac	*yahipxat*	——
mantle containing the visceral mass (squid)	*ilít*	'its head'
mantle, hump-shaped, containing the visceral mass (octopus)	*itni*	'its umbilical hernia'
pen, internal (squid) (also given to the internal shell of sea hare)	*iixquij* (sometimes called *itac* 'its bone')	'its cartilage'
raspers (squid)	*inooslca*	'its claws'
suction cups (octopus), suckers (squid)	*yatopyas*	'with which it attaches (with suction)'

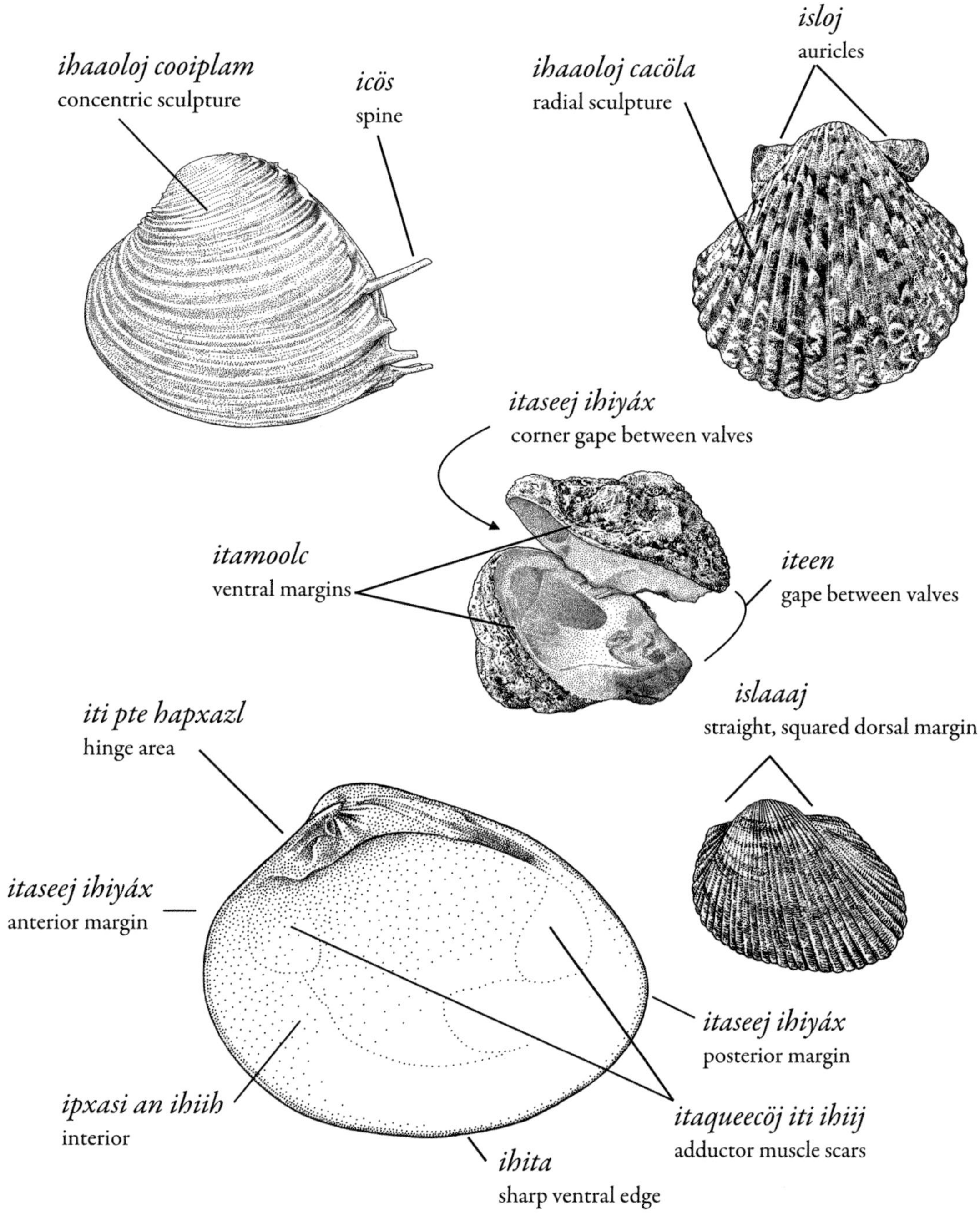

Figure 3.3. Anatomy of the bivalve shell.

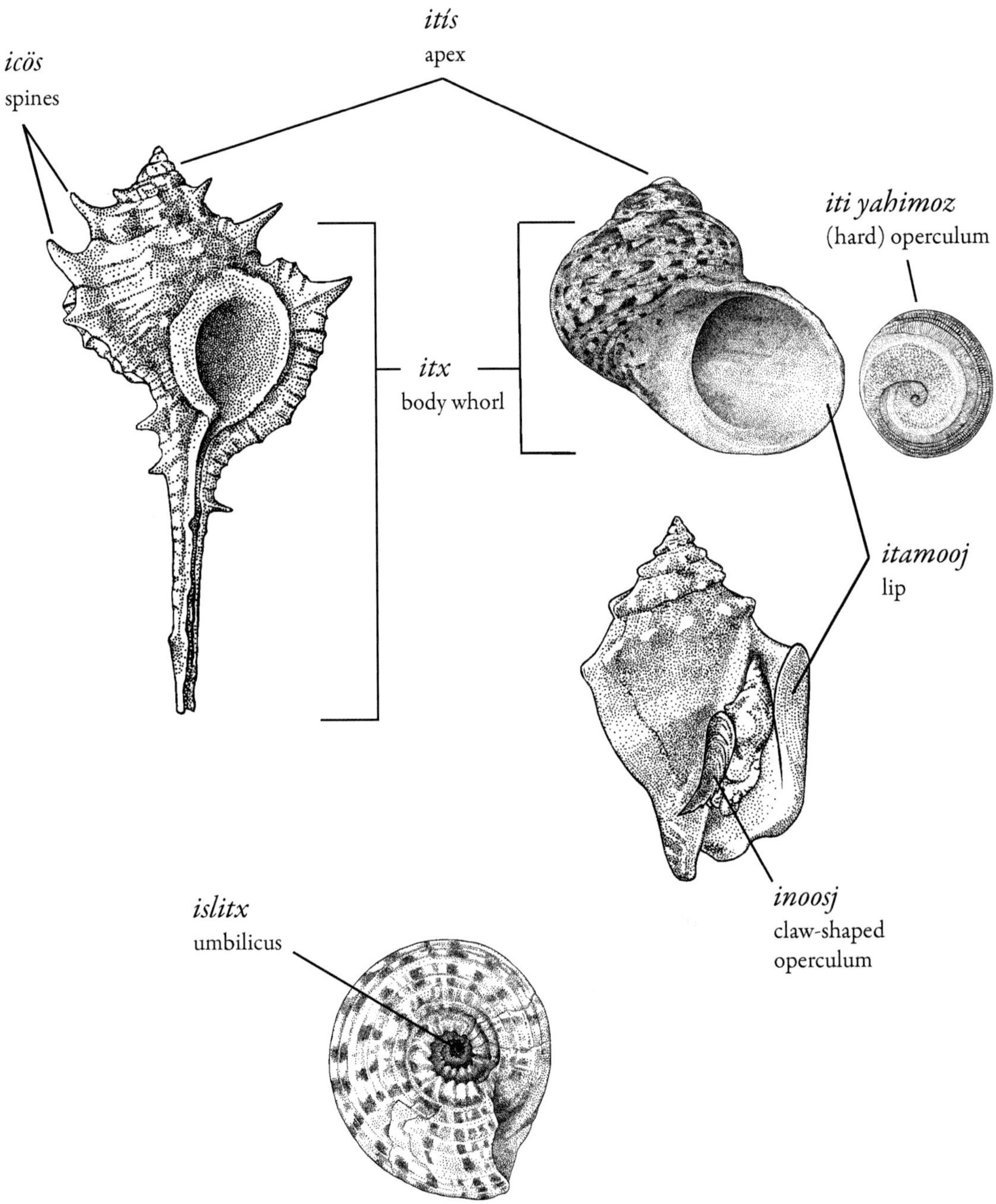

Figure 3.4. Anatomy of the gastropod shell.

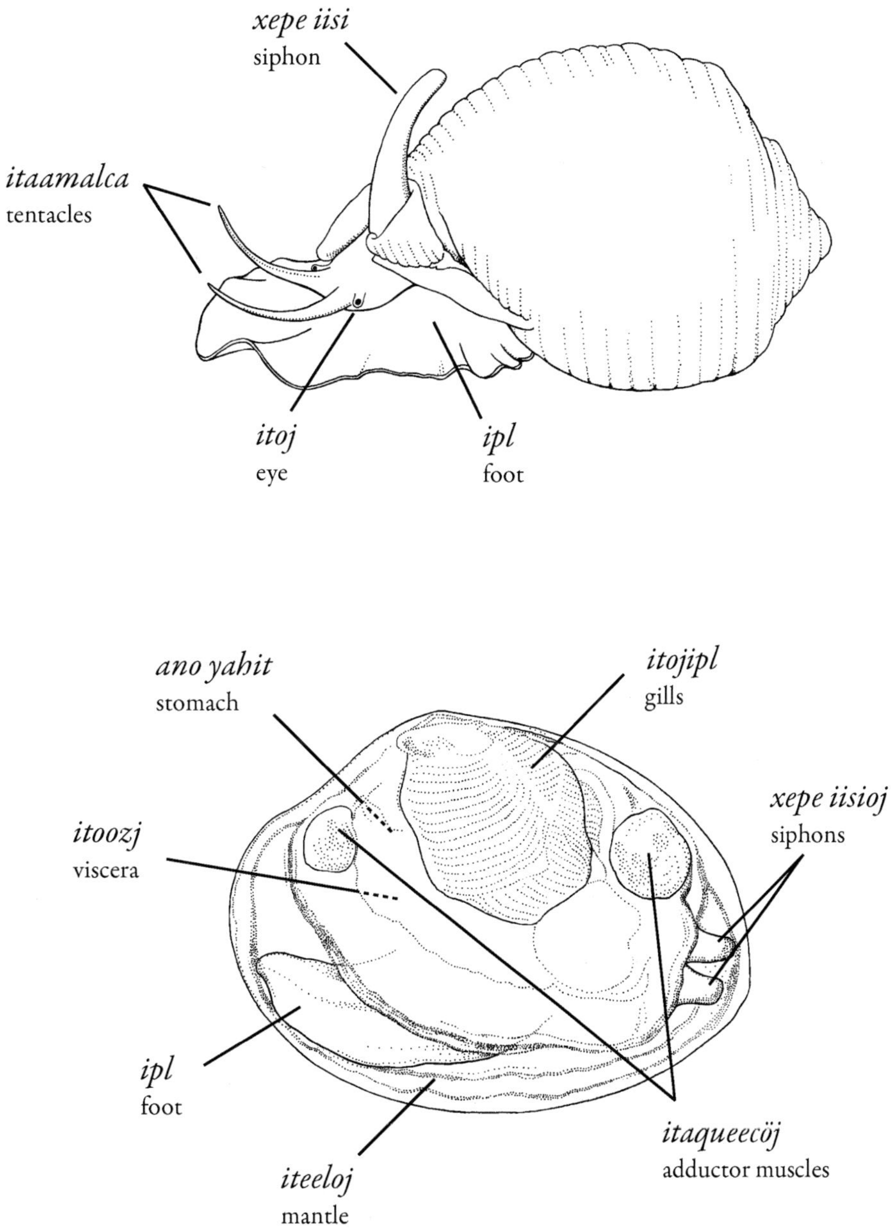

Figure 3.5. Anatomy of the internal animal.

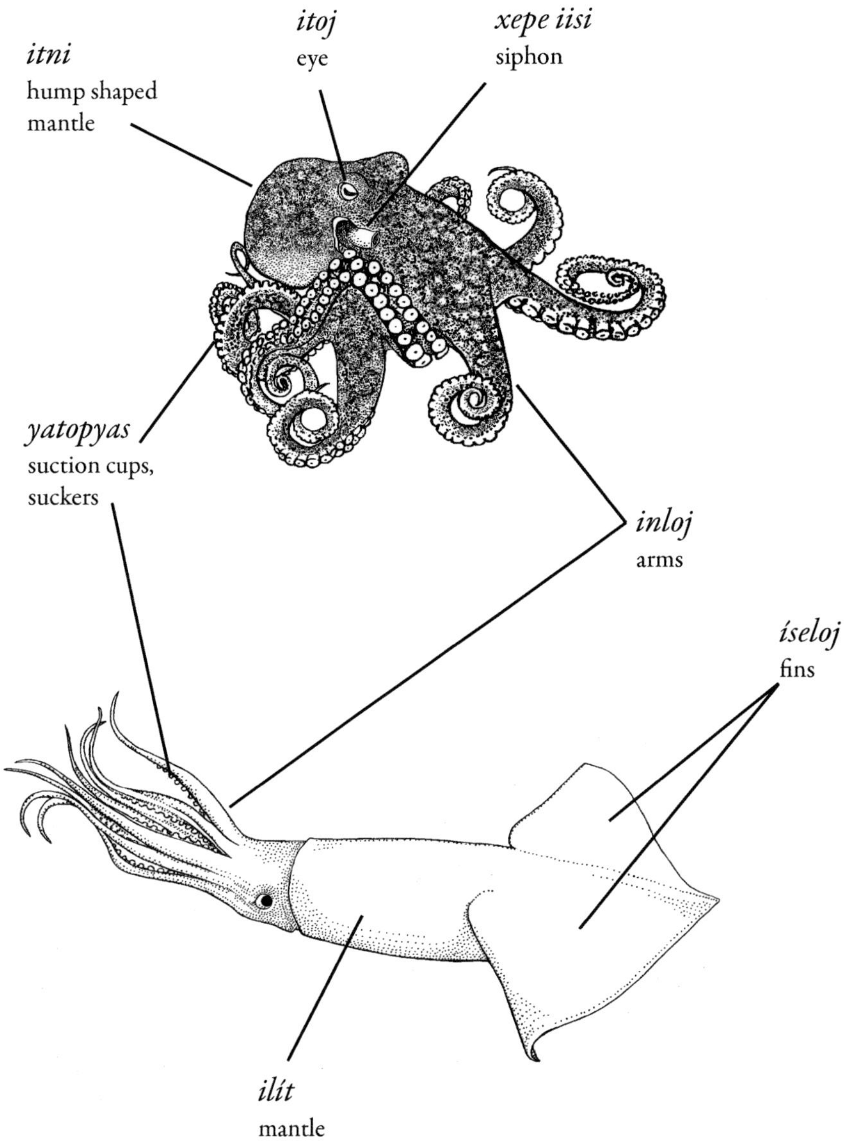

Figure 3.6. Anatomy of the cephalopods

The opercula of most snails are called *iti yahimoz* '[a hard thing] with which it covers itself'. The narrow, claw-shaped opercula of other snails, such as that of the *Strombus gracilior*, are described by the name *inoosj* 'its claw'.

Shapes and textures for shells utilize terms that are widely used and understood.[28]

Mythology and Folklore

The Seri origin story is somewhat variable, and although the central theme is fairly consistent, the order of events and minor participants vary with the person recounting it. While the main personages are generally known and named, there is some variation in their roles as well.[29]

Figure 3.7. Chavela Torres Molina, one of the *hant iiha cöhacomxoj*, loved to share the ancient knowledge of the *Comcaac*. Punta Chueca, 2008.

Hant Caai (the land maker) was the creator of the world. After the earth was formed, another prominent figure—to some people synonymous with *Hant Caai*—appeared on the scene. His name, *Hant Iiha Quimx*, alludes to his being the one who told others about the ancient things (literally, the possessions of the Earth). A third figure whose name is well-recalled is *Cmaacoj Cmasol* (yellow old man), who at a specific and known location on the northern end of Isla Tiburón, instructed the Seris as to what should be considered taboo and how cultural events were to be celebrated. He is most often credited with naming the plants and animals, although folklore accounts describing how animals were named often refer to *Hant Iiha Quimx* as being the name giver.[30]

The oral transmission of traditional knowledge is explained as a process begun by *Hant Iiha Quimx,* who gave his knowledge to others, the *hant iiha quimxoj* (those who tell about the ancient things). The *hant iiha quimxoj* in turn passed it on to the *hant iiha cöhacomxoj* (those who have been informed about the ancient things). Today a few Seris are recognized as *hant iiha cöhacomxoj* and, having the greatest understanding of the world of the *Comcaac,* they are entrusted to successfully pass the knowledge on to future generations.

An ancient and well-known song recounts the Earth's creation:

Hant Quij Cöipaxi Hac

Ha mat cmá tpa xi ma ya
Hant cmaa tpaxi ma
Za hi ti pá xi
Zaah iti ipaxi
Ha ma tí ya xa no sín la xói ton te ya
Hant quih ihiyáx com ano siinla xo, siitom
Ha ma ti ca hé ya
Hant Caai
Ha ma ti ca zé ne i ma ya
Hant Quizin
Tá ya za mó ya
Taa yazám
Ha ma ti xa cá no gui
Hant ixacaanoj
Ha mi mí xó tin
Hamiime ixootni
Hán ti ha quím xoée nec te
Hant iiha quimxoj oeenec.

About the Earth's Creation

When the Earth was newly created,
On the day it was created,
The noise would sound to the ends of
the Earth [and] it would speak.
Earth Maker.
Land Dryer.
[The Earth] took a step.
[The Earth] resounded with a boom.
[The sound] touched infinity.
[The story is] carried by "Those who tell
about ancient things."

As summarized by María Luisa Astorga, *Hant Iiha Quimx* (or *Hant Caai*) was present. The sea creatures were also there. The whale dove deep to the ocean floor and was tired. He slept. The sharks also descended; they were hungry and ate there. The male sea turtle swam down and picked up some earth and brought it to the surface—just a tiny bit of earth held beneath a claw of his flipper. He gave it to *Hant Iiha Quimx*, who blew it into the air, and the dry land appeared, separated from the sea.

In one version, *Hant Iiha Quimx* then sent out the daddy longlegs (Phalangiidae) to walk on the newly created Earth to see if it were solid, since his legs were long and light. *Hant Iiha Quimx* also sent the prince cone snail to test out the firmness of the land, since it could so lightly glide over its surface.

Before the animals were named, they were people. *Hant Iiha Quimx* asked each person what animal they wanted to become.[31] Some, including those to be mollusks, participated in their own naming process by comments they made. For example, in one of several versions of a well-known conversation, the differently shaped valves of the *Euvola vogdesi* scallop decided their names when one said, "You be the convex one, and I'll be the flat one." In the account about the giant spindle shell, *Fusinus dupetitthouarsi*, he felt overlooked and was tired of waiting to be named. He offhandedly commented, "Well, I guess I'll just stand over here," and he was given that phrase as one of his names.

Prominent in folklore and legends are the *xiica coosyat* 'singing things', considered to be Giants. The name is used when referring to unknown beings who inhabited the area in the past, who some say were Seris. Giants' speech, besides having a special vocabulary, is described as having a song-like quality. The Giants are described as very large, with huge heads and massive bones. For armor they hung rocks, such as grinding mortars, over their chests and backs and wore stone or shell ornaments in their noses and ears. Objects of especially large size, such as strangely formed rocks or prehistoric artifacts, are attributed to the Giants.[32]

The origin accounts include a Catastrophe, described as a cataclysmic event involving a great flood, fire, and earthquakes. Peculiar geologic formations are explained as having been Giants who turned into these at the time of the flood. For example, oddly shaped and weathered igneous peaks near Bahía de Kino had been a Giant and his wives. As they tried to escape the flood they were overtaken and became the peaks. Another transformation involves a mountain north of Desemboque on which fossil shells are abundant. The Giants climbing the mountain to escape the floodwaters became the shells. A final example is of a small rock island (Morrito de Turner) near Isla Dátil (Isla Turners), named *Hast Xnoois* 'eelgrass seed rock', that was once the young daughter of a Giant. When someone chided her she became sad and ran away, taking with her the baby of an older sister because the baby cried as she was leaving. When the Catastrophe occurred, the girl turned into this tiny island, and the baby became the ***teexoj*** (thorny oysters) found growing there. The account includes the report of a huge octopus lurking near the island.[33]

The Seri oral tradition includes a significant corpus of songs relating to the sea. Many refer to sea winds and travel, marine mammals, sea birds, fish and sea turtles, or smaller marine life. Some songs poignantly illustrate the somewhat tumultuous connection of the people with an ever-changing sea.

Ernestina Morales sang about some of the sea creatures:

Hant Quixoaa Cöicoos

Hai ní ji yái qui je me yí qui ja no mi ta
Hai iinj quih yaai iiqui xeemej, iiqui xaanim
Hai ní gui yái qui je me gui ta
Hai iinj quih yaai iiqui xeemej
Hi ya la co jo ma he ya
Hiyal haaco xooma
Hai ní gui yái qui je me gui ta
Hai iinj quih yaai iiqui xeemej.

Song about the Hermit Crab

The wind slowly veered away; it had
been blowing very hard.
The wind slowly veered away.
My companion is powerful!
The wind slowly veered away.

Ernestina explained the song:

> The nerite snails, the small swimming crab, the hermit crab, and other creatures were in the ocean. The strong, cold sea wind blew for eight days. The sea was churning, and the creatures were all going to die. The shell house of the hermit crab broke. He crawled out of it; he was tired. The old man ghost shrimp was in his home in the ocean floor. He was a shaman. He left his home, went to the hermit crab, put his arm around him and led him to shore. He made a fire and they sat by it. They sang this song, the wind stopped and the ocean became calm.

Figure 3.8. An amulet made of a ball of *csipx* and an Osprey talon. Collected by Gwyneth Harrington, 1941. Arizona State Museum, catalog no. E-980.

Beliefs and Practices

The supernatural played a central role in everyday Seri life, evidenced in attempts to appease malevolent spirits and seek protection from them, in seeking power to cure illness or curse enemies, or in soliciting good luck for hunting, fishing and harvesting.[34]

An early note made by Edward Moser accompanying a place name on Isla Tiburón reads: "*Ikx yáX* [*Iicj Yax*—CMM] The woman had the power of all of the shell fish [*sic*] in sea, she came paddling the balso [*sic*] with the man—she said the balsa was singing. They left from *ʔast án* [*Hast Án*—CMM]." Such brief references record a relationship with the sea and provide a tantalizing glimpse at a past world.

Some interactions involved mollusks, perhaps most obviously as amulets worn by young children.

Amulets

The youngest child of a pregnant woman, known as the *caal*, was singled out as needing special protection from the *Coen*, a spirit having the shape of a fat woman who lived in the camp and specifically targeted the

Figure 3.9. A child, Fernando Montaño, wearing a necklace of *Olivella dama* shells and a crab claw amulet. (The young man on the left is José Ángel Montaño.) Photograph by William Neil Smith, courtesy of University of Arizona Libraries, Special Collections. MS316_B27_f5_18.

Figure 3.10. Casimiro Comito (left) with his younger brother, who is wearing a necklace on which are strung *Zonaria annettae* shells and other items. Tecomate, Isla Tiburón, 1951. Photograph by William Neil Smith, courtesy of University of Arizona Libraries, Special Collections. MS316_B32_slb5_K.-209–8.

Figure 3.11. A child wearing a necklace of shell amulets. Bahía de Kino, 1935. Photograph by Laurence M. Huey, courtesy of University of Arizona Libraries, Special Collections. MS241_B4_f2_N-9308.

caal with hunger and general weakness. As one woman recently explained, the *caal* was especially vulnerable because the pregnant mother lacked adequate milk to nourish the child, and as milk was not commercially available, the child was anemic.[35]

As special protection, the *caal* wore amulets strung on a necklace. The amulets were such things as a ball of *csipx* formed from the lac of the creosotebush or brittlebush (figure 3.8),[36] the child's own dried umbilical stump contained in a tiny cloth bag, the stuffed scrotum of the small Harris's antelope squirrel (*Ammospermophilus harrisii*), dried octopus suction cups, or other uncommon objects such as the talon of an Osprey (*Pandion haliaetus*), claw of a badger (*Taxidea taxus*), or rattlesnake (*Crotalus* sp.) vertebrae.

Shells were often strung as amulets. Especially remembered was the uncommon and somewhat unusual shell of the Gulf sundial. Cowries, such as the easily found measled cowrie (*Zonaria annettae*) and the more scarce little deer cowrie (*Macrocypraea cervinetta*)—both strikingly polished shells—were used more commonly.

Not only worn by children, amulets were used by anyone. A necklace, called *hee yahaaho* 'antelope jackrabbit's trail', made of bleached *Olivella dama* shells strung with blackened stem sections of *haat* (*Jatropha cuneata*, limberbush) or *najcaazc* (*Asclepias subulata*, a milkweed) was worn for protection against rattlesnakes when walking in the desert (figure 3.12).[37] The name *hee yahaaho* evokes the droppings left by the jackrabbit (*Lepus alleni*), marking its trail. A similar strand could be worn around the ankle for the same purpose.

An obscure use for the pen shell byssal fibers, and possibly those of the mussels as well, was to wear them as some kind of amulet. The fibers were either entwined or formed into small bundles and hung on a necklace.

Roberto Herrera said that his wife recalled her mother had a saying when telling her to do a chore—to do it immediately so that "she wouldn't lose her ***seeten*** *imas*." He imagined it to be some kind of charm made from the byssal fibers.

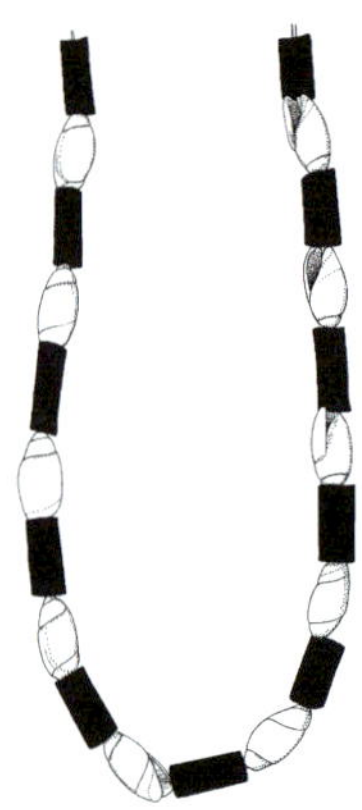

Figure 3.12. Detail of a necklace made with bleached *Olivella dama* shells strung with blackened plant stem sections.

Burial Practices

W J McGee wrote of shells used in mortuary practices, basing his conclusions on his excavation of a woman's grave at a camp he identified as Agua Amarillo [*sic*].[38] Among the items he collected from the burial is a pottery bowl containing shells and other items (figure 3.13). McGee speculated that the bowl contained "mortuary food" and suggested that the items were placed in the grave to provide the deceased with food for her journey.[39] He further proposed that another, unidentified shell was a "mortuary cup" (see figure 4.47, p. 112). McGee's interpretations remain uncorroborated by any researcher or Seri.[40] Indeed, several of the items in the bowl appear to be *Simomactra dolabriformis* shells; Seris who have seen McGee's illustration note that the shells were of a kind used as tools by a potter, adding that when the woman died her personal possessions would have been buried with her. Such a custom was still occasionally practiced in the 1980s, but has since been discontinued.[41]

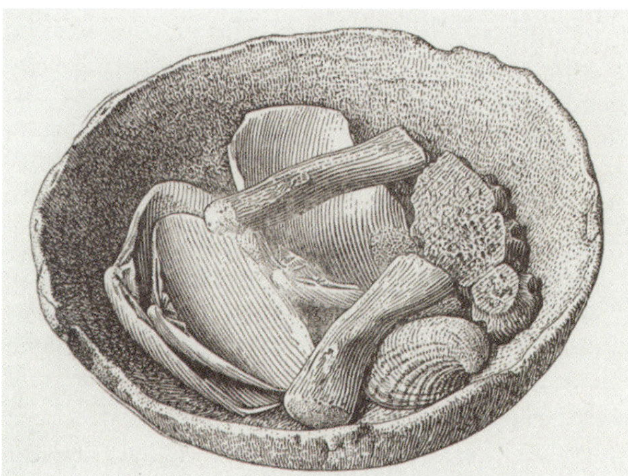

Figure 3.13. Illustration from McGee (1898:291, figure 41) of a pottery bowl containing putative "mortuary food."

McGee also mentioned seeing great numbers of shells, mostly *Arca* spp., covering a purported burial area on Isla Tiburón, and containing human excrement. He collected several such shells,[42] again speculating about their role in the deceased's journey. I have been unable to corroborate any such practice with the Seris.[43]

Before the current practice of capping a grave with cement, graves were commonly covered with such things as ashes, rocks, shells, cactus sections, or thorny brush to discourage scavengers (figure 3.14).

Figure 3.14. The grave of Santo Blanco (d. 1939), Bahía de Kino, covered by beach rubble including shells of clams and *Turritella* spp. The remains of two crepe paper-covered wire wreaths can be seen on the cross. Bahía de Kino, 1945. Photograph by William Neil Smith, courtesy of University of Arizona Libraries, Special Collections. MS316_B31_slb2_R.-108–8.

[redacted]

Figure 3.15. A wire wreath labeled a "corona," measuring 43 cm in diameter, decorated with *Oliva spicata* and *Anomia peruviana* shells. Collected in Desemboque in 1963 by Margarita Nolasco. Instituto Nacional de Antropología e Historia collections, catalog no. 35024(64)30.08b3–95.

A wire wreath strung with shells was collected in Desemboque in 1963 (figure 3.15).[44] The use of such decorations is a more recent custom, as Seri graves were traditionally undecorated. One person noted, "they would just bury the body and put brush on the top, and that was all." According to several people who saw a photograph of the object, the wreath was probably modeled after a crepe paper-covered wire wreath common on non-Seri Mexican graves. Rosa

Barnett said that in the recent past such wreaths were made with shells, adding that flower shapes made of jingle shells (*Anomia peruviana*) were sometimes used on the wire hoop.

Vision Circles

In the past, vision quests were undertaken by young people seeking spirit power to become shamans.[45] The solitary quest lasted three or four days and was usually done in a cave or brush hut away from camp, walking along the ocean shore, or in a vision circle most often constructed of stones, but that could also be made with shells.[46] According to María Luisa Astorga, the shells used were large murexes, *Chione californiensis* or "whatever there were a lot of around." She recalled as a child seeing such a circle north of El Desemboque, at *Hast Itliitxo* (Las Cuevitas), and being told by her father that it was a vision circle. The circle was formed with rocks and had an inner ring of *Chione californiensis* shells.

Stopping the Wind

When the fierce sea wind known as *hai xepe imac quiij* blew for many days, the people were unable to fish or easily collect shellfish. Various practices were used to stop the wind. According to Juanita Herrera, a person could draw eight crosses in the sand at the high tide line and then throw either four or eight heated sections of organ pipe cactus (*Stenocereus thurberi*) into the ocean. Bundles of leafy twigs from the elephant tree, brittlebush (*Encelia farinosa*), or *Frankenia palmeri*, a desert shrub, could be waved at the clouds. One might burn seaweed and then throw it into the sea, or make a fire on an exposed rocky intertidal area.[47]

A practice recorded by Edward Moser was still in use by older Seris in the 1950s. A live sea creature, such as an octopus, mussel, fish, or crab, was put into a fire. Before it was dead, it was removed and waved in the wind while speaking to the animal, and the following was recited: "*Hai quih inseemetim xo! Hai quih inseemetim xo!*" "Stop the wind! Stop the wind!" The animal would then be put back into the fire or, according to some, it could then be thrown into the sea. As one consultant graphically explained, the writhing of the creature in the fire is like the turbulence of the wind and ocean waves. When the creature (referred to specifically here as an octopus) died, it was quiet and peaceful—then the ocean would also be calm (see *Octopus fitchi*).

Soliciting Good Luck

Shells, most often of Venus clams, were inserted into rock crevices, trees, or columnar cactuses such as the cardón (*Pachycereus pringlei*) as an offering to solicit good luck from the spirit of the area or plant.[48]

When fishing by boat, sometimes a sea creature was placed in the boat for luck in bringing in a good supply of fish. An orange-colored starfish such as *Phataria unifascialis* was sometimes used as well. When a person rowing the boat was tired and sleepy, he would pick up the starfish and recover his strength. Such a starfish could also be kept in a house for good luck. Live sea turtle barnacles (*Chelonibia testudinaria*) were also placed in boats to attract the turtles to the boat to be easily harpooned.

A similar custom was to place *Nerita* spp. snails in a wooden boat for good luck. In groups of three or four, the live snails were placed inside the prow and stern and beside the ribs of the boat. Because the boats were not watertight, seawater often seeped into the boat; the snails would reproduce, and soon a colony would be established. This practice was said to protect the boat in stormy seas and bring good luck in fishing or hunting sea turtles. As one person remarked, "There was not one boat around without these snails in it!"

Food Avoidance (*Hapáx*)

In the process of completing the Seri dictionary, previously unrecorded words sometimes came to light. One such word was *hapáx*, which has the implied meaning that a food item was to be avoided, usually by children or women of childbearing age. There is little consensus about what things were *hapáx*, and the term seems to have almost disappeared, with only a few older people recalling it as applying to a specific food. As explained by María Luisa Astorga, in the past the concept that a certain food item was *hapáx* was generally known and the practice observed. She added that today there really is not any food considered *hapáx*, but instead what one eats is a matter of personal choice.[49]

In a practice still followed by many, when a girl began her menses she was to abstain from eating meat containing blood, including fish and sea turtle, for four days. Each month thereafter the practice continued and on through her childbearing years.[50] According to some, shellfish containing red-colored blood, such as an *Arca* sp. or *Carditamera affinis* clam,[51] were included. The large white-shelled *Dosinia ponderosa* clam was included by some as well; the clam's name, ***halít cahooxp*** 'what causes white hair', reflects the common belief that eating a lot of these clams would turn one's hair prematurely white. According to one woman, following puberty, women should not eat *Leukoma grata* clams with purple-hued beaks, because of the color's similarity to blood. A few people include the larger murexes, *Chicoreus erythrostomus* and *Hexaplex nigritus,* as being *hapáx*. One person mentioned *Codakia distinguenda*, a large clam with a pinkish-red interior margin, as to be avoided as well.

A woman of childbearing age was cautioned against stepping on or handling the sea hare *Aplysia californica* because of the violet ink released by the animal when disturbed. Since the color is similar to blood, it was said that touching the animal would cause heavy menstrual flow or hemorrhaging during childbirth. A woman was also cautioned against eating the peritoneum of a deer or other animal, since this would later block her birth canal, preventing birth.

Prepubescent boys were cautioned against eating the *Turbo fluctuosus* snail in great quantity, since doing so might cause their foreskin to retract and the head of their penis to be exposed. A boy was also advised against eating the tendons in the lower leg of a deer or its fatty bone marrow. If he ate such parts, as an adult his footsteps would make noise when he hunted, and any deer would be frightened and run away.

Finally, in a practice generally followed today, the gills of mussels are removed after cooking, as eating them is said to cause blindness (see Mytilidae).

Trade and Commerce

The archaeological record provides evidence of extensive trade of shells from the Gulf of California. However, there is no indication from the Seris today that shells had significant historical trade value or were considered prestige items.

A few sketchy oral accounts mention that shells were sometimes traded to people on non-Seri ranches within or near the Seri territory, and even as distant as the current state capital, Hermosillo, as early as the late nineteenth century. Most certainly, live shellfish were not included in trade to the latter, as the long distance and small quantity that could be carried would make this highly unprofitable (see Food).

Figure 3.16. An unusual photograph showing Seri women and children in what is probably Hermosillo, most likely before 1950. Photographer unknown, Moser collection.

The shell-related trade with nonnative peoples included pearls. In the past, according to Juanita Herrera, pearls from pearl oysters as well as those from mussels and pen shells were traded in Hermosillo.[52] However, this was probably done on a very small scale, since the Seris only collected pearls from oysters that were easily found on or near the shore (see Pteriidae).

The development of Bahía de Kino (also known as Bahía Kino) that began in the mid-1920s with the establishment of the Kino Bay Club, a hunting and fishing club started by an American, opened for the Seris a new venue of easy contact with the outside world. As more outsiders visited or settled there, the people took advantage of the opportunity to trade or sell their baskets and other crafts, as well as food they harvested from the sea.

Some individuals who frequented the area are remembered by name today. According to Raquel Moreno, a non-Seri Mexican man known as *Cahtxima Caacoj* 'Large Rich [Man]' bought shellfish from the Seris. In an anecdote humorously recounted by older Seris, an early fish buyer—some say a Japanese man known as *Cocsar Cahtxima* 'Rich Mexican'—supplied the people with water and firewood without charge. When buying their fish, he paid them with *Chione californiensis* shells, which could then be redeemed for other supplies he brought. The arrangement ended quickly.

Figure 3.17. Seris selling *Crassostrea* spp. oysters. Bahía de Kino, 1928. The oysters were probably gathered in the nearby Estero Santa Cruz. Photograph by Carlos Ronstadt. Arizona Historical Society, Ronstadt Collection, Indians-Seri, #49623.

Figure 3.18. A wood-burning truck procured by Roberto Thomson Encinas, dubbed *Trooqui Treen* 'Train Truck' by the Seris. Estero Santa Rosa, 1930. Photograph by Alfred L. Kroeber, courtesy of the Phoebe A. Hearst Museum of Anthropology and the Regents of the University of California, catalog no. 15–8729.

One man who traded clothing for empty shells was the American playwright Porter Emerson Browne, called *Cmaacoj Haari* 'Old Man *Haari*' by the Seris, named for his greeting of "Howdy, howdy!"[53]

With the onset of the commercial fishing industry in Bahía de Kino in the late 1920s, the Seris stepped into the then unfamiliar money-based economy and began selling to the outside world what was previously their own rich food supply from fishing, hunting, and gathering (figure 3.17). The Mexican government gave assistance through the contact of Roberto Thomson Encinas, a special friend of the people whose family ranched in the area. A federal school was established in Bahía de Kino, and a twelve-ton boat and a truck were supplied to the Seris for transporting fish and firewood (figure 3.18).[54] This increased involvement in a radically different economy introduced new factors and challenges into the isolated, traditional hunter-gatherer Seri society.[55]

In the following years, great quantities of fish, sea turtles, and shellfish were sold to commercial fish buyers through the Seri fishing cooperative, begun in Bahía de Kino in 1938 by a non-Seri Mexican fish buyer and moved to Desemboque in 1941.[56] In the early 1960s the people also began living in Punta Chueca, selling their catch directly to non-Seri Mexican buyers.[57]

Several older Seris recalled harvesting boatloads of the large *Crassostrea corteziensis* oysters in the mid-1940s from Estero Santa Cruz (see Ostreidae).[58] The oysters were sold in Bahía de Kino to individual buyers, who then took them by truck to Hermosillo. Other shellfish sold were *Crassostrea columbiensis* and *Myrakeena angelica* oysters, *Modiolus capax* mussels, larger murexes, and *Strombus* spp. snails. Venus clams from Estero Santa Cruz, and later, from an estero on Isla Tiburón were sold as well. Juanita Herrera also remembered an American man buying empty murex shells after the Seris had consumed the meat, sometime before the mid-twentieth century. This was probably an isolated incident.[59]

Xavier Moreno told me that for a period of seven years, beginning in the early 1980s, he and a few other men free-dove for thorny oysters (*Spondylus limbatus*) in rocky areas north of Desemboque, selling the raw adductor muscles to individual non-Seri buyers in the village, who then transported the meat for sale elsewhere. Xavier confirmed that previously the Seris had not dived for the purpose of collecting shellfish or any other food, nor to harvest pearl oysters.

Around 1973 the people began selling large quantities of meat (the adductor muscles) from pen shells they harvested in the shallow Infiernillo Channel between Isla Tiburón and the mainland, using a hookah apparatus. The system consists of an air compressor in a boat connected to a long hose that feeds air to a diver working on the sea floor. Still using the same system, the pen shell fishery continues on a large scale today, the meat sold to individual non-Seri buyers (see Pinnidae).[60]

Figure 3.19. Discarded empty pen shells at *Xana* (Campo Almon) in 1968. These pen shells were said to have been harvested at low tide, as the Seris were not yet involved in hookah diving. The meat was sold to non-Seris. Photograph courtesy of Richard S. White.

In addition to the extensive sale of pen shell meat, other shellfish are sold to individual buyers (both Seri and non-Seri) on a smaller scale and apparently only sporadically. These include *Spondylus limbatus*, *Strombus galeatus*, the two largest murexes, mussels, smaller rock oysters, and some of the Venus clams. Octopus meat is sold raw and whole, while the shellfish are usually boiled and the meat removed from the shell for sale. Since roasting the meat under a brush fire is time-consuming, such a cooking method is used only if the meat is for personal consumption (see Roasting under a Brush Fire, p. 51).

According to several people, in recent years pearls from the pearl oyster *Pinctada mazatlanica* were taken on a small scale by divers harvesting pen shells in the Infiernillo Channel and the pearls sold to individuals outside the community.

In the second half of the twentieth century, tourists and wholesalers visiting the two Seri villages provided an extensive market for the large quantity of shell jewelry made by the people. In recent years, that market has declined as fewer tourists visit the area, and today the jewelry (mostly necklaces) is often sold by the Seris in Bahía de Kino, Hermosillo, and even tourist-populated areas farther away.

Food

Just as the ocean tides in the Gulf of California followed a dependable rhythm, so did the world of the people who hunted, fished, and gathered the sea's bounty. In a land considered harsh by outsiders, the rhythms of the sea and the desert were well understood by those who lived there, their lives intimately joined to their environment.

As a constant supply of fresh water adequate for a large population was not available in this part of the Gulf coast, the Seris travelled to where water could be found, often digging for it in shallow holes. The people moved together in small groups, sharing the food they hunted and gathered; when the food or water resources in the area were diminished, they moved on.[61] Although closely tied to their places of birth, called *ihiizitim*, the Seris were not limited to these ancestral areas and moved beyond them—especially the men when hunting or fishing.

Historically, these smaller groups fit into six different larger groups or bands.[62] Although nearly all used the resources from both the desert and sea,[63] each band was known to be most knowledgeable about the resources found in its traditional area. Some people really knew the sea, and were expert at hunting sea turtles or adept at spearing fish; others were the inland people like the *heeno Comcaac*, who knew the desert and mountains well and were the expert hunters of land animals.

According to María Luisa Astorga, even while living away from shore, the people retained strong ties to the ocean. She explained that when the spring sun warmed the land, signs of new life showed in the desert. When the buds of the *hantipz íteja* (hedgehog cactus, *Echinocereus* spp.) opened into vivid pink blossoms, the people knew that it was time to go down to the sea, because the sea turtles were rich with fat and eggs.

As the inland mesquites leafed out in green, the ocean mirrored the desert's new life, becoming lush with seaweed and eelgrass (*Zostera marina*). The *Leukoma grata* clams would be fat, and the triggerfish (*Balistes polylepis*) would grow fat from eating them, searching for them heads downward on the ocean floor. The desert people knew that the time for good eating from the sea had arrived.[64]

Figure 3.20. A sign of the rich cactus fruit harvest to come; *Colpomenia tuberculata* growing on intertidal rocks. Desemboque 2007.

The ocean gave its clues about the desert as well. As the round, hollow green *xpeetc* seaweed (*Colpomenia tuberculata*) began to form on the intertidal rocks, so the people would know that the columnar cactuses would soon be budding, heralding the time of cactus fruit harvest. The mesquite flowers would form their pods in a promise of rich harvests, and the people would head inland in the heat of early summer, making their camps in the mesquite forests and cactus areas. In describing such anticipation, María Luisa laughed as she recalled people playfully sipping the liquid inside the small round sacs of *xpeetc* and draping bits of seaweed over their ears like earrings, in joyful anticipation of the coming bounty.

Then, as the seasons changed and the *hai xepe imac quiij* (a strong northwest sea wind) blew hard and cold against the shore, the people moved into the more protected areas near the Infiernillo Channel between Isla Tiburón and the mainland, where winter dormant sea turtles could be hunted on the shallow channel floors[65] and to the esteros where fish, crabs, clams, and oysters were plentiful.[66]

Even as the people moved with the rhythm of the seasons, there remained a constant supply of food lying under their feet—food that did not move far or quickly, nor fluctuate greatly with seasons or climate. That food was portable and could be kept alive for days; even a hungry child could find it on his own.[67]

The mollusk-rich shores of the Gulf of California stretched in a seemingly endless line through esteros and around islands, providing easily accessed food.[68]

Not just emergency food, shellfish were an important part of the Seri diet, eaten "for variety" even when other meat was available. For instance, when a fiesta was being held, different people were responsible for providing deer meat or fish, sea turtle, mussels, and clams. All foods were enjoyed.

Although many species were considered edible, some were especially favored. To any Seri asked, it is clear which of the mollusks were considered "real" food: mussels, oysters, and most Venus clams (especially *Leukoma grata*). Common in the area, Venus clams were easily harvested. A walk in the desert through old camps provides evidence of their importance, as their shells are found in great abundance, often in small heaps, remnants of meals shared.[69]

Some mollusks were eaten but were harder to harvest in quantity or did not occur as commonly. While some murexes found at low tide were heavily eaten, the large *Strombus galeatus*, because it was from deeper water, was not consumed as often. Such shellfish found as solitary specimens were collected together with other mollusks.

Large snails, such as the turban snails (Turbinidae) were commonly gathered at low tide for food. Others, such as the turret snails (Turritellidae), worm snails (Vermetidae), cones (Conidae), cowries (Cypraeidae), olives (Olividae and Olivellidae), or smaller snails were not eaten. When asked about eating such snails, one woman retorted, "Is the flesh seen?," indicating that their small size was not worth the bother.

The pen shells, although providing a good amount of meat in their large adductor muscles, grow partially buried and sturdily anchored in the low intertidal zone or in deeper water, and because their spiny shells are sharp, they are difficult to harvest. Consequently, the pen shells were often avoided.

The two Gulf pearl oysters are found in somewhat deeper water, so were not commonly eaten, except when high waves occasionally deposited live ones onshore and they could be easily collected. Similarly, the large and heavy-shelled thorny oyster, though not easily accessed, was eaten when found.

Some mollusks were consumed almost exclusively by adults—they were considered *hapáx* and to be avoided by younger people (see Beliefs and Practices, p. 41). Others, such as the button-sized clam, *Heterodonax pacificus,* were sometimes eaten by children "just for play." Some were occasionally eaten but were not considered good food: a small sharp-edged and bitter-tasting growth form of the rock oyster *Saccostrea palmula* was said to cause headaches and dizziness when eaten.

Neither the California sea hare (*Aplysia californica*) nor the much smaller and less common nudibranchs were eaten. However, recalled as food by a few older people today are a marine slug (*Onchidella binneyi*) and a glass-hair chiton (*Acanthochitona exquisita*), both of which were eaten raw, or cooked by boiling. Other chitons and the limpets were consumed as well, the latter said to be eaten as a last resort "when there was nothing else left to eat."

Octopuses were eaten, with the exception of the tiny *Octopus fitchi* and the *Octopus alecto*, the latter considered strange because of the way its long, thin arms easily detach. Although squids were not hunted nor often found, some people did say they had eaten them; however, people sometimes commented that they didn't like the meat because it smelled bad or irritated the throat. One woman added that she liked the broth only after the meat was cooked with tomatoes, garlic, onions, and oregano (*Lippia palmeri,* a native shrub).

In describing different mollusks with regard to taste or texture, attributes such as sweet (*coaatjö*), bitter (*cacat*), fat (*iiha capii*), mushy (*ctapjö*), rubbery (*coqueht*), or tough (*ctoozi*) are used. The expression *hateen hai cöimahoiitot* 'it knocks the air out of the mouth' describes an odd flavor or sensation, said to be similar to what happens when chewing gum with a strong mint flavor, and is associated with eating limpets, some of the mussels, and the large squid.

When people describe unpleasant physiological effects caused by eating certain shellfish, the expression *hapos iyahaaxat* 'it affects (irritates) the throat' is used, most often when referring to the limpets, large squid, and *Chama buddiana* clams. Others say they eat these without any such effect. Several Seris noted that some people experience headaches and stomach pain when eating too much of certain shellfish, such as the largest *Modiolus capax* mussels.

The Seris are aware of paralytic shellfish poisoning. Two anecdotes describe a historic event in which a group of people died from eating shellfish contaminated by *xepe án iheel* 'red (state) of the sea area' (red tide) (see *Leukoma grata*). Raquel Moreno also recounted a similar event in which a group of people was said to have been affected by eating such food and were "like they were drunk," but recovered. Similar toxic events are known as *xepe án imasol* 'yellow (state) of the sea area' and *xepe án ihamoc* 'darkness (state) of the sea area', one of which is said to have been responsible for the large pearl oyster die-off that took place north of Desemboque in 1956 (see Preface).[70]

Older people comment that the Giants ate anything from the sea. When talking about food today, one woman said that while everything in the ocean is edible and nothing is known to be poisonous,[71] there are things that are not eaten. For instance, when describing octopuses, she said that she did not eat the uncommon *Octopus alecto*, an octopus with easily detached arms: "Known things are eaten; we don't eat things we don't know." She then continued by saying that rather than having a young person eat something new or strange, one could ask a *yaxi imxizi* "[one] whose death is not mourned" to try it first—if he did not die, the food was okay to eat.[72]

The Seris harvest intertidal shellfish for their own consumption; however, with food easily purchased in stores today, it is not as commonly eaten as in the past.[73] When I asked various people about which shellfish they had recently eaten, the most common reply was "mussels." Others mentioned *Leukoma grata* and other Venus clams, the *Turbo fluctuosus* snail, a few larger clams such as *Dosinia ponderosa* and *Laevicardium elatum*, and oysters. A few mentioned octopus. Only one woman said that she had eaten *Arca pacifica*, a clam consumed in the past, as evidenced by its common occurrence in shell middens.

According to Efraín Estrella, desert plants are healthy when growing near a decomposed cow or horse. In the same way, clams will be fat at a shore were a beached whale decomposes. Mollusks found buried, such as the Venus clams, are described as being fat but sandy, while rock snails and oysters do not contain sand. And when talking about eating the animal, some people said they ate the whole thing, while others first discarded any sand-filled parts, especially those of the larger clams such as *Laevicardium elatum*. One woman commented that she did not like to eat that clam because "its intestines are too big." It all seems to be a matter of personal preference.

Cooking Methods

Mollusks were sometimes eaten raw (*quis*) both at the beach where found and after being carried back into a camp. Any found dead were discarded. Unless consumed raw, the animal was usually not removed from the shell prior to cooking.

Although men sometimes cooked meat or baked shellfish when away from camp, usually women prepared food. According to María Luisa Astorga, an archaic phrase for food is *ziix cmaam ocaa* 'thing that a woman looks for'.

The method used to cook the food depended on size, consistency of the meat, and availability of water, among other things. Different kinds of shellfish gathered at the same time were often cooked together (see Species Accounts for details).

Boiling

Shellfish could be boiled (*hapaznij*) in either fresh- or saltwater; when in the former, the broth could be consumed.[74] Seawater could be used for seasoning when no salt was available; it was sprinkled on meat or a small amount added to fresh water in frybread dough. Smaller clams or snails were boiled in the shell or eaten raw because they were too small for other cooking methods (figure 3.21). Sometimes the meat was removed after cooking and reheated with other ingredients such as vegetables or rice; today vegetable oil has replaced the highly favored sea turtle oil used in the past.

Figure 3.21. *Leukoma grata* clams being cooked. Desemboque, 2013.

Before metal vessels were available, the Seris used clay pots for cooking. Such foods as *xnoois haaztoj* (eelgrass flour gruel) and *haas cooil hapamjöc* (mashed green mesquite pods) as well as deer meat or seafood, could be cooked in such vessels. Efraín Estrella recalled seeing swimming crabs boiled in clay pots, and noted that, "the clay vessels made by the people long ago were thin, hard, and well made—now they aren't like that." Very large, thin-walled vessels were used to transport and store water, while the bowl-shaped, wide-necked *cojít* vessels were used for cooking purposes.

Roasting under a Brush Fire

A cooking method often used in the past involved roasting larger clams or snails in their shells on the ground, directly under a brush fire; such food was called *hamaai*. Sea turtles were also cooked in this way.[75] Plants used for the fire, called *hehe iti icamaai* 'plants that one roasts with', were the bushy and abundant *seepol* (*Frankenia palmeri*) or *hatajixp* (*Atriplex* sp.). When cooked by this method, the live snails or clams were carefully placed close together in rows in the sand, either hinge up or hinge down. A large quantity could be cooked in this way; one person indicated an area of around ten square feet. When the brush burned down and the shells were open, the meat was ready to eat (see Mytilidae). The meat, left in the half-shell, could be placed in a container such as a basket or sea turtle shell and easily carried (figure 3.22). If necessary, the meat could be rinsed in seawater or freshwater to remove any sand or ashes.

If one were only cooking a larger mollusk, such as a large murex or conch, and one did not have the time to collect brush, the whole animal could just be pushed into the coals of a fire, where it would be cooked quickly.

Figure 3.23. Children roasting mussels on the beach. Punta Chueca, November 2005.

Cooking with Coals

Another cooking method was to bake (*hapát*) the shellfish on hot sand beneath coals.[76] This method was used if one wanted to cook food quickly, since a fire that was already made could be used. The coals were scraped aside and a hole dug in the hot sand. The shellfish were placed in the hole and covered with the coals. When the fire cooled, the food was ready to eat. When water was available, the food might be rinsed off.

Figure 3.22. Freshly roasted mussels, placed in the shell of a sea turtle. Desemboque, 2013.

Figure 3.24. A wind-exposed hearth area on a dune north of Desemboque, 2007. The shells are primarily *Leukoma grata*, *Carditamera affinis* and *Tegula rugosa*.

Food was sometimes cooked on a grill or directly on live coals (*hapasni*). Deer or other meat could also be cooked on a wooden skewer, called an *hamquee*; food cooked by this method was called *hameequet*. Neither was a common way to cook shellfish.

Drying and Storing

Venison and sea turtle meat was often dried for later use. The meat of octopuses and squids was dried as well: octopuses were cooked, sectioned, and dried; large squids, an uncommon food, were sometimes filleted raw, salted, and dried. The meat of both was later rehydrated by boiling.

Other mollusks were not commonly preserved by drying; however, as recalled by one person, the meat of the ***haxöl*** clam was sometimes cooked, removed from the shell, dried, and later rehydrated and cooked (see *Leukoma grata*).

Oral history relates that a strange group of people from Baja California, thought to be Seris, sometimes visited the people on Isla Tiburón. The island people were struck by one of the visiting group's customs, that of transporting cooked, dried scallop meat for eating on their ocean voyages (see *Euvola vogdesi*).

Gathering Methods

There was no clear division of labor for harvesting mollusks; they were gathered by men, women, and children. However, shellfish found nearer camp and close to shore were most often gathered by women and children. The harvesting of mollusks found farther away or requiring heavier work involved men as well. For instance, only men harvested mussels in deeper water by spearing clumps of them from a boat, and before trucks or cars were available, octopuses hunted from remote beaches in larger quantities for sale were transported in boats piloted by men.

When collecting mollusks for use in shell jewelry, only women, girls, and very young boys participated. Live mollusks or their shells found in beach drift were collected. One person remarked strongly that older boys and men would not do such collecting.[77]

When women foraged for shellfish or shells within view of their camp, they often were accompanied only by their young children (figure 3.26). However, when going farther away, such as in the desert in quest of plants or firewood, they went as a group. And, according to René Montaño, they often sang in order to prevent the emergence of dangerous supernatural creatures living underground.

Large quantities of shellfish were harvested for sale during the early to mid-twentieth century. Men, women, and children participated. Handmade wooden boats piloted by men ferried people into esteros to harvest clams and oysters since the boats provided easy access to remote areas and could carry large loads. Clams buried in sand, such as the Venus clams or cockles (Cardiidae), were collected by digging with a sharpened stick or large shells. One woman recalled using a metal rake to harvest the clams (see *Chionista fluctifraga*).

Figure 3.25. Gathering live snails on Isla Tiburón, 2010. Photograph courtesy of Lynn Johnson.

Figure 3.26. Rosa Flores digging up *Leukoma grata* clams. Desemboque, 1946. Photograph by William Neil Smith.

In the past, an area could be claimed by a person in his *ihiizitim* (traditional home area or birthplace) so that no one else could use it. One consultant explained that if he wanted to keep such a place from being used by others, he might say that something dangerous, like a large snake, mountain lion, or ghost, had been seen there.[78] The claimed area might be a good place to harvest desert fruit or plants, or it might be a beach or area in the ocean where there was an abundance of fish, sea turtles, or shellfish. Several people recalled one man claiming a mussel bed near Punta Chueca in the mid-twentieth century (see *Modiolus capax*). Since such a custom was no longer in common practice, the occasion stood out as memorable.

Hunting and fishing were considered male activities,[79] and in general, shellfish were not harvested by men while involved in such work. According to María Luisa Astorga, men only ate the meat and fish they procured while away from camp. Her husband, Efraín Estrella, concurred, saying that there was little time to cook shellfish. However, he recalled eating clams raw in an estero while on a fishing trip, but said that shellfish were not gathered at such times to take back to his family. Another man, however, said that if there were nothing else to eat while away from home, he would often eat shellfish, even taking the time to roast the food under a brush fire. He then added that sometimes he did collect shellfish to take back to his home and family.

Figure 3.27. Cleotilde Morales and her granddaughter Claudia digging up *Heterodonax pacificus* clams near *Xoop Cacöla quih Inoohcö*, 2010.

Table 3.6. Seri word or expression used for collecting mollusks

Term	Type of mollusk
quiptx, to pick up from the surface	***xiica cooxp*** (*Olivella* spp.) and other small snails on sand or rock surfaces, as well as shells in beach drift
quip, to dig up	***haxöl*** (*Leukoma grata*) and other buried clams
capóc, to pry loose	***stacj*** (*Saccostrea palmula*) and other attached oysters
an hant cöcoiitim, to dive down into	***seeten*** (Pinnidae), from deeper water
caai, to detach	***satoj*** (Mytilidae) and other attached mollusks (including larger snails) that are at least partially visible and do not have to be removed by digging
csaii, to pry loose (mussels) with a spear	***satoj hasaiitaj*** (*Modiolus capax*) mussels in clumps from deeper water
quicö, to kill	***hapaj***, octopus
caahit, to fish with hook and line (although this can generally mean to fish)	***pajaas***, squid

Since the people did not dive for food, most mollusks were collected at low tide, with the exception of mussels taken by spearing from a boat. Harvesting mollusks involved various methods, from surface gathering to the more difficult task of collecting those found submerged. Specific terms were used for each type of harvesting.[80] All these terms, with the exception of *csaii* 'to pry up (mussels) with a spear', are also used for collecting objects other than mollusks.

Clams found in soft sand could easily be collected with bare hands (figure 3.27). Those found in harder substrates were dug up using sturdy shells such as the giant cockle ***xtiip*** or a wooden prying stick called *hehe hapxeezc* 'wedge-shaped stick'. Since metal has become available, metal implements or large spoons have often been used.

Oysters found attached to mangrove roots or rocks were broken off with sharp rocks or large knives.

Octopuses, usually the common *Octopus bimaculatus* and *O. hubbsorum*, were hunted at low tide. In the recent past, a metal bar was used to pry up large rocks. After they were killed, the octopuses were strung on a loop of wire drawn through their muscle for easy carrying. Presently, octopuses are captured by using a metal rod with a hooked end and are carried in a mesh bag or plastic bucket.

Containers and Transport

In the past, women gathered shellfish and transported them in baskets placed on headrings and balanced on their heads.[81] These shallow, bowl-shaped baskets were made without elaborate designs, as they were used for work rather than for sale. Clay vessels or handsewn cloth bags were used as well. Later, in the early to mid-twentieth century, containers were made from commercial five-gallon tins into which a wire or wooden handle was fixed for carrying. The clay vessels, metal containers, or bags could be hand-carried, transported by women on their heads, or suspended from the ends of a wooden pole or carrying yoke called a *peen*, used by men (figure 3.28). When they were to be used with a *peen*, the clay vessels were suspended in net bags made from mesquite root twine or cotton cord.

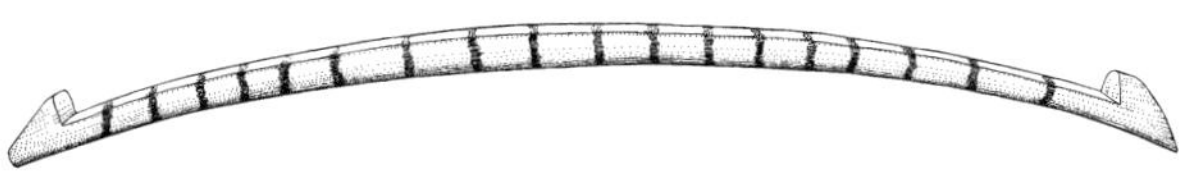

Figure 3.28. A *peen* (carrying yoke) made from mesquite wood, painted with a design to identify it to its owner.

Some shelled mollusks could be kept alive for several days.[82] They were not washed, but kept together in some of the moist sand in which they were found, wrapped in a damp cloth or covered with moist seaweed and taken back to camp. Oysters, mussels, small clams, and some snails were transported for later consumption in this way.

Angelita Torres recalled a time from her childhood on Isla Tiburón:

> We were at *Hajháx* [Tecomate], at *Xneelcam* [an inland camp]. My father gathered lots of ***haxöl*** [*Leukoma grata*], ***quiit*** [*Carditamera affinis*], and ***cotopis*** [*Turbo fluctuosus*], and he carried them [back]. He got to the desert camp where we were and we cooked them in a metal pot together with sea turtle oil (because we didn't have any 'white fat' [lard]) and oregano. We were poor; that is where we lived, in the desert near *Hajháx*. We wanted to drink coffee, but it was bitter. We used *xpajoocsim* [*Batis maritima*] root to make it sweet.[83] That is what we drank. We were poor.
>
> We carried the live clams tied up in cloth. We [also] carried them to *Siimen Hax* [a camp at a water hole on the east side of Sierra Kunkaak], far away. It took two days to walk there. We kept the clams in the wet sand that we found them in because when you wash them off they soon die. My father carried them in a sack made from old blue [denim] pants.
>
> ***Xiica cooxp*** [*Olivella dama*] were like that; there were lots of them. [Carried] like that, they didn't die. Two or three days would go by, and if they weren't washed off but still had seawater on them, they didn't die. Just like that, they were there. Then, a long time later, maybe three days, they were cooked [to remove the animal before stringing the

Figure 3.29. Jesús Morales carrying metal containers of *Crassostrea* spp. oysters for sale. He is not using a carrying yoke, but rather a simple pole. Bahía de Kino, 1929. Photograph by Carlos Ronstadt. Arizona Historical Society, Ronstadt Collection, Indians-Seri, #49624.

shell] and died on the fire. "Poor things, they were killed!"

You could keep [shellfish] in seaweed. ***Xnoocat*** [large murexes] and ***satoj*** [mussels] from *Ziipxöl Iifa*, *Xoop Cacöla quih Inoohcö*, and *Quipcö Coospoj* [seaside camps on the mainland shore of the Infiernillo Channel] were kept like that.

Utensils and Tools

In a world without china, plastic, or easily obtained metal, a supply of vessels, utensils, and tools usually lay within easy reach. In his report of visits to the Seri area in 1894–95, W J McGee elaborately observed:

> The marine shells applied industrially comprise the prevailing local genera, *Cardium*, *Mactra*, *Arca*, *Chama*, and others. They are used ordinarily as drinking-cups, dishes, dippers, receptacles for fats and facepaints, and as small utensils generally; and they are used nearly as commonly for scraping skins, severing animal and plant tissues, digging graves and waterholes, propelling balsas, and especially for scraping reeds and sticks and okatilla [*sic*] stems in the manufacture of arrows, harpoons, bows, balsas, and jacal-frames—indeed, the seashell is the Seri familiar, the ever-present handmate and helper, the homologue of the Anglo-Saxon Jack with his hundred word-compounds, a half-personified reflex of habitual action and thought. . . . The abundance of abandoned shells about the rancherias and camp sites, and over all Seriland for that matter, indicates that the objects are discarded as easily as they are found along the prolific shores.[84]

Without a doubt, the most extensively used shell was that of the giant cockle *Laevicardium elatum*. Large, sturdy, and portable, these became bowls and cups,[85] pigment containers, digging tools, and dippers (figure 3.30). They could be transported or buried and kept for a return visit to a desert camp.

Shell spoons were plentiful. In fact, sometimes shells such as those of mussels were carried in quantity in pockets; as Eva López commented: "When one broke, another was just taken out and used!" Even the Giants were said to have conveyed the shells of the largest mussels, hung over their ears as if on hooks.

Figure 3.30. Sara Villalobos and her children eating what is probably a gruel made from toasted and ground seeds of *ziim* (*Amaranthus* sp.) from a basket, using shell utensils. Bahía de Kino, 1929. Photograph by Edward H. Davis, courtesy of the National Museum of the American Indian, Smithsonian Institution (N24008).

The shells of other clams served as spoons as well.[86] The word *quihaxöl* 'to have an ***haxöl*** (*Leukoma grata*)' (in this case used as a general term for clam) means to use a shell for a spoon. A modern spoon is called *eenm haxöl* 'metal ***haxöl***'.

Twine made from the bark of mesquite roots was prepared by pounding the bark with a rock or the edge of a large shell in order to soften and separate the fibers so they could be twisted into twine.[87] One shell mentioned today as being durable enough for such use was the *Glycymeris gigantea* shell, considered sturdier than that of the *Laevicardium elatum*.

Any large or medium-sized shells such as those of pearl oysters, mussels,[88] scallops, and even larger limpets served as pigment containers.[89] The most commonly used paints were red from a pulverized ochre mineral or the blood-red sap of the elephant tree; a muted blue pigment made from natural ingredients or the intense blue commercial bluing powder; white made from gypsum; and yellow from various plants (figure 3.31).[90]

Before metal was available, razor-sharp shells were used to butcher deer (figure 3.33), scrape meat from the plastron of a newly roasted sea turtle, and clean an animal hide in the tanning process.[91] Such sharp shells were also used to cleanly slice through the umbilical cord of a newborn.[92]

Figure 3.31. A woman, identified as María Juana Necia, applying face paint contained in *Laevicardium elatum* shells. She is holding the shell containing the sap from the elephant tree (*Bursera microphylla*). Note the three different pigments. Tecomate, Isla Tiburón, 1946. Photograph by William Neil Smith, courtesy of University of Arizona Libraries, Special Collections. MS316_B31_slb1_R.-123–16

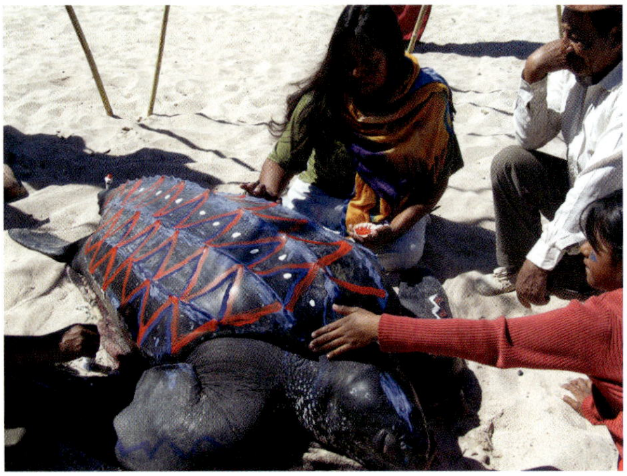

Figure 3.32. María del Carmen Hoeffer painting designs on a leatherback sea turtle (*Dermochelys coriacea*) during the four-day fiesta that is celebrated when such a turtle is found. She is holding a *Nodipecten subnodosus* shell containing the red pigment. Other shells used during that fiesta were those of *Laevicardium elatum* and *Dosinia ponderosa*. Desemboque, March 11, 2011. Photograph courtesy of Mayra Estrella Astorga.

Figure 3.33. José Torres using a mussel shell to butcher a deer. Isla Tiburón, 1936. Photograph by Edward H. Davis, courtesy of the National Museum of the American Indian, Smithsonian Institution (N20240).

Shells played an important part in the manufacture of clay vessels. Most often used was the shell of the *Simomactra dolabriformis* clam, a potter's special possession. The shell was used to thin the unfired clay to achieve strikingly thin-walled vessels,[93] as well as smoothly polish the exterior of the vessel or other clay objects, such as figurines (see figure 4.48, p. 113).

The narrow, elongate *Tagelus* shells were also used in working clay and were described as being used to shape a narrow area of a clay vessel.

The mouths of large clay vessels used for seed storage were covered with pottery, a flat rock or a large shell such as *Laevicardium elatum,* and sealed with *csipx* (a creosotebush lac).[94]

The pointed shell of a turret snail (*Turritella* spp.) could be used as an auger to make holes in soft material such as a deerskin.

A *Strombus galeatus* shell was used as a "trumpet" to summon people or warn them of danger. Such a shell was used during the last half of the twentieth century by the Seri Apostolic church in Desemboque to announce its meetings. A hole made in the shell's apex was blown into, creating a far-reaching and easily heard resonant sound.

Shells were used to scrape the flesh of a barrel cactus (*Ferocactus tiburonensis* and related species) in order to obtain liquid for drinking when water was scarce (figure 3.35).[95] Named for this, and most commonly used, was the shell of the ***siml ihataxoj*** 'barrel cactus scraper' (*Trachycardium procerum* [and possibly *T. consors*]), although other shells could be used as well.

Figure 3.34. *Simomactra dolabriformis* and *Tagelus* sp. shells, possibly cached, in a desert camp near Bahía de Kino, 2009.

Figure 3.35. Alberto Molina "Vaquero" (left) and Guadalupe Astorga (right) scraping barrel cactus flesh to obtain juice for drinking, 1929. The event may have been staged since the photograph shows the toxic cactus *Ferocactus emoryi*, which was not used (Richard S. Felger, personal communication). Photograph by Edward H. Davis, courtesy of the National Museum of the American Indian, Smithsonian Institution (N24032).

Older men and women smoked tobacco in pipes made of clay, a section of reedgrass (*xapij*, *Phragmites australis*), or even a piece of pelican bone. Sometimes a section of a large worm snail shell, called *hapis ano cöihiip* 'what tobacco is in' was used. Of five native plants that were used for smoking, the tobacco especially favored was *hapis áa* (*Nicotiana obtusifolia* [= *N. trigonophylla*]) from an area north of Desemboque called *Hapis Iihom* 'where tobacco is'. The leaves were dried and stuffed into the hollow tubes with large bird quills. Also smoked was the *haaxat ihixcoocoj*, the leaf gall on the creosotebush.[96]

The two valves of a clam held closed together were sometimes used as tweezers for plucking whiskers. This past use of clam shells was noteworthy enough that some men who did this are remembered by name today. Rather than clam shells, a few older men still use the folded lid of a metal can.

Finally, because of their abundance and availability, shells were used as hygienic implements for toilet use (see *Arca pacifica*).

According to the Seris, shells were not made into fishing hooks, something supported by their absence in the region's archaeological record.[97]

Adornment

Historic Use of Shells for Personal Adornment

The earliest recorded description of the Seri people that mentions the use of shells for personal adornment was made in 1678 by Father Ortiz Zapata, a Jesuit missionary. While meeting with some Seris regarding a mission settlement at the site of Pópulo, Sonora, Ortiz Zapata noted that two Seri men showed him some pearls that they reportedly obtained from other Seris who were "well acquainted with the sea"; one of the men wore two strung mother-of-pearl shells around his neck.[98]

In 1692 a Jesuit missionary living at the Pópulo mission, Father Adamo Gilg, wrote a letter in which he gave a detailed description of Seri personal adornment:

> Now, I will describe my Seris' manners and customs. Many of them die soon after birth because they not only pierce the ears and nasal septum of their new

born babies, but they also sharply slash their faces about the eyes and mouth with thorns, so that after the wounds have ceased to bleed little black dots or flecks remain from the scars, which are their most beautiful adornment. In their pierced ears and noses, they put either a precious stone or a pretty little piece of shell, but adults, instead of that, stick a skewer under the nose.[99]

In the center nasal cartilage hangs a blue stone or a skewer and in the ear lobes either shells or ribbons of different colors. . . . At the neck hangs a beautifully polished round shell, together with all sorts of grains, fruits, beads, and other sorts of childish playthings.[100]

Figure 3.36. Drawing by Father Adamo Gilg in 1692. Note the ear and nose ornaments as well as necklaces, bracelets, and anklets. Original in the Central Jesuit Archives in Rome (Archivum Romanum Societatis Jesu, Boh. 108).

Gilg provided a treasure in a drawing he made with a map and enclosed with the letter, showing two Seri men, a woman carrying a baby on a cradleboard on her head, and a child, walking through the desert. This early illustration provides details of the Seris' attire, possessions and personal adornment, and the first depiction of the Seris using shells (figure 3.36).

In describing Seris living at Pópulo in 1749, Father Tomás Miranda, also a Jesuit missionary, noted:

> The Indians are tall and very dark and take great pains with their appearance. They fill their hair with wildflowers, wear eight or ten earrings in each ear and a blue stone in their nose, and sport necklaces and bracelets of different colors and shells.[101]

A great number of prehistoric ornaments found in the region are of shell manufacture.[102] The people call the common shell disc beads *haapxij*, a term possibly derived from ***haapx*** (*Tegula rugosa*).[103] More unusual are pendants or beads of other shapes, such as the tubular beads formed from short sections of worm snail shells. Turquoise disc beads are called *yom*, a term also given to naturally perforated shell 'beads' common in beach drift.[104]

Prehistoric artifacts made with shell or rock inlay are unknown in the area, and the Seris today do not know of such shell use historically, nor have they described any etching process they used for shells in the past.[105]

The Seris did not use pearls, precious stones, or metals of their own manufacture for personal adornment, although they did incorporate found metal or glass objects, and clay beads that they made. Most often, however, they took what was provided by the world around them, using an assortment ranging from plant parts—seeds, flowers, sectioned stems, and roots—to snake vertebrae, crab claws, seaweed, and even sections of lobster antennules.[106] Octopus suction cups, round and containing a natural center hole for stringing, were occasionally strung, as they were easily removed from the cooked animal. However, shells were without a doubt the most popular material for use as jewelry, since they were sturdy, abundant, and easily gathered and prepared for stringing. Shells or other items were strung on sinew or prepared plant fibers, such as fiber from mesquite roots. Later, cotton string or thread was used, and today most people use nylon fishing line.

It is clear from early photographs and collections of historic Seri adornment that the most commonly used shell was that of the *Olivella dama* snail. Two other shells often strung with it were *Trivia solandri* and *Anomia peruviana*.

Necklaces were worn by anyone regardless of sex or age. Men often sported necklaces (figures 3.37 and 3.38); very commonly used were pendants of shell or stone hung on a string.

Long strings of clay beads, called *hatleen oo hapácatx* 'what is put on the upper chest', were worn crisscrossed under the arms and over the chests of unmarried men to show that they were eligible for marriage.[107] Strands of shells were worn in such a way as well (figure 3.39).

Figure 3.37. Chico Romero wearing a necklace made of alternating natural and blackened plant stem sections, 1924. Photograph by Edward H. Davis, courtesy of the National Museum of the American Indian, Smithsonian Institution (N23817).

Figure 3.38. Chico Romero wearing a strung pendant, possibly a shell, in 1930. Photograph by Alfred L. Kroeber, courtesy of the Phoebe A. Hearst Museum of Anthropology and the Regents of the University of California, catalog no. 15–8742.

Figure 3.39. Jesús Montaño (1eft) wears strands of *Olivella dama* shells strung together with blackened plant stem sections. José Ángel Montaño is on the right. Isla Tiburón, 1949. Photograph by William Neil Smith, courtesy of University of Arizona Libraries, Special Collections. MS316_B31_slb4_R.-185–4.

A necklace named *hee yahaaho*, 'antelope jackrabbit's trail', was made with bleached *Olivella dama* shells alternating with the small black seeds of the *xneeejam is hayaa* (*Viscainoa geniculata*) shrub or blackened sections of limberbush or milkweed stems. Such a necklace was worn as protection from rattlesnakes while walking in the desert. A similar strand worn as an anklet and used for the same purpose was called *hapij hacaaix* 'what is put on the lower leg'.

Hatbands, *haaonam ihaafin* 'what goes around a hat', were sometimes made of *Trivia solandri* shells strung with *xneeejam is hayaa* seeds. Hats were also decorated with shells sewn directly onto the hatband.[108]

A belt, *hatj ipaxquim* 'what is put (a long thing) around the waist', worn for adornment rather than support and used by either sex, was sometimes made of *Trivia solandri* shells, strung side-to-side.[109]

Earrings, *hasitcoj*, made from small *Codakia distinguenda* shells, were considered especially attractive because of the shells' red interior rim. Shells of the globular bubble snail (*Bulla gouldiana*) were strung on threads and hung from the ears, where it was said they would make a pretty sound as the wind blew them together. Other shells such as *Olivella dama* and *Trivia solandri* were used as earrings as well.

Sometimes shells were used to decorate objects. A doll made from the cranium of a pelican was called *toohit imcám* 'what eats but doesn't get full', so named because sand fed into a hole ("mouth") would sift through the cavity and come out the opposite end. Small *Olivella* spp. and *Trivia solandri* shells were used as earrings on the doll (figure 3.40).

Figure 3.40. A doll made from the cranium of a pelican wrapped with cloth, with earrings made of small *Trivia solandri* shells strung together with *Viscainoa geniculata* seeds. Made by María Luisa Chilión around 1970.

A unique object in the Arizona State Museum collections is the dried plastron of a juvenile hawksbill sea turtle (*Eretmochelys imbricata*), with head, tail, and flippers still attached (see figure 5.9, p. 146). Strung near each front flipper is a *Zonaria annettae* cowrie, giving the appearance of earrings. The museum record notes the object as a fetish, something corroborated by several Seris who have seen the photograph, although one person suggested it might have been used as a doll.

Shells were not used as buttons.[110] When I asked Lorenzo Herrera whether shells were used to decorate clothing, he recalled his grandfather telling him that long ago they were sometimes sewn onto clothes with sinew.[111] Lorenzo added that sea turtle barnacles (*Chelonibia testudinaria*), with their distinctive star-like pattern, were well-cleaned to whiten them, and sewn as decorations onto the muslin *coton*, an early type of shirt. The barnacles also decorated deerskin arrow quivers. Another person, Chavela Torres, told me that long ago deerskin clothing was ornamented with *Heterodonax pacificus* and *Theodoxus luteofasciatus* shells.

Present-Day Personal Adornment

Early photographs show the use of necklaces made with shells as well as commercial beads, and in later photographs from the early to mid-twentieth century it is clear that the people often wore the necklaces they made. However, in today's tourist market, priority is usually given to sales. Sometimes, though, a special shell is kept and worn. In 2007, Juanita Herrera wore a shell pendant she found in an abandoned camp, made from an unusually thick *Anomia peruviana* shell (see figure 4.23, p. 97).

As was previously noted, the most commonly used shells in the past were those of *Olivella dama, Anomia peruviana*, and *Trivia solandri*. Since the mid-twentieth century many other shells, mostly snails, have been strung on necklaces for the tourist market.[112]

More recently, less traditionally used shells have been strung, such as the round and flat valves of the *Euvola vogdesi* scallop. Occasionally strung are such things as the opercula of the murex and strombus snails.[113] Since many people now own vehicles, beaches that were rarely visited are easily reached, making more unusual shells accessible, and so a greater variety of shells is used today (appendix 5). The hardness of a shell no doubt determines its suitability for stringing, as some shells cannot be easily perforated by hand.

Figure 3.41. Necklaces made for the tourist market.

The Seris enjoy bright, vibrant colors. Small colorful cloth bags containing a fragrant herb such as desert lavender (*Hyptis albida*), *xeescl*, are sometimes sewn onto shell or seed necklaces, and were traditionally used as amulets. Since the 1970s, snail shells colored with vivid commercial dyes have been used in necklaces as well.

The *Olivella dama* snails were often bleached by toasting them in hot sand; such bleached shells are referred to as *hacaaat* (see Olivellidae). In a modern twist, they are further whitened by soaking them in a commercial bleach solution. Today, other snail shells, commonly those of the nerites, are bleached in this way as well.

Small snails containing hermit crabs and gathered from rocky shores are spread out in the sun to dry. The dead animals are then removed with a needle (figure 3.42). When the snails are gathered live, they are boiled, and the dead animals are removed in the same way. The limpets used in necklaces are prepared by pouring hot water over the shells; the soft parts are then easily slipped out.

María Luisa Astorga described a method she recently used to clean large quantities of tiny snails: she sprinkled the dead snails with sugar and buried them. After about a week she unearthed them; they had been completely cleaned by ants. After a quick rinse in water with a bit of bleach they were odor-free.

Figure 3.42. A needle is used to remove the dead animal from a dove snail. Desemboque, 2006.

Figure 3.43. Using nail clippers to remove the apex of a *Mitrella ocellata* snail in preparation for stringing. Desemboque, 2006.

In preparation for stringing, the apex of the snail is removed by gently tapping it with a rock; it is then smoothed by rubbing the tip against a rock or sandpaper. However, a nail clipper is used to remove the apex of tiny snails (figure 3.43). The shells of chitons, limpets, or larger snails are punctured by placing them on a board and piercing them with a sharpened metal awl. The shells are then strung on nylon fishing line.

Sharing of shells among women is common. If a large quantity is gathered, or unusual shells are collected from a distant or special beach, they are often shared with relatives or friends.

Although historically the people did not use a shell inlay technique, a few resourceful artisans have learned the process, setting pieces of shell into ironwood or other material using a commercial resin matrix. The shell most often used is that of *Pinctada mazatlanica*, as it is the thicker-shelled of the two Gulf pearl oysters, and is easily worked with a hacksaw and file. Occasionally a nonnative shell such as abalone is used as inlay or shaped into pendants.

A recent and unusual use of the *Lottia* spp. shells is as overlay on small carved wooden sea turtles, in imitation of a turtle shell (figures 3.44 and 3.45). Another recent use of the *Lottia dalliana* limpet involves cutting the shell and gluing the pieces into the shape of a sea turtle onto fragrant herb-filled cloth bags which are then fastened onto a necklace.

Figure 3.44. Pendant carved from *Bursera hindsiana* wood overlaid with *Lottia dalliana* shells. Made by Francisca Romero, purchased by Carolyn O'Meara in 2007. Photograph courtesy of Scott Gable.

Figure 3.45. Earrings carved from *Bursera hindsiana* wood overlaid with *Lottia stanfordiana* shells. Made by Francisca Romero, purchased by Carolyn O'Meara in 2007. Photograph courtesy of Scott Gable.

Other Decorative Uses for Shells

Few images from the past show the use of shells to decorate living spaces. In what is more than mere decorative use, one photograph taken in the 1940s shows fetishes hung in a brush shelter on Isla Tiburón, together with strings of white *Olivella dama* snails strung with blackened sections of plant stems (figure 3.46).

Figure 3.46. Strands of *Olivella dama* shells and wooden santos made by Ramón Montaño, in a brush shelter. Isla Tiburón, 1947. Photograph by William Neil Smith, courtesy of University of Arizona Libraries, Special Collections. MS316_B29a_f10_25–8.

Today one often sees houses or porches decorated with strung shells (figure 3.47). An unusual object in the Arizona State Museum Seri collection, composed of strings of shells, is identified by a few people today as a decoration hung from windows and doorways of non-Seri houses (figure 3.48). The record describing the item notes that "according to Mr. and Mrs. Moser these ornaments are made to be hung in doorways to keep insects out of the house. Made for Mexican

Figure 3.47. Modern shell decorations made with beach-worn *Turbo fluctuosus* shells. Desemboque, 2008.

Figure 3.49. A rock sculpture in the form of a snail shell made by Miguel Barnett, purchased by Irving Davis in Desemboque, 1987. Given to the author by Imogene Davis.

Figure 3.48. A house decoration made with *Trivia solandri*, *Olivella dama*, and *Zonaria annettae* shells. Collected by James Manson, mid-twentieth century. Arizona State Museum, catalog no. E-8518.

people in the area." One Seri woman recalled seeing strings of white *Olivella dama* snails hanging in windows and remarked that they were "really pretty."

Ironwood or rocks are sometimes carved into shell shapes for sale to tourists (figure 3.49).

Medicinal Use

Desert plants were the source of most medicinal substances.[114] However, some marine animals, including a few invertebrates, were also used. A tea made from the starfish *Luidia phragma* was drunk to stop postpartum hemorrhaging. Another starfish, *Phataria unifascialis*, was toasted and ground with salt and beef kidney fat, and the paste rubbed onto a swollen area to stop the swelling. The fireworm *Eurythoe complanata* was used along with a grass (*Cenchrus*) to stop menstrual flow.[115] A tea made from the spines of a sea urchin (*Echinometra vanbrunti*) was said to stop hemorrhaging during pregnancy.

Among the mollusks, a few were considered medicinal. The tiny *Octopus fitchi* was dried and prepared as a tea and drunk to give strength and endurance; it was also given to a horse "to make him run fast."[116]

A snail having the name *iquihimz iic cöihiipe* 'ringworm medicine' (*Mexacanthina lugubris angelica* and possibly *Plicopurpura pansa*) was used to treat ringworm. A secretion from the live snail was applied to the affected area and was said to be effective in treating other skin conditions as well.

Shells of the pearl oysters *Pinctada mazatlanica* and *Pteria sterna* were toasted, ground to a fine powder, and applied to one's face to dry pimples. The same powder was applied to the umbilical stump of a newborn to help it heal and dry more quickly. Other shells, such as those of the mussel *Modiolus capax* and the turban shell *Turbo fluctuosus,* were used in this way as well.[117]

It was said that a person could lose weight by eating the meat of the diminutive *Olivella dama* snail. Something that caused weight loss was called *cateexotol* 'what causes to be thin'.

According to Victoria Astorga, *haxölipj* (a shell gravel found on some beaches), was boiled to make a tea to treat stomach ailments (figure 3.50). If a miscarriage were imminent, one drank the liquid to stop the bleeding.

Figure 3.50. Shell gravel on shore at Bahía de Kino, 2009.

Recreation

The rich desert and sea world provided a wonderful array of entertainment opportunities. Very young boys easily learned the skills of the men, filling small toy boats with minnows and tiny crabs taken from tide pools with miniature spears, or shooting objects—or animals—with small bows and arrows. Girls helped harvest cactus fruit using miniature fruit-gathering poles, learned to make baskets and pottery, and played with dolls made from sea turtle or deer bones, shells, or even crab claws that they placed in tiny houses fashioned of sticks and turtle bones and furnished with miniature shell vessels (figure 3.51).

Toys and Games

Shells of larger snails were sometimes used as dolls. Because of its elongate shape, the *Fusinus dupetitthouarsi* shell was used as a male doll, while the short, round shape of the *Strombus galeatus,* and others, suggested a female shape.

A flying toy used by boys was made by tightly inserting the quill-end of a large bird feather into a

Figure 3.51. Dollhouses made with sea turtle bones and furnished with shell dishes. The bone dolls, usually dressed in cloth, are here unadorned flipper bones. Desemboque, 1971. Photograph courtesy of James Hills.

hole made in the apex of a large olive or cone snail. Tossed by the feather or thrown as a dart, the toy was said to really fly. The game was called *ziic ina xpa̲leemelc ano cöhazám* 'bird feather stuck into a ***xpa̲leemelc*** shell' (see figure 5.17, p. 162).

Gwyneth Harrington recorded that "the use of rattles of shell and shells as playthings was noticeable among the children."[118] An object she collected and documented as a child's rattle is made of *Macrocypraea cervinetta* and *Zonaria annettae* cowrie shells, and the flat valve of *Euvola vogdesi* (see figure 5.7, p. 145).

The people entertained themselves with games of chance, and men often bet with their material possessions—or, long ago, their wives. Running games where objects were tossed or kicked were popular.[119] Some entertainment involved races, as described in one event when a group of both men and women raced from an inland mountain camp to a shore miles away "just for fun." Especially popular were contests of strength or skill where men participated in balsa races, or foot races while carrying large clay vessels filled with water hanging from shoulder yokes. Other competition involved arrow or rifle shooting matches.[120]

In one contest of skill the shell of a large turret snail was flipped into sand to see who could make it enter point-down. The shells were also used as playing pieces in a gambling game, where several of the shells were flipped off of one's hand in the same way that flat pieces of cane were thrown.

A game called *xtapacaj caahit* 'what makes the *xtapacaj* (turret snail) make a sound' was played by two men (although some said it was played by women as well).[121] A hollow section of a large cane stem was held in one hand, while a sand-filled turret snail shell was flipped into the open end of the tube so that the shell's point entered the hole. The players took turns; the first to get ten hits in a row was the winner. Edward H. Davis describes such a game in detail (figure 3.52).[122] In the game he witnessed the counting was different, although the manner of playing the game was the same. The shell and cane tube Davis purchased are in the National Museum of the American Indian collections (see figure 5.4, p. 135).

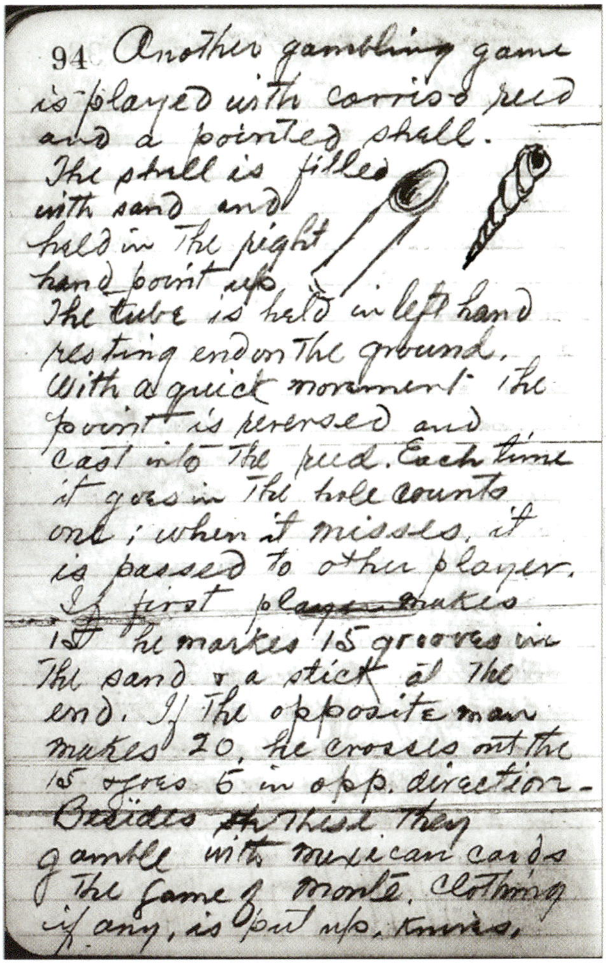

94 Another gambling game
is played with carriso reed
and a pointed shell.
The shell is filled
with sand and
held in the right
hand point up.
The tube is held in left hand
resting end on the ground.
With a quick movement the
point is reversed and
cast into the reed. Each time
it goes in the hole counts
one; when it misses, it
is passed to other player.
If he marks 15 grooves in
the sand & a stick at the
end. If the opposite man
makes 20, he crosses out the
15 & goes 6 in opp. direction.
they
gamble with Mexican cards
the game of monte. Clothing
if any, is put up, knives,

Figure 3.52. A page from the 1929 notebook of Edward H. Davis, detailing a game played with the shell of a turret snail. Cornell University Library (Division of Rare and Manuscript Collections).

Cleotilde Morales described a game she saw played as a child. A deer or cow bone was used instead of a section of cane; the turret snail shell was flipped using a finger on each hand, aimed at a hole in the bone, which was on the ground. The winner was the one who could make the shell's tip enter the hole.

Boys played games of competition using the sturdy *Chione californiensis* shells. In one game the shells were collected into piles onshore, and a shell was thrown simultaneously into the ocean by each boy to see whose shell entered the water last, and thus, was thrown the farthest. Another game involved throwing the shell into a strong headwind. The wind would return the hard-to-see shell at a high velocity, and the boys would dodge it. The game was rather dangerous, as suggested by its name, *iti icamaptx*, roughly translated as 'head injury game'. In another game, boys would throw the shells high into a cardón cactus to see who could make the greatest number lodge into the cactus flesh, again dodging the shells that ricocheted off the cactus (figure 3.53).

According to José Juan Moreno, the game of checkers, *raama* (from Spanish *damas*), was introduced to the Seris by non-Seri Mexicans in the 1920s. Santo Blanco, a Seri man who spoke Spanish and

Figure 3.53. *Chione californiensis* shells lodged in a cardón cactus near Desemboque, 2007. Note the shells around the cactus base.

interacted easily with outsiders, learned the game and taught others to play. A handmade painted wooden game board was used. Objects such as buttons, shark vertebrae, stones, or shells, including broken pieces of pearl oyster shells, were used as game pieces, called *iixaca* 'pets' (figure 3.54). Commonly used were *Turbo fluctuosus* opercula.

Figure 3.54. Boys playing a game of checkers at Tecomate, Isla Tiburón, 1951. The game pieces appear to be pebbles and *Leukoma grata* shells. Photograph by William Neil Smith, courtesy of University of Arizona Libraries, Special Collections. MS316_B31_slb1_R.-130–4.

Musical Instruments

Music was an important part of life, often performed by a singer and a single dancer whose rhythmic steps resonated on a board or large sea turtle shell placed over a hollow in the ground. Most often the singer, usually a man, accompanied himself with a handheld rattle. Pebbles or small pieces of broken shells were used as noisemakers in rattles made from cans with sticks inserted as a handle (figure 3.55).

Figure 3.55. A rattle made from an evaporated milk can, collected by Gwyneth Harrington in 1941. Arizona State Museum, catalog no. E-933.

Sometimes the dancer wore cocoon leg rattles, a practice learned from neighboring peoples.[123] Empty cocoons of the *Rothschildia cincta* moth into which noisemaking pebbles or even shell fragments were inserted, were sewn together into strands and wrapped around a dancer's lower leg and ankles (figure 3.56). Edward H. Davis noted seeing a shell "used for a cup" (probably *Laevicardium elatum*) broken and the small pieces inserted in the empty cocoons.

Figure 3.56. Rattles made with cocoons of the *Rothschildia cincta* moth.

Whole shells were apparently not used as musical rattles or noisemakers, with the exception of a large *Pinna rugosa* shell used as a musical rasp, and a stick rhythmically stroked against it produced a sound.[124] Although apparently not common, such use was confirmed by one consultant.

A likely incidental use of a shell was also recorded by Davis on Isla Tiburón. He described a traditional one-string violin being played and noted that the shell of a small, round clam was used as a bridge under the string.[125] In most cases a bridge was carved from *Bursera hindsiana* (the same wood used to make the violin) or ironwood.

Figure 3.57. A one-string violin played by Francisco Barnett. Punta Chueca, 2010. Photograph courtesy of Lynn Johnson.

Bait

Shellfish were used as bait for fishing.[126] The meat of abundant and easily obtained intertidal snails such as *Tegula rugosa*, *Turbo fluctuosus,* and *Nerita scabricosta* was used to bait small hooks. The snails were also crushed and cast into the ocean to attract fish for spearing.

The meat of octopuses was cut into pieces and used to bait hooks. On larger hooks used in deep water, the raw octopus was strung with the mantle covering the visceral mass removed (see Octopodidae). As Lorenzo Herrera explained, the fish would not approach the bait if a complete octopus were used. Fish such as *caanj* (Gulf grouper, *Mycteroperca jordani*), *zixcám coospoj* (spotted cabrilla, *Epinephelus analogus*), and *zixcám caacoj* (giant sea bass, *Stereolepis gigas*) were said to be especially attracted to octopus meat.

When no other bait was available, the sea hare *Aplysia californica* could be used. One man mentioned that even a lizard could serve as bait, adding that he once saw his father use small sections of barrel cactus flesh to catch a triggerfish.[127] In such cases, once a fish was caught, that fish was the supply of further bait.

Although considered good food, lobsters (*ptcamn*) were not commonly eaten, as they were only found in some areas. Lorenzo Herrera recalled living on the south shore of Isla Tiburón at *Hehe Hasoaaj quih An Ihiip* for two years around 1950, where the people were catching lobsters in traps supplied by an American, and selling the lobsters to non-Seri Mexicans. They found that the bait that worked best in the traps was meat of the *Turbo fluctuosus* snail.

Place Names

In the well-traveled Seri world, names were given to mountains and camps, to areas in the sea or on land where a resource could be found. Even trees were given personal names. Still recalled by name today are certain huge mesquite trees known for their large, sweet beans that grew in the extensive wooded areas inland from Bahía de Kino. Their names exist only in oral accounts since the trees were long ago cut down by outsiders for the commercial charcoal market.

Especially important were sources of fresh water—and virtually any place to be referenced or visited.[128]

Names might be simple, such as *Xnit*, a primary name. Many are descriptive, evoking strong images perhaps based on topographical features, such as an area of sprawling, jagged red hills named *Coiiz Cheel* 'red spider'. Others anchor an event in space, such as *Quimoontaj Iihca* 'where those who danced the victory dance sat', a place where an important war victory was celebrated. Or *Xapij An Ihapancojc* 'where they ran inside the reedgrass', named for an event long forgotten. Still others refer to something apparently mundane, such as *Comcaii Inoosj An Ihatootoj*, 'where the ***comcaii inoosj*** (snails) were baked'. Without a doubt, however, the best-known place name referencing mollusks is *Haxöl Iihom* 'place of the ***haxöl***' (more literally, 'where the ***haxöl*** lie'), Desemboque.

Among the hundreds of recorded place names are a number referencing mollusks; see those listed in table 3.7.

Though we surely have recorded only a small part of the many place names that the Seris have used over the centuries, the hundreds of names we do have give us a clear and penetrating look at a past world and how the people thought about that world. The place names that refer to mollusks are one facet of that larger picture.

It is obvious from what has been presented in the preceding pages how very central mollusks were in the lives of the people who lived along the shores of the Gulf. The sections that follow lay out in greater detail the ways in which mollusks touched so many aspects of life.

Figure 3.58. Desemboque, *Haxöl Iihom* 'place of the ***haxöl***'. March 2013.

Table 3.7. Seri place names referencing mollusks

Comcaii Inoosj An Ihatootoj, 'where the ***comcaii inoosj*** (*Nerita scabricosta*) were baked'	a seaside overnight camp north of Bahía de Kino
Copas Caaizi Iti Yaii, 'where the people who gathered ***copas*** (*Pinctada mazatlanica*) lived'	a shore area on Isla Patos, said to have been used by divers harvesting ***copas*** pearl oysters long ago, possibly when the pearling industry was active in past centuries
Cotopis Inaail quih Yeen, 'face of the ***cotopis*** (*Turbo fluctuosus*) shell'	an area by *Hasteemla* (a mountain south of Desemboque) where ***cotopis*** snails were eaten
Haanc, possibly an archaic plural form of ***haan*** (*Chionista fluctifraga*)	Punta San Miguel, on the east coast of Isla Tiburón
The following place names incorporating the name ***haan*** are located near *Haanc*	
Haanc Icaheme, '*Haanc* camp'	a well-used camp
Haanc Iicot, 'with *Haanc*'	a place in the Infiernillo Channel near *Haanc Iyat*
Haanc Iyat, '*Haanc* point'	a sand point where many ***haan*** clams are found
Haanc Pnaacoj, '*Haanc* mangrove [area]'	a mangrove area
Haanc Xtaasi, '*Haanc* estero'	an estero
Haanc Xtaasi quih It, 'base of the *Haanc* estero'	a place near *Haanc Xtaasi*
Haanc Yaiij, '*Haanc* shoal'	a shoal
Haitepl	an archaic name for *Hast Quipac* (Campo Dólar), as well as the name of an unidentified shelled mollusk
Hant Xpeecol, 'land of the ***xpeecol*** (unidentified) shell' (a variation of the name is ***Xpeecot***)	a camp on the edge of Estero Santa Cruz
Hapaj Itni, 'octopus umbilical hernia'	a mountain near Desemboque with a shape similar to the hump-shaped mantle containing the visceral mass of an octopus
Haxöl Hast, '***haxöl*** (*Leukoma grata*) mountain'	a seaside mountain north of Bahía de Kino near a bay where large ***haxöl*** are found
Haxöl Hast quih It, 'base of *Haxöl Hast*'	a camp at the base of *Haxöl Hast*
Haxöl Iihom, 'place of the ***haxöl***'	Desemboque, named for the clam beds there
Haxölinaail iti Yahimoz Iti Icaheme, 'camp of the shell's operculum'	a camp
Haxöl Quipcö, '***haxöl*** mound'	a raised rocky intertidal area near Punta Chueca, where ***haxöl*** clams are abundant
Haxt Án, 'area of the ***haxt*** (*Crassostrea corteziensis*)'	an area inside Estero Santa Cruz where ***haxt*** oysters were harvested
Iimox Itleen oo Quicatx, 'who wears ***iimox*** (*Spondylus limbatus*) over his chest'	a camp name; the name of nearby mountain with a strip of red rock that gives the appearance of a string of beads worn by men crossed over their chests
Satoj Hasaiitaj, '***satoj*** (*Modiolus capax*) pried up by spear'	a submerged area between Estero Sargento and Isla Tiburón, where clumps of mussels were harvested by spear
Satoj Iicot Iime, 'living place in the middle of the ***satoj*** (*Modiolus capax*)'	a place in the sea between mussel beds, where sea turtles were hunted

Seeten Ctamcö (plural of ***seeten ctam***, *Pinna rugosa*)	a place in the sea where many *Pinna rugosa* were found
Teexoj Tis ma Ihacanaj quih Iyat, 'point where the raw ***teexoj*** (*Spondylus limbatus*) were chewed'	a rocky point near Puerto Libertad where raw ***teexoj*** meat was eaten by non-Seri fishermen
Xnoocat Coos, 'the singing *xnoocat*'	a desert dune near Desemboque where the ***Xnoocat*** *Coos* 'singing ***xnoocat***' song was sung (see *Chicoreus erythrostomus*)
Xojmás Queaaf, '***xojmás*** (terrestrial snail) having a waistcord'	a mountain on Isla Tiburón around which are found live ***xojmás***
Xojmás Queaaf quih Hapx Ihiip, 'where *Xojmás Queaaf* emerges'	a place in the sea from which *Xojmás Queaaf* can be seen
Xtoozajö Cöhanípatim, 'where the ***xtoozajö*** (Vermetidae) struck it'	an area on the eastern shore of Isla Tiburón that was given this aptly descriptive name because eroding cemented beach deposits contain masses of horizontal ***xtoozajö*** shells
Yom Chast, 'rocky *yom* (either a turquoise bead or a sea-worn pierced apex of a snail shaped like a bead)'	a mountain north of Puerto Libertad
Zaah Cooil, 'blue/green sun'	a small estero near the larger Estero Sargento, said to contain bluish shells, perhaps referring to ***zaah*** (*Megapitaria aurantiaca*)

PART III

SPECIES ACCOUNTS

The mollusks supplied important food that was easily accessed at most times, and their sturdy and variously shaped shells provided the basic things the people needed in their everyday lives. And they were not only of utilitarian value; the shells were talked about and enjoyed.

The species presented here are representative of the larger group undoubtedly used for millennia by other Gulf peoples; one can only imagine their stories. The Seri names given in this book were shared with me by many people as I was growing up and later when I was researching for this book. A great many names are no longer used, and what was previously common knowledge is disappearing.

Where a name is obviously the most commonly used, I mark it with a diamond (♦); where there is no clear preference, no name is marked. When it is fairly obvious that a name was somewhat improvised and was otherwise unknown, the name is marked with an empty diamond (◊).

went along slowly in a boat, thrusting his spear down into the mussel beds. When the spear hit a clump, the man would twist the spear to uproot it, and the mussels were then brought up into the boat. Large quantities of these mussels, called ***satoj*** *hasaiitaj* '***satoj*** brought up with a spear' were harvested in this way.

Figure 4.1. Clumped *Modiolus capax* mussels tossed upon a beach. Desemboque, 2007.

BIVALVIA

The bivalves were heavily consumed. They were easily found and gathered in the intertidal area, whether rocky or sandy; on estero sand flats exposed at low tide, the clams were plentiful and their meat easily prepared for eating. Their largest shells were important tools and vessels, as easily used as they were found on the ocean shore, and their smallest shells were simple to incorporate in shell jewelry.

There is no common Seri term for "bivalve." Of the species presented here, more than sixty are recognized by Seri names, some with multiple names. A few unnamed species are included as well, for the record.

Mytilidae (mussels)

The mussels are included under the phrase *ziix* ***saacj*** *hant cöquiih* 'thing grouped with ***saacj***' (see Classification and Naming).

Because of their flavor and good quantity of meat, mussels were a favorite food of the Seris and reportedly of the Giants as well—"they were *real* food."

Mussels were usually gathered at low tide. When many had been taken from an area and the supply depleted, the people would then harvest clumps of mussels by boat from deeper water. Such clumped mussels are called ***satoj*** *ptiiqui cosiiija* '***satoj*** lying together'. A long, single-pronged barbed spear, *hacaaiz icosaii* 'spear for prying mussels loose' was used to harvest them. As the mussels were often not visible from the surface, a man who collected them (called a *csaii*)

Seri folklore details a well-known conversation referring to a time when the mussels were being harvested, when the mussel confronted the jingle shell about why it did not help protect the mussels (see Anomiidae).

Juvenile mussels were called *l̲oomz*, a word almost forgotten today. Lorenzo Herrera described them:

> *L̲oomz* are baby mussels. When the Ancestors collected mussels, they would go in a boat over the mussel bed and collect [them]. When only medium-sized mussels could be found, they would put them into the boat and take them instead [of the larger ones]. Then the women and men on shore would meet the boat as it arrived and say in jest, "These people have just brought *l̲oomz*. Do they think they are food?!" [1]

The variety of terms used to describe types or size of mussels is perhaps an indication of the importance of this mollusk (table 4.1).

Table 4.1. Seri names for different kinds of mussels

satoj	*Modiolus capax*, the common mussel, usually those of moderate size
hant quiitij iti ***satoj***	'***satoj*** on the steep slope (shoals)' mussels found growing on underwater shoals
loomz	juvenile or very small mussels
satoj *hasaiitaj*	'***satoj*** brought up with a spear' clumps of mussels growing in deeper channels, harvested by spear from a boat
satoj *ptiiqui cosiiija*	'***satoj*** lying together' mussels growing clumped together
xiica hant cooyoj	'standing things' mussels growing singly on the ocean floor, rather than in clumps
saacj	very large, heavy-shelled mussels, found only in deeper water; probably *Modiolus capax*
satoj aapa, 'huge or true ***satoj***'	undetermined species, described by some as small, with inflated valves and a very curved foot; others say it is possibly *Mytilus californianus*, *Mytella charruana*, or a large *Modiolus capax*
oot isatoj, 'the coyote's ***satoj***'	*Modiolus tumbezensis*, *Mytilus californianus*, *Entodesma pictum* (a bivalve that is not a mussel)

Mussels were harvested for sale to non-Seri commercial fish buyers. Juanita Herrera recalled the Seris selling mussels to Jesús Solórzano at *Quipcö Quih An Icaheme* (Paredones), a camp on the mainland side of the Infiernillo Channel. She explained that the people cooked the mussels and removed their meat, which Solórzano then pickled in vinegar and trucked to Hermosillo.[2]

Cooking

Although eating mussels raw is not common today, several people commented that they did like eating them that way. José Ángel Montaño recalled his father enjoying them, and added that in the past other shellfish were eaten raw as well.

When cooked, mussels were either boiled in the shell or, more commonly, roasted under a brush fire. When roasted, the shellfish were placed carefully in rows, and not piled haphazardly. However, when clumps of mussels were cooked, the individual mussels were not separated, but a fire was built over the whole clump. As described by Cleotilde Morales: "When [the mussels] were cooked, they were all there still clumped together, with their mouths opened. When [the shells] were cool, then [the mussels] were eaten."

Angelita Torres recalled times when mussels were cooked and eaten:

> We were at the *Xepe Coosot* [the Infiernillo Channel]—at *Ziipxöl Iifa*, *Xoop Cacöla Inoohcö*, *Quipcö Coospoj*, or at *Xana*—and my father got some ***satoj*** [*hasaiitaj*]. He got them from the sea floor, far away, by spearing them from a boat. They were growing together in clumps, called *ptiiqui cosiiija*. He speared [the clumps] and filled the boat with them, and brought them to us. We covered them with fresh green eelgrass and threw seawater on them. When we were ready [to eat them] we uncovered them, cleaned them, and then we roasted them or [pan] cooked the meat. They don't die if you cover them with seaweed [or eelgrass]. They last a long time that way—three or four days—and they don't die. You cover them like that, with eelgrass.
> [Another time], when I was young, we were at *Xana*. Three or four old people were there, and they roasted ***satoj*** [under a brush fire]. They sat around the cooking area and ate and were so happy. They ate that really good food. That is what it was like. In the camp there would be two or three families there, at the *Xepe Coosot*. It's not like that today. The old people . . . now they don't do that anymore.

Figure 4.2. Removing the gills from a cooked mussel. The roasted mussels, still in their shells, are contained in the shell of a sea turtle. Desemboque, 2013.

When pan-cooked, the mussels were often flavored with leaves of oregano, which was used to flavor other meat as well. By whatever means the mussels were cooked, the whole animal was eaten except for the *itojipl* (ctenidia, or gills), which were easily removed after cooking, in a practice that continues today (figure 4.2). Eating the gills was said to cause eye discharge and blindness, and if a pregnant woman ate them, her baby would be born blind. The name for eyelashes is *itojipz*; both names contain *itoj* 'its eyes', and that of the gills, *ipl* 'its tongue' (although a tongue is never mentioned when discussing the name). The names and gestures used when the information is given suggest a connection; however, it remains unclear.

I was told that for some individuals, eating too many mussels results in what was described as (using a modern diagnosis) "elevated blood pressure." Such a person is described as ***satoj*** *oiicöt* '[who] is killed by mussels'.[3] Several singled out the large ***saacj*** mussels as causing headaches in some people.

Use of the Shell

Mussel shells were used as utensils for eating and drinking. When a sea turtle was roasted under a brush fire, the cooked broth was spooned up using these shells.

In speaking of the shell, Eva López described its use:

> We really like [mussels]. We ate them—they are real food. [The shells] were also kept by the Ancestors. They kept and carried them. When a *moosni* [sea turtle] was roasted [under a brush fire], men and women all had these. The *moosni* plastron was removed, and they all would come around and drink the broth. There was enough for everyone. The plastron was scraped and then they *really* went for it. The broth was enough for everyone. Little children and everyone carried these [shells]. That is what these were kept for. They kept them in their pockets. When one would break, they would just take out another. They thought [of them as if] they were spoons.

Figure 4.3. William Neil Smith and a child eating roasted sea turtle meat from its carapace. *Hoona*, 1960. Note the use of a metal spoon and fork instead of shells. The unusual photograph is included here to show a sea turtle being consumed in this way. Photograph by Gordon M. Daetz, courtesy of University of Arizona Libraries, Special Collections. MS316_B27_f8.

The shell could be used as a scraper (*ziix ipexöp*). Sea turtle skin and air bladders of large fish were scraped in preparation for toasting (see *Arca pacifica* and *Dosinia ponderosa*). Such smooth-edged shells could also be used to form clay vessels, in the same way as the *Simomactra dolabriformis* shell. Victoria Astorga kept a mussel shell for such use.

Sharp as a razor, the shell (which could be sharpened against a rock) was used to cut the umbilical cord of a newborn. To demonstrate, Sara Villalobos held up a taut thread for me to cut—the shell cut well.

The shell could be used as a receptacle for pigments, although other larger shells were more commonly used.

Mussel shells, as well as those of the ***cotopis*** (*Turbo fluctuosus*) snail, were toasted, ground into a powder and applied to a baby's umbilical stump to help it heal and dry.[4] Pearl oyster shells could be used in this way as well.

Pearls found in mussels are described as being dark, bluish, and pretty. They are uncommon and sold as curiosities.

Unlike those of the pen shells, the byssal fibers of the mussels are not used in making dolls, since they are not as dense or attractive as the former (see Pinnidae). Those of both bivalves are referred to as *imas* 'its pubic hair', *ilít* 'its [head] hair', or *itaa* 'its—' (possibly an archaic word for hair).

Seeding Mussels

In two instances of which I am aware, mussels were taken from one area and seeded in another. This was called *iti yapatjc* 'spreading out'.[5]

According to Rosa Flores, when her children were small (c.1940), the Astorga family harvested a boatload of ***satoj*** from an area near Estero Sargento and brought them to Desemboque, where the large ones were eaten. The children took the small mussels and at low tide placed them in sandy areas in front of Desemboque. Years later as Rosa gathered ***satoj*** from the area, she said that they came from the ones they had planted.

A second instance of seeding mussels is recounted from the mid-1900s, when a man claimed to own a submerged bed of mussels north of Punta Chueca. It is said that he would throw small mussels into the water there, so they would grow. The man became quite indignant when anyone else would harvest from the area. Since the idea of claiming ownership of an area is unusual today, his reaction is rather humorously recounted.

Leiosolenus spatiosus

stac ano coom 'what lies in coral'

Beach drift corals from commercial shrimp boat dredging or brought up by divers often contain these mussels. The fragile shell is covered by a thin, light-colored chalky incrustation.

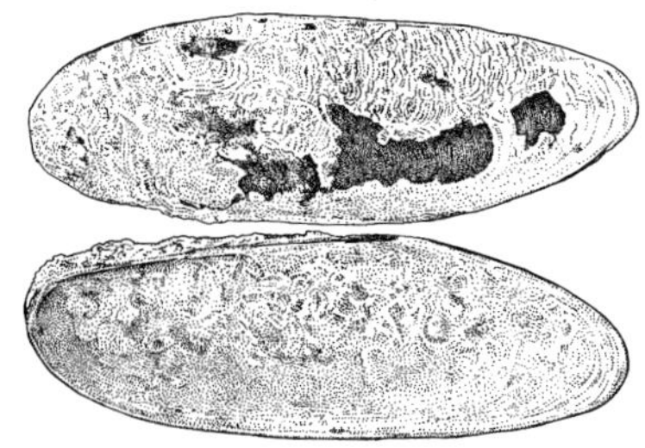

L 36 mm

Lithophaga aristata

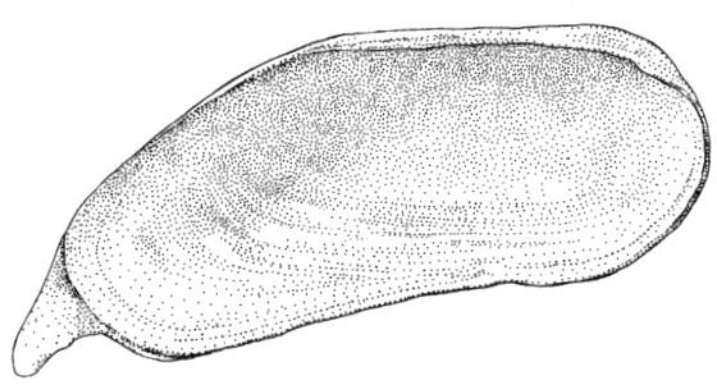

L 12 mm

This small, fragile-shelled boring mussel has distinctive incrustations extending forceps-like beyond its shell.

One afternoon I found María Luisa Astorga picking apart the shell of a large cup-and-saucer snail (*Crucibulum scutellatum*). She said she had seen the holes bored into such shells and always wondered what they contained, so decided to find out. She carefully removed this tiny mussel shell, calling it a ***satoj*** *heeque* 'small ***satoj***', probably a spontaneous descriptive name, and said that she had not seen one before.

Lithophaga attenuata

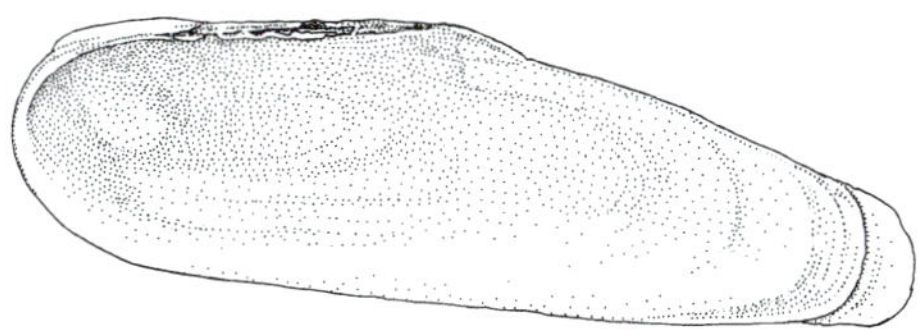

L 29 mm

This small rock- or shell-boring mussel has a very fragile shell covered by an incrustation extending beyond the ends of the shell. The shell given to me had bored into a *Chama buddiana* shell that was attached to an *Atrina tuberculosa* pen shell, brought up by divers near Isla Tiburón.

Modiolus capax

satoj

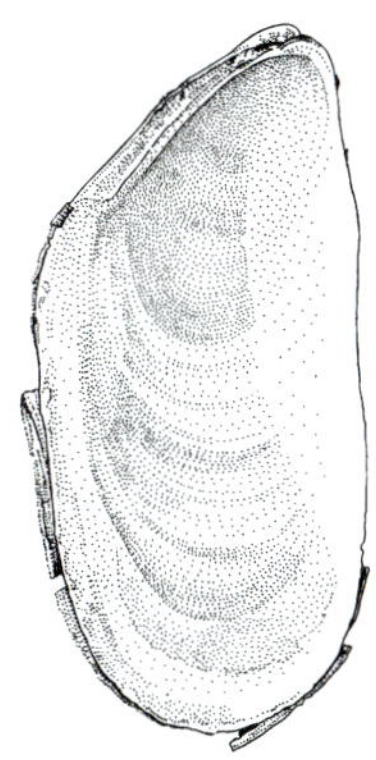

L 90 mm

This is the most common mussel in the region, growing in extensive beds along some shores. Its shell is sometimes marked in purple or orange and, when fresh, covered by a fibrous periostracum.

Satoj is a primary name. These mussels were heavily consumed. Place names referencing mussels pinpoint areas in the sea where they were abundant and easily harvested. *Satoj Hasaiitaj* is a submerged area between Estero Sargento and *Hant Cmaa Coocp* (a spit of land on the northeast shore of Isla Tiburón) where clumps of the mussels are found. *Satoj Iicot Iime* 'place (where the sea turtles live) between the mussels' is an area in the sea where sea turtles were easily hunted among the mussel beds.

In April 2006, I was shown how shellfish were traditionally roasted under brush (figures 4.4a–4.4g). Cleotilde Morales and her husband, Alfredo López, brought a half-full bucket of ***satoj*** and ***stacj*** (*Saccostrea palmula*) that they had collected from the rocky intertidal area at Punta Chueca. Cleotilde and I walked down to the beach to an area of coarse sand. She smoothed the sand and began placing the shellfish on the sand in rows about an inch apart, completing an area of about two feet square. The mussels were placed hinge-side down, while the oysters were laid flat.

Figure 4.4a–g. Cleotilde Morales roasting mussels and oysters. From left to right: a) placing the shellfish in rows, b) gathering brush, c) placing brush on the shellfish, d) lighting the fire, e) roasting the shellfish, f) removing the shellfish, g) the shellfish, ready to eat. Punta Chueca, April 2006.

After the mussels and oysters were in place, we searched nearby bushes for dry branches. Cleotilde broke off the smaller, bushy branches and placed them in small bunches on the shellfish, finishing the pile with larger branches. When the pile was about a foot high and completely covered the shellfish, she set it on fire. After about ten minutes of a good fire, the wood was completely burned and all that remained were ashes. Cleotilde then carefully removed the shellfish one at a time, using the end of a stick since they were too hot to handle, and placed them on the ground, taking care not to get sand in them. When they were cool enough to remove the meat, we ate them. The meat was delicious, with a somewhat smoky flavor.

Digitized portions of films made by William Neil Smith on Isla Tiburón in the early 1950s illustrate traditional Seri life (see Preliminaries). Included is a brief section showing women roasting mussels on a beach. After the fire burns down, the placement of the mussels in rows is clearly seen. When cool enough to handle, each mussel is tapped and brushed lightly to remove sand and ashes, and the meat removed and placed in a container rather than being consumed from the shell.

Modiolus capax

saacj

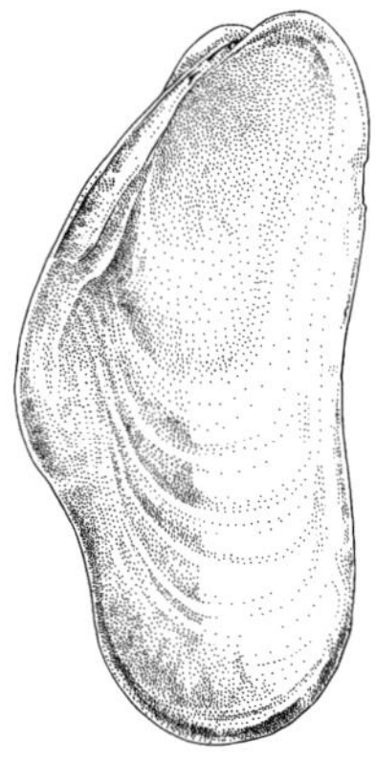

L 170 mm

Not commonly found alive, nor are its shells found in beach drift, this very large, robustly shelled mussel grows only in deeper water in the Infiernillo Channel, where it is occasionally brought up by divers harvesting pen shells. According to the Seris, it does not occur with the smaller ***satoj*** mussels, which grow in shallow water and can be found in the low intertidal area.[6] The shell's exterior is a deep purple and pink beneath a fibrous periostracum; its white interior is blotched with purple.

Saacj is a primary name. Although other names are given to mussels (table 4.1, p. 76) most people did not suggest such names for this one. However, one account of a name change due to name taboo does in fact relate some of the names to it. As related by Eva López, a tall boy was named *Saacj*. He died, and the name of the ***saacj*** mussel was changed to ***satoj aapa*** 'huge or true ***satoj***'.[7] And, according to María Luisa Astorga, *xiica hant cooyoj* 'standing things' then began to be used for this mussel as well.

The mussel's meat is substantial and described as sweet-tasting. Although considered to be good food, some people say that eating a large quantity can cause headaches. One woman said that she did not like the meat, because it "disintegrated" (was mushy) in her mouth.

The Giants were said to have used this large shell as a spoon and carried it hung over one ear, as if on a hook.[8] Eva López chuckled as she related that when someone sees an especially curved footprint in the sand, they might quip, "Oh, look how that person's footprint looks like a ***saacj***!"[9]

Modiolus tumbezensis

oot isatoj 'the coyote's ***satoj***'

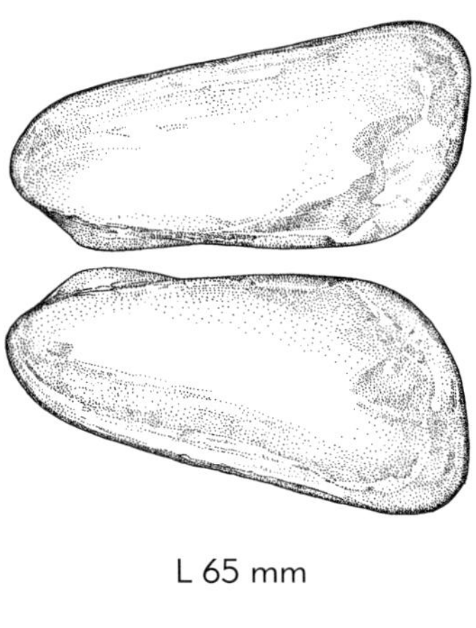

L 65 mm

The delicate shell of this small mussel has an iridescent interior, described as resembling a rainbow. Its difference in size and shape from the more common mussel, ***satoj*** (*Modiolus capax*), provides it the name ***oot isatoj*** 'the coyote's ***satoj***'.

According to some people the meat of this mussel was not eaten. Juanita Herrera recounted an incident when she was young, and her family was eating a bunch of cooked mussels; she took one from the pile, and her mother told her quickly not to eat it—that it was an ***oot isatoj*** and would be bitter, have an odd taste, and irritate her throat. Other people, however, said that these mussels were eaten. Victoria Astorga confirmed that she had eaten them, commenting that she came from a family familiar with the sea and ate *lots* of different seafood.

Mytella charruana

satoj aapa 'huge or true ***satoj***'

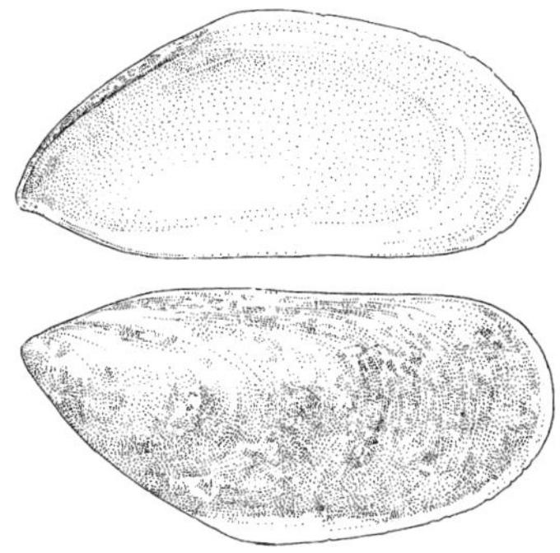

L 81 mm

The shell of this uncommon mussel is somewhat fragile, with a greenish periostracum. The shell was collected and named by Alfredo López, who found it in beach drift on the rocky north shore of Isla Tiburón. Alfredo described the mussel as occurring in that area rather than in the mussel beds of the Infiernillo Channel.

Mytella guyanensis

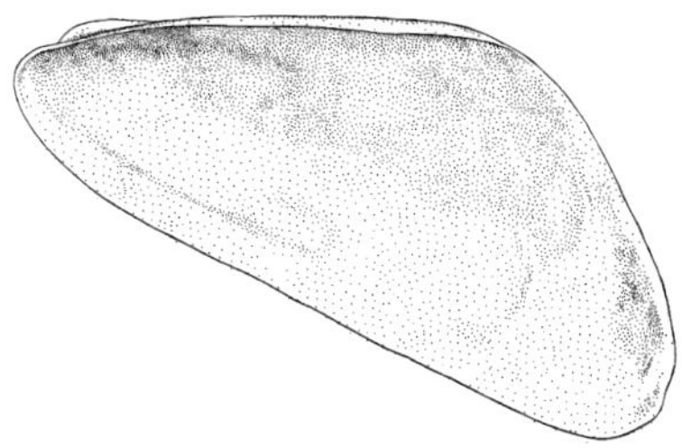

L 80 mm

Apparently uncommon, this fragile-shelled mussel is only known from several shells found near Estero Santa Cruz and is unrecognized by name today.

Mytilus californianus

oot isatoj 'the coyote's ***satoj***'
satoj aapa 'huge or true ***satoj***'

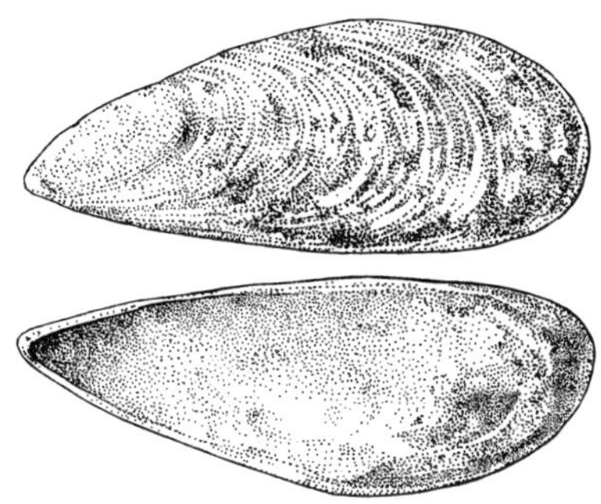

L approx. 130 mm

A single valve of this mussel was collected on Isla Tiburón by Edward H. Davis in 1936 and is now in the National Museum of the American Indian, labeled as a knife.[10] When she saw a photograph of the shell, María Luisa Astorga identified it as ***oot isatoj*** 'the coyote's ***satoj***', and said that it was not often seen, adding that such a mussel shell was particularly sturdy and sharp-edged. On a later trip I made to Desemboque, she gave me a similar shell brought from Ensenada, Baja California, by her sister Amalia. Other people have more readily given the name ***oot isatoj*** to *Modiolus tumbezensis,* which is relatively common in the Infiernillo Channel.

Davis mentions a mussel identified as *Mytilus californianus* being used as a knife, adding that it was "ground to a keen edge on a stone."[11]

The name ***satoj aapa*** 'huge or true ***satoj***' was given to this mussel as well as to *Mytella charruana.*

Arcidae (ark clams)

Wide in their distribution and their shells common in middens and ancient camps, the ark clams were heavily consumed in the past but are not eaten by most people today. In fact, a remark sometimes made is that they are enjoyed by non-Seri Mexicans, who purchase them commercially outside the region. Some of the clams are said to contain a red-colored blood, so were considered *hapáx,* a food to be avoided by some people (see Beliefs and Practices, p. 41).

Various Seri names are supplied for the ark clams; the name ***isliicot itaail*** is the one most clearly recognized and commonly known for *Arca pacifica*.[12] Other species are named as well, but with less confidence, and the names vary in form. It is indeed easy to conclude that although probably valued in the past, the clams have for some time not been as important.

The dark fibrous periostracum that covers the ark clams is called *iteems* 'its whiskers'. Some species are anchored by a solid byssus that does not have a name, unlike the fibrous byssus of the mussels and pen shells, which is recognized by name.

Anadara formosa

The white shell of this clam has around thirty-six ribs and is covered by a rather heavy, bristly periostracum. Living offshore, the clam was not recalled as being food.

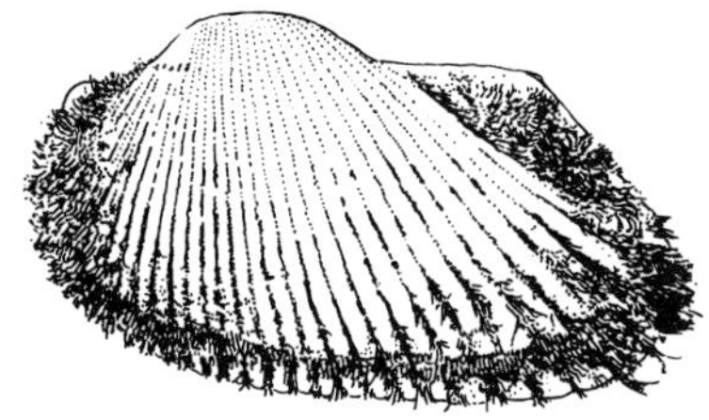

L 80 mm

Anadara grandis

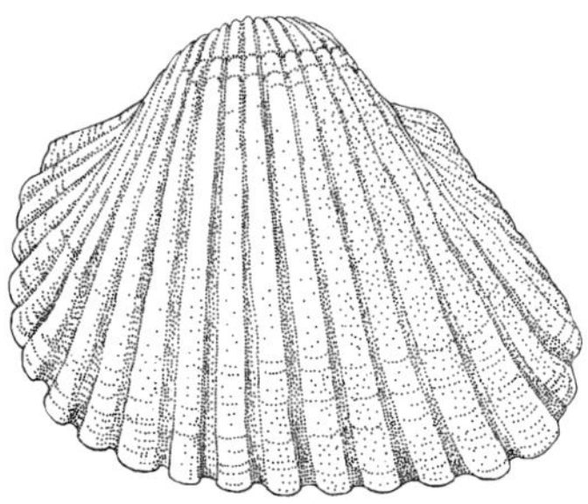

L 65 mm

The large, heavy white shell of this ark clam has about twenty-six ribs.

The clam was eaten in the past; its shell, together with those of oysters, is common in old camps surrounding some of the esteros in the area.

Anadara multicostata

quiimosim xepenoyaafc 'what the beggar pounds in the sea'
xnahaail

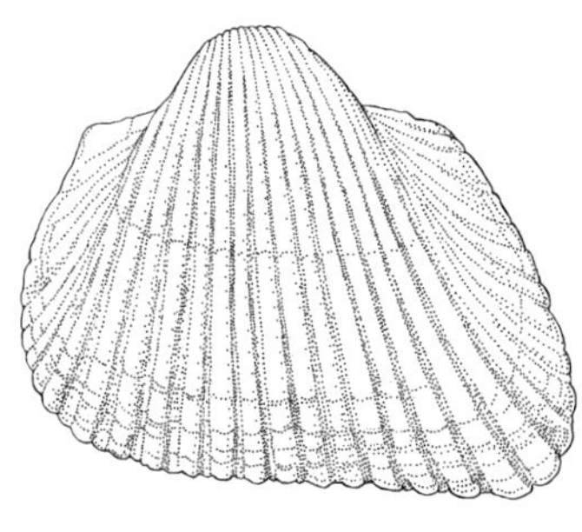

L 97 mm

The heavy, white shell of this large ark clam has about thirty-five ribs. It is found on sand flats at very low tide. The clam was said to have been eaten in the past.

The name ***quiimosim xepenoyaafc*** 'what the beggar pounds in the sea', also given to *Cardites laticostatus*, describes a person begging unsuccessfully for food, who then stands in the ocean striking the clams together in order to break them open to remove their meat—a method commonly used when eating smaller clams raw at the beach.

The name ***xnahaail*** is probably a variant of ***xomcahaail*** and ***xnahooil***, given to other ark clams. There is no common use known for the clam's good-sized shell. However, Ramón López (b. 1941) recalled playing with the larger, heavier ark shells as a boy when living at *Hast Quipac* (Campo Dólar).[13] The boys, armed with long sturdy sticks, would take turns pushing a shell along the ground with the stick inserted into its beak and quickly flip the shell into the air for all to chase.

Anadara tuberculosa

xnahaail

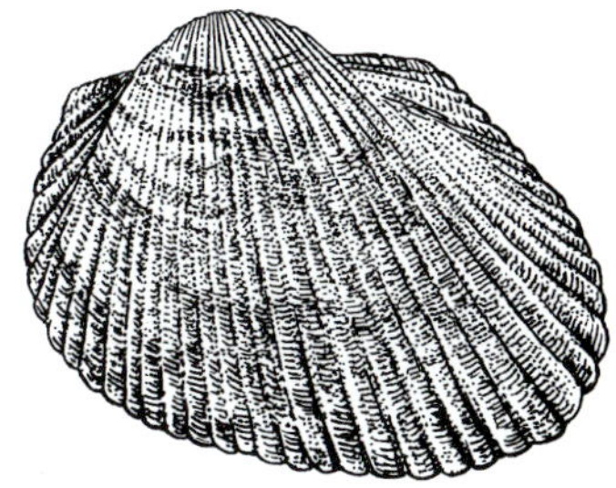

L 60 mm

This smaller ark clam is common on some estero mud flats. Its somewhat delicate shell has about thirty-seven ribs and is covered by a fine, dark periostracum.

The name ***xnahaail*** is probably a variant of ***xnahooil*** and ***xomcahaail***, given to other ark clams.

Arca pacifica

♦ ***isliicot itaail*** (var. ***isliicot ihaail***) 'wide-based upper back'
xnahooil
xomcahaail

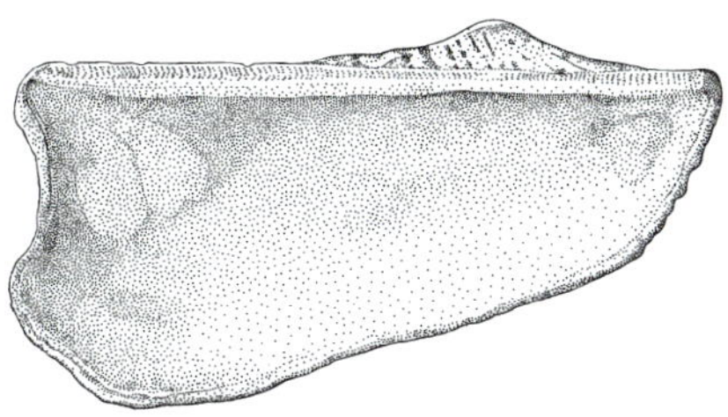

L 75 mm

This clam has a sturdy shell with V-shaped grooves on its broad, flat dorsal surface. Unlike some of the other ark clams, its blood is not red, so the clam was not considered *hapáx* and therefore something to be avoided as food.

The clam's best known name is ***isliicot itaail*** 'wide-based upper back', a name derived from *isliicot* 'its upper back', *it* 'base', and *-aail* 'wide'. The name is based on a comment made by the person who became this clam when living things received their names at the world's creation and reflects the shape of the shell when the valves are closed—the broad flat area between the beaks looks like a person's upper back between the shoulders.

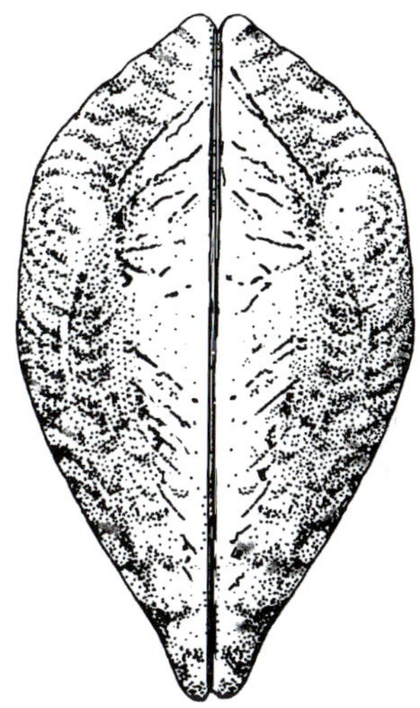

Figure 4.5. An *Arca pacifica* clam showing its "broad upper back."

As Eva López recounted:

> An old man with a wide upper back arrived and sat down. He said, "I think that I, the old man with the wide upper back, will sit here and listen to what is being said by *Hant Iiha Quimx* [the one who tells about the ancient things]." So *Hant Iiha Quimx* said, "Then that will be your name, ***isliicot itaail***."[14]

The names ***xnahooil*** and ***xomcahaail*** were provided by two different individuals. A variant, ***xnahaail***, is given to other ark clams. The names are possibly based on the words *xnaai* 'south wind', *caail* 'wide', and *cooil* 'blue/green' or *ooil* 'crest (of bird)'. However interesting, one would be only guessing as to their meaning, since it is unclear today.

The shell was used as a scraper; the deep valve was easily grasped and used to scrape in a motion directed away from one's body (figure 4.6). The shells of the *Arca pacifica*, mussels (*Modiolus capax*) and large, edge-chipped *Dosinia ponderosa* were used to scrape clean the air bladders of large fish or the soft skin of sea turtles, which were then toasted and eaten as cracklings. It is said the food was delicious when dipped in sea turtle oil.

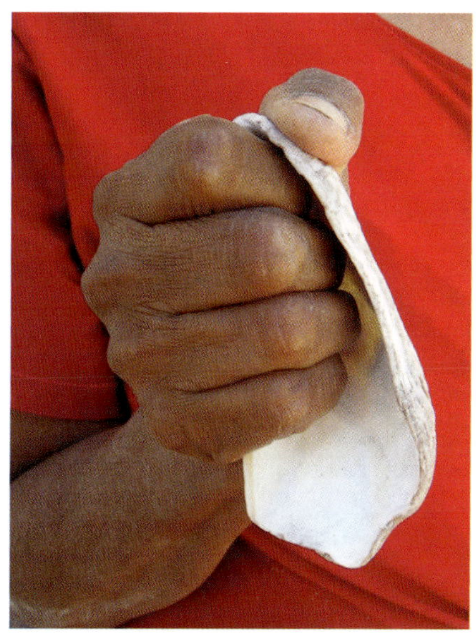

Figure 4.6. An *Arca pacifica* shell grasped for use as a scraper. Desemboque, 2007.

W J McGee speculated erroneously that these shells were used as repositories for human excrement and placed on graves with some sort of mortuary significance (see Beliefs and Practices, p. 41).[15]

Barbatia reeveana

xatoscö ihaxöl 'the Eared Grebe's ***haxöl***'
xomcahaail

L 42 mm

The white shell of this small ark clam, covered with a fibrous dark periostracum, has a somewhat variable shape due to distortion from growing between rocks.

The name ***xatoscö ihaxöl*** 'the Eared Grebe's ***haxöl***' suggests some relationship between the clam and the Eared Grebe (*Podiceps nigricollis*), but the connection is unclear. The clam's other name, ***xomcahaail***, is similar to names given to other ark clams.

The clam was eaten; Angelita Torres recalled eating many as a child on Isla Tiburón.

Glycymerididae (bittersweet clams)

Glycymeris gigantea

naapxa yaat 'what the Turkey Vulture baked in the coals'
satim
saxáp
seeex 'will grunt'
xpanooil (var. ***xpanoois***)

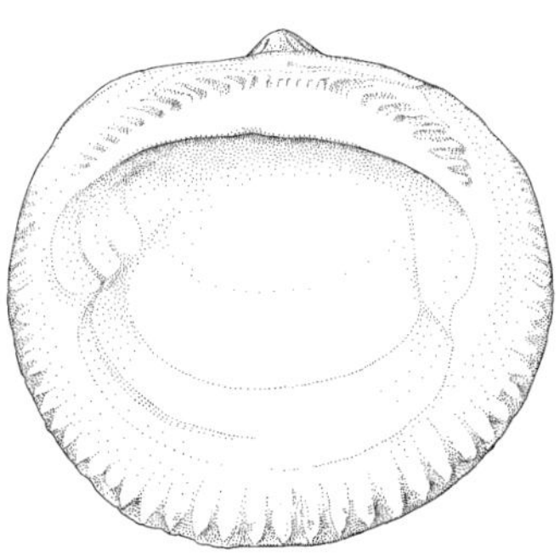

L 67 mm

The robust shell of this large clam features prominent teeth and is marked on its exterior with brown zigzag designs. As the clam is not often collected live since it lives mostly offshore, its shells are usually found in beach drift. Larger shells are common in some middens. According to Keen, this species, one of the largest of its family, is endemic to the Gulf of California.[16]

For such a distinctive and common shell it is somewhat surprising that it was not recognized by name by many people and when named, there was much variety.

The name ***naapxa yaat*** means literally 'what the Turkey Vulture baked in the coals', from *naapxa* (*Cathartes aura*, Turkey Vulture) and the verb *caat* 'to bake in the coals'. The name is given to three different things: a well-used hearth having a thick and bowl-shaped dirt base baked hard by the fire; sea salt that has been compressed into a round flat shape for easy transport and storage; and this clam.

Satim and ***saxáp*** are primary names.

Seeex is the future tense of *queeex* 'to grunt'. According to Eva López:

> The Name Giver was teaching, giving the names of things. This [person] came along and said, "I think I will sit down here and grunt." The Name Giver said, "Well then, that will be your name: ***seeex***."

The name ***xpanooil*** and a variant ***xpanoois*** are compounds incorporating *xepe* 'sea', but any further analysis would be speculative.

The clams are not recalled as a common food. Several people commented that the clam was eaten by the Ancestors as well as the Giants, but noted that eating it was said to irritate the throat, causing coughing. Sara

Villalobos, when talking about the clam, laughed as she described the Giants: "[The Giants] ate them wherever they were; they lived where we live. It is said that the Giants were blind. They had big red pop-eyes. They used their own urine as eye drops!"

The shell was sometimes used as a utensil for eating, and one person specifically mentioned its use for eating cooked sea turtle blood. The shell, sturdy and not easily broken, could also be used as a tool to scrape the flesh of the barrel cactus to obtain its liquid for drinking.

Bracelets and nose ornaments (called *ziix coosyat ineemj* 'Giant's nose ornament') made of this shell have been found at prehistoric sites in the region. The people today do not personally recollect such use and comment that the objects were worn by the Giants.[17] Intriguingly, among a collection made by William N. Smith on Isla Tiburón in the mid-twentieth century are twelve handmade wooden boxes used to store personal items (see Preliminaries): five of the boxes each contain several unfinished *Glycymeris gigantea* shell bracelets of non-Seri prehistoric manufacture.

Noetiidae

Arcopsis solida

This small clam grows attached by a byssus to the underside of intertidal rocks. Its white shell is finely ribbed.

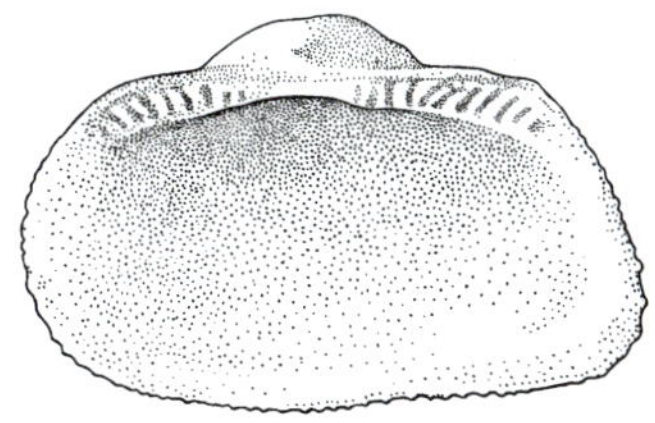

L 14 mm

Pteriidae (pearl oysters)

The Gulf of California figures prominently in the history of the early exploration of Mexico; tales of great riches in pearls kept by native peoples fueled the fires of such exploration. The pearling industry in the Gulf lasted for several centuries and was so extensive that by the nineteenth century the pearl oyster beds were depleted.[18]

There is no historical record or oral account of the Seris using pearls for adornment, and the people today confirm that they were collected only for trade or sale to non-Seris.[19] Since the Seris did not dive for pearls,[20] they were usually harvested when a strong storm and heavy seas deposited oysters (possibly having been affected by a natural toxic contamination) on shore, as happened in 1956 when great numbers of *Pteria sterna* oysters were cast onto a long stretch of beach near Desemboque.

Figure 4.7. Pearls from the Gulf, collected by Edward H. Davis in 1922. National Museum of the American Indian, Smithsonian Institution (#112407.000). Photograph by the author.

The pearls were collected and kept in small, hollow stem sections of reedgrass (*xapij*, *Phragmites australis*) called *xapij hapazix xiica iictoj ano yaii* 'cut *xapij* in which are pearls'. A pearl, called *ziix iiquet* 'child of the thing', was considered bad to possess, as pearls were said to belong to the *xpacaao* 'who travels the sea', a mermaid-like creature who wore pearls as jewelry.[21] According to one consultant, however, pearls were kept by shamans as a type of protection. In 1678, in what may be the earliest ethnographic note about the Seri knowledge of mollusks, Father Ortiz Zapata recorded what a Seri man told him about the pearl oyster after showing him some pearls: "and said that inside of these, from [an oyster], had been those granules that become the animals they gather in the sea."[22] This is interesting to note, as the Seri names for the oyster and the pearl perhaps reflect such a belief, yet one that the Seris do not maintain today.

According to Juanita Herrera, the meat of the pearl oyster is free of sand. She recalled that when she was young, her mother cooked some pearl oysters and afterward found a grape-sized pearl inside one; it was not affected by the cooking, and her mother sold it to a Mexican fish buyer, Antonio Topete.[23] Juanita added that the people would take pearls from pearl oysters, mussels, and pen shells, and sell them in Hermosillo, the state capital.

The pearl oysters were not a major source of food. Those eaten were the ones collected at very low tide or found cast on shore still alive. The oysters could be eaten raw or roasted under a brush fire.

The shells of pearl oysters were not used for decoration or personal adornment, and although prehistoric worked shells are found in the region, the Seris have no knowledge of their use, and consider them as having been used by the Giants. On the other hand, the shells themselves were sometimes used as spoons or containers in which to mix pigments for face painting.

The shells were used medicinally. They were roasted, ground into a fine powder, and applied to the face to heal skin eruptions. The powder was also applied to the umbilical stump of a newborn to help it heal and drop off more quickly. Shells of mussels and *Turbo fluctuosus* snails, as well as sun-bleached deer antlers, were also prepared and used in this way.

Pinctada mazatlanica

caaitmoj
casopaj
♦ ***copas***
xiica quiictoj 'pregnant things'
ziix iiquet inaail 'the pearl's shell'

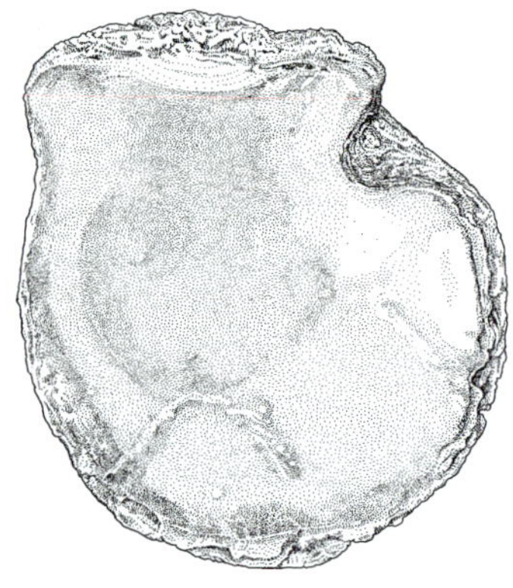

L 128 mm

This large oyster lives attached by a byssus to rocks or other hard substrate. Its shell is heavy and thick, with a striking pearly-white interior. One woman, noting the fact that this oyster is less common than *Pteria sterna*, quipped: "There aren't any now because the Seris ate them all!"

A ***caaitmoj*** is a deer with five or more points on each antler; the name derives from *cooitom* 'five'.[24] The connection to the oyster's name, if any, is unclear.

Casopaj and ***copas*** are primary names. ***Xiica quiictoj*** 'pregnant things' is the name given to the pearl oysters because they contain pearls.[25]

According to many Seris the original name for this oyster was ***copas***. In explaining the different names, Eva López related that long ago a child was named *Copas* because his pre-emerged teeth resembled little pearls beneath his gums. He died, and because of name taboo, his name was no longer spoken. The oyster was then renamed with the descriptive phrase, ***ziix iiquet inaail*** 'the pearl's shell'. However, another person suggested that juveniles of both species of pearl oysters are called ***copas***, but when large enough to contain pearls, they take on the latter name.

Modified fragments, including perforated and smoothed shell discs made from this shell have been found in old campsites. Recently, a few Seri carvers have used the shell for inlay in items made from ironwood or other materials.

Figure 4.8. An *Oenothera arizonica* plant near Desemboque, 2006.

The Arizona evening primrose (*Oenothera arizonica*) is called *hantoosinaj cooxp* 'white *hantoosinaj* (*Abronia villosa*)'. María Luisa Astorga provided another name for the plant: *hant copas* '***copas*** of the land', because the flower's white petals resemble ***copas*** shells.

One almost forgotten place name refers to this oyster. *Copas Caaizi Iti Yaii* 'where the ***copas*** gatherers lived' is a shore area on Isla Patos, said to have been used by divers harvesting pearl oysters. As this was the then named San Xavier region where placers (beds of pearl oysters known to be good sources of natural pearls) were exploited in the eighteenth century,[26] and since the Seris did not dive, the place name possibly references pearlers working the area in the past.

Pteria sterna

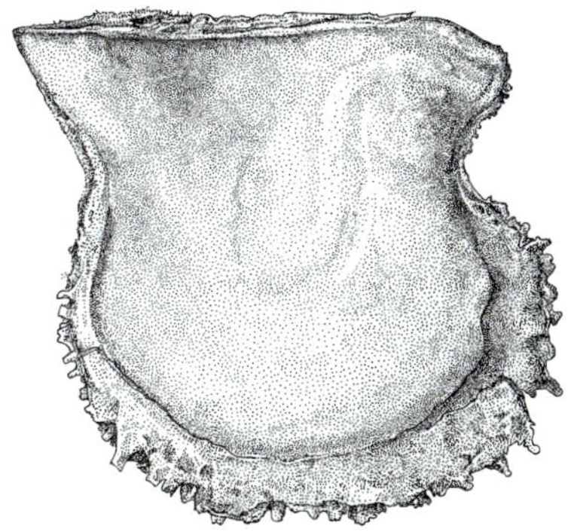

L 90 mm

soocajam
xiica quiictoj 'pregnant things'
♦ ***ziix iiquet inaail*** 'the pearl's shell'

The more common of the two pearl oysters in the region, this oyster is often found clumped and attached by a byssus to sea fans or corals in beach drift from bycatch of commercial shrimp trawlers. The smaller and more thin-shelled of the two pearl oysters, its shell has a nacreous purple-blue interior.

Soocajam is a primary name. ***Xiica quiictoj*** 'pregnant things' is the name given to the pearl oysters because of their bearing pearls. ***Ziix iiquet inaail*** 'the pearl's shell' is the oyster's most common name. Some people give the name to *Pinctada mazatlanica* as well, as both oysters produce pearls.

The oyster's shell was often used as a receptacle for face paint (figure 4.9).

Figure 4.9. A *Pteria sterna* shell containing pigment made with commercial bluing. Collected by James Manson in the mid-twentieth century. Arizona State Museum, catalog no. E-8816.

Isognomonidae (purse oysters)

Isognomon janus

♦ ***copas hapapl*** 'strung ***copas***'
oot icopas 'the coyote's ***copas***'

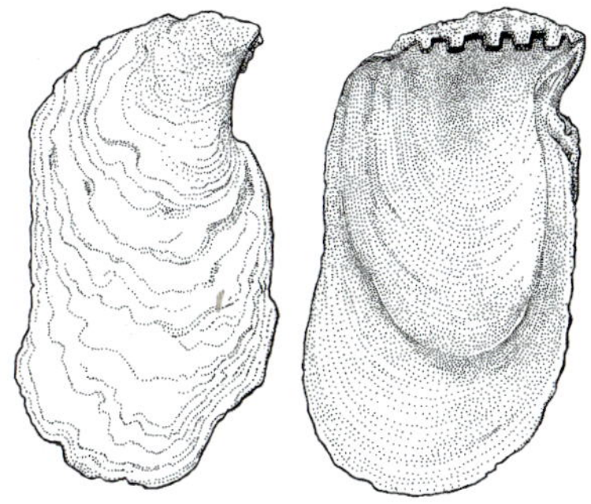

L 16 mm

This very common small oyster grows attached by its byssus, in clusters beneath intertidal rocks.

The oyster is called ***copas hapapl*** 'strung ***copas***' because the shape of its shell and nacreous interior are similar to the pearl oyster ***copas*** (*Pinctada mazatlanica*), while being small enough to be strung whole on necklaces.[27] In addition to being similar to the ***copas*** shell, the small size and odd shape suggest it belongs to the coyote, hence the name ***oot icopas***.

Gathered alive at low tide, this tiny oyster was not eaten. The shell's rough, ridged exterior is described as resembling an old woman's fingernail; in preparation for stringing, most of its sharp, ventral scaly edge is removed, leaving the pearly interior intact (figure 4.10).

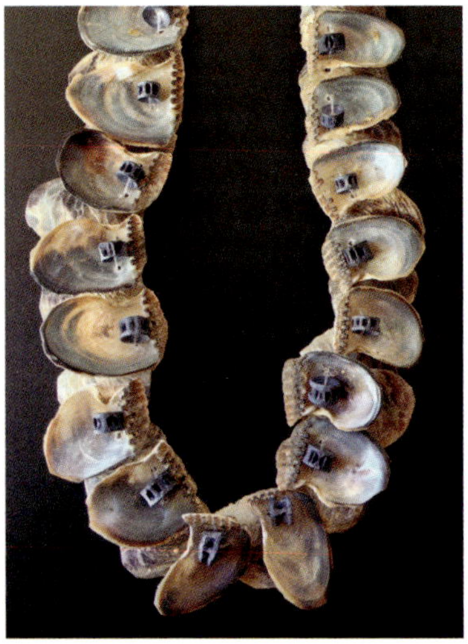

Figure 4.10. Detail of a necklace of trimmed *Isognomon recognitus* shells strung together with dyed shark vertebrae.

Isognomon recognitus

♦ ***copas hapapl*** 'strung ***copas***'
oot icopas 'the coyote's ***copas***'

This oyster is very similar to the *Isognomon janus*, but is a darker purple. Often found growing together in the same intertidal habitat, the shells of both are prepared for stringing in the same manner.

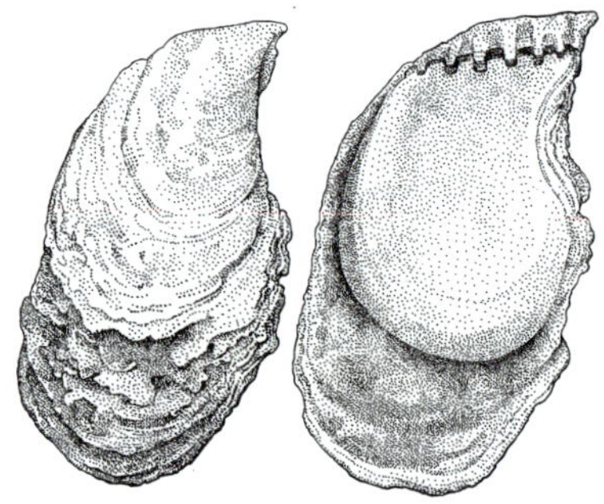

L 17 mm

Pinnidae (pen shells)

All pen shells are named ***seeten*** and are generically grouped as *ziix* ***seeten*** *hant cöquiih* 'thing grouped with ***seeten***'. Although their adductor muscles are noted as being of different shapes or sizes, "their insides are the same" (see Classification and Naming).

The two most common species are individually known as ***seeten ctam*** 'male ***seeten***' (*Pinna rugosa*) and ***seeten cmaam*** 'female ***seeten***' (*Atrina tuberculosa*). As the people are aware that there are both males and females of each species, the names do not refer to the sex of the animal, but rather to the shape of its shell.[28] The names suggest typical sexually specific shapes: the "male" shape is described as elongate, while the "female" is short and round (see Names and Naming, p. 25).

Three other Gulf pen shells have recently acquired names. There seems to be a lack of consensus in correlating the names with species, however. One name appears to have originated with the shell's local Spanish name, *marimacha*, probably resulting from contact with non-Seri fishermen after commercial harvesting began (see *Atrina maura*).

The pen shells were not a major source of food in the past since they grow firmly anchored and at least partially buried, making them somewhat difficult to harvest. Their spiny shells are sharp and can easily cut, so they were generally not worth collecting. Although known to be edible and even considered to be good food by some, only those found in the intertidal area were consumed.[29]

According to one person, a small pen shell is referred to as *imaii* '[the one] that doesn't wake up'. As was explained, just as a sleeping person is not disturbed, so a pen shell of this size is left alone until it is large enough to harvest.

Both a word and a phrase are used to refer to the harvesting of pen shells. The term *caai*, also used for harvesting cactus fruit, implies harvesting something that is anchored. A phrase used for gathering something by diving, probably more recently applied to the pen shells since Seris were not traditionally divers, is *an hant cöcoiitim* 'to go down [toward]' (see table 3.6, p. 53).

Figure 4.11. Cleotilde Morales removing the adductor muscles from newly harvested pen shells, Isla Tiburón, March 2009.

Presently the pen shells are harvested commercially by both Seris and non-Seris by hookah diving from a boat, primarily in the Infiernillo Channel.[30] The whole pen shell is uprooted using a metal tool with a widened triangular tip and called *seeten ipápotim* 'with which one extracts ***seeten***' (figures 4.12 and 4.13). The complete animal is removed from the shell while in the boat and the shells thrown back into the sea; the edible parts are then separated in the boat or onshore. The large adductor muscle, *itaqueecöj*, brings the higher price, while the mantle (*iteeloj* 'its edges') and the foot with its byssal gland (*iti cöihiti* 'what it connects with') are sold for less (figure 4.14).[31] Often the latter parts are consumed locally rather than sold, unless large quantities are harvested.

Figure 4.12. A metal tool used to uproot pen shells. Desemboque, 2013.

Figure 4.13. A freshly harvested *Pinna rugosa*, still held by the extracting tool. Isla Tiburón, March 2009.

Though not heavily consumed in the past, when the pen shells were eaten, they were eaten raw or boiled after being removed from the shell, or the live animal in the shell was roasted under a brush fire. One person said the mantle could be cooked in or under hot coals. When referring to the raw meat, Cleotilde Morales said that "[the people] would eat the meat of the mantle and muscle—it was eaten raw by people while in the ocean—that's how they would eat it."

Juanita Herrera recalled that when first offered the meat, she did not want to eat it; when she later tried a spoonful of soup made with it, she was surprised and really liked it. She explained to me: "You might think they were eaten in the past, but they weren't." Juanita said that she now prefers eating the adductor muscle (locally referred to in Spanish as *callo de hacha*) raw, with onion and lemon juice, as "It isn't tough like that—when it is cooked it is tough." Her son, René Montaño, commented that he liked eating the mantle the best, dipped in egg and fried.

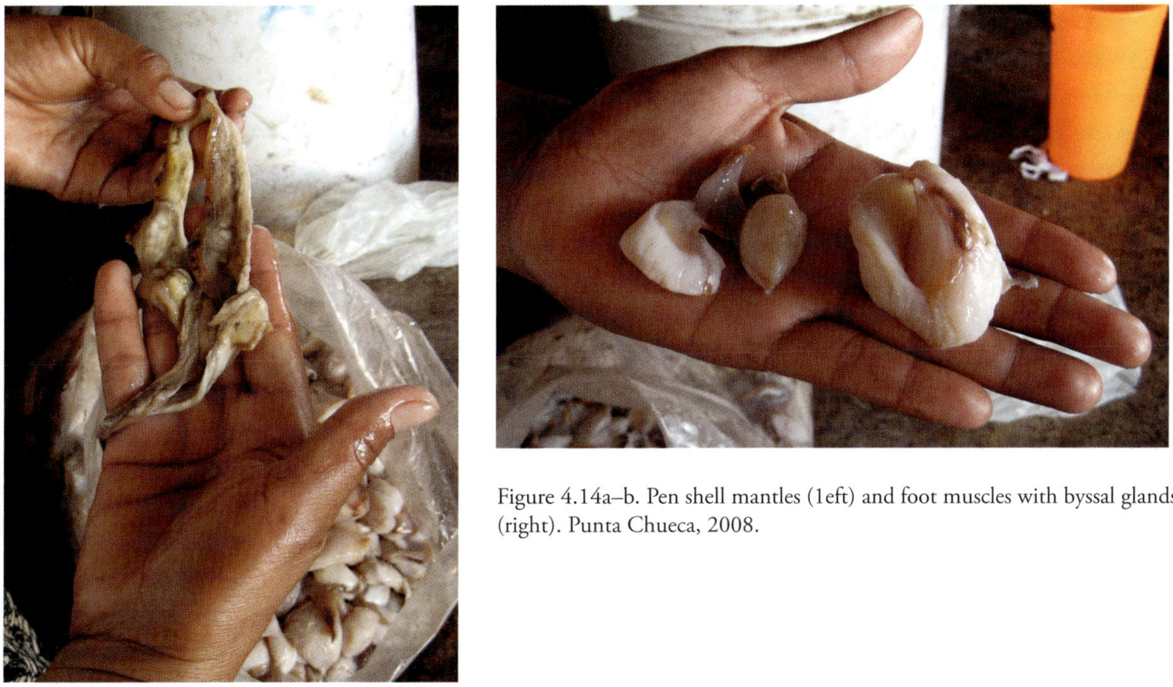

Figure 4.14a–b. Pen shell mantles (left) and foot muscles with byssal glands (right). Punta Chueca, 2008.

Figure 4.15. A doll with pen shell byssal fiber hair, made in the 1970s.

The pearls from the pen shells were sometimes sold to outsiders,[32] together with those from pearl oysters and mussels, as the Seris did not personally use pearls (see Pteriidae).

W J McGee reported that according to his Seri interpreter, Mashém, women paddled balsas lying prone, using large shells.[33] Edward H. Davis mentions women paddling balsas using "large hatchet clam shells" when navigating the short distance to Isla Alcatraz for a pelican hunt.[34] Angelita Torres confirmed such use of the pen shell. Davis further recorded that the shell of a large *Pinna rugosa* was sometimes used as a musical rasp, as a stick or other object stroked against its rough surface produced a rasping sound.[35] This use was also confirmed by a Seri woman.

Although rather fragile, the large, flat shells could be used as a surface on which to place meat or other solid foods.

The name for the byssal fibers of the pen shell, ***seeten*** *itaa* '—of the ***seeten***' uses an archaic word that also occurs in the name for corn silk, *hapxöl itaa*. Another name is ***seeten*** *imas* '***seeten***'s pubic hair'. Since the name is considered somewhat vulgar, a more commonly used name is ***seeten*** *ilít* '***seeten***'s hair'. These fine, shiny golden-brown fibers are sometimes sewn as hair to the head of a cloth doll (figure 4.15).

An intriguing early note by Edward Moser gives an archaic use for byssal fibers, in which Roberto Herrera recounted something he was told by his wife, Ramona Casanova:

> *Setoj immaas* old, old use. Even Ramona's mother hadn't seen it—but she talked of it. Seems to be old custom—of rolling a piece of twine of *séttooj* hair. Had some sort of good luck use. Ramona's mother said that Ramona should take away the fire ashes in the afternoon and make a new fire so that she would not "lose her *setoj ímmaas*." R. T. [Roberto Herrera] still thinks that it was a sort of woven fabric made of *setoj ímmaas* twine. Chico [Romero] and Vichi [Jesús Morales] don't know about it.[36]

María Luisa Astorga recalled that her grandmother referred to such an item, where small bundles of the fibers were hung on a necklace as amulets.

Atrina maura

oot iseeten 'the coyote's ***seeten***'
seeten cmaam ctam 'female-male ***seeten***'
seeten iizax coocp 'quarter-moon ***seeten***'
seeten inaail cquihjö 'red-shelled ***seeten***'

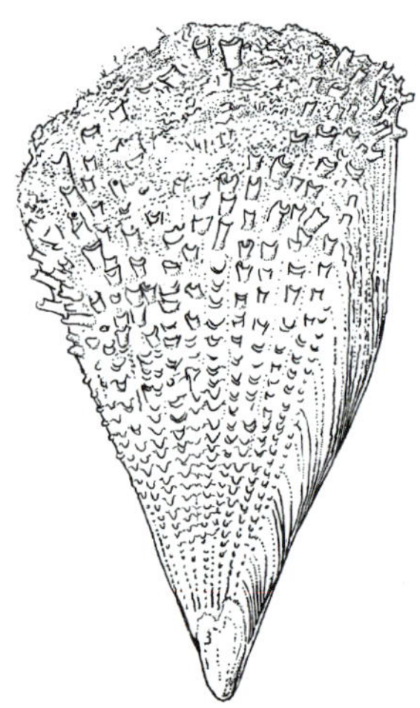

L 235 mm

This pen shell is not common in the region.[37] Its reddish-hued shell is shorter than that of *Pinna rugosa* and is somewhat less rounded than that of *Atrina tuberculosa*. The spines are slender and almost tubular in shape.

The name ***oot iseeten*** 'the coyote's ***seeten***' is sometimes given to this shell because it is unlike either *A. tuberculosa* or *P. rugosa* and is less common. The name is also given to *Atrina oldroydii*.

Since this pen shell shares characteristics of both ***seeten cmaam*** (*A. tuberculosa*) and ***seeten ctam*** (*P. rugosa*), both names occur in its name ***seeten cmaam ctam*** 'female-male ***seeten***', which is probably a calque or direct translation of the local Spanish name, *callo marimacha*, an offensive term for lesbian.

This pen shell is locally known in Spanish as *callo media luna* 'half-moon pen shell'. The Seri name ***seeten iizax coocp*** 'quarter (waxing) moon ***seeten***' is a new name, possibly a calque of the Spanish name.

Another name, ***seeten inaail cquihjö*** 'red-shelled ***seeten***' describes the shell's reddish hue.

Atrina oldroydii

oot iseeten 'the coyote's ***seeten***'
seeten comihj 'smooth ***seeten***'
seeten cooil 'blue/green ***seeten***'

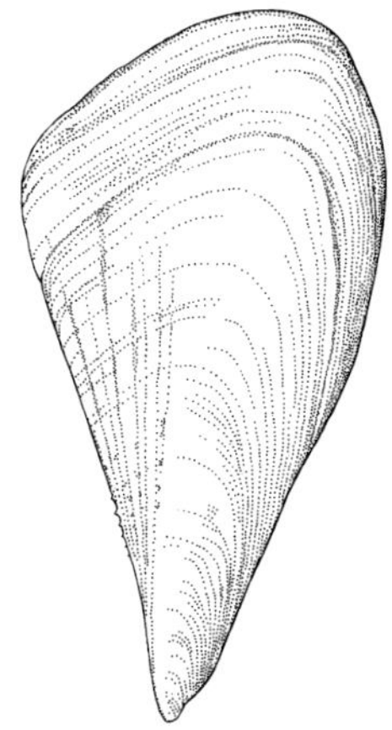

L 195 mm

This uncommon pen shell is generally unknown to the Seris. Its smooth-surfaced shell is a dark, bluish gray.

A live specimen collected off the northeast coast of Isla Tiburón was identified as ***oot iseeten*** 'the coyote's ***seeten***' because it was so strange. At that time, I was told by the man who found it that it was only the second he had ever seen in the area. When I showed the shell to Alfredo López, he confirmed the name ***oot iseeten*** and described the different kinds of pen shells. He added that he had seen such a shell on the west coast of Baja California, as well as several around Isla Tiburón.

The descriptive names ***seeten comihj*** 'smooth ***seeten***' (given because of the shell's lack of large spines) and ***seeten cooil*** 'blue ***seeten***' (reflecting its color) were more recently supplied by Xavier Moreno. He described the pen shell's muscle as being good-sized and buff-colored.

In 2010, a large bed was discovered by non-Seri Mexicans in deep water near Bahía de Kino, and was commercially harvested.

Atrina tuberculosa

seeten cmaam 'female ***seeten***'

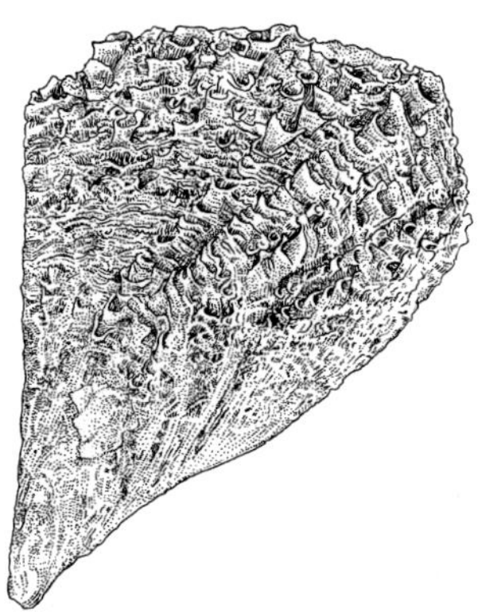

L 120 mm

This smaller pen shell lives almost completely buried, with only a small portion of the upper shell visible. Its shell has rows of dense, broad tubular spines.

The name ***seeten cmaam*** 'female ***seeten***' describes the shell's more blunted, round "female" shape. The pen shell is known locally in Spanish as *callo riñon* 'kidney-shaped *callo* (adductor muscle)'.

Pinna rugosa

seeten ctam 'male ***seeten***'

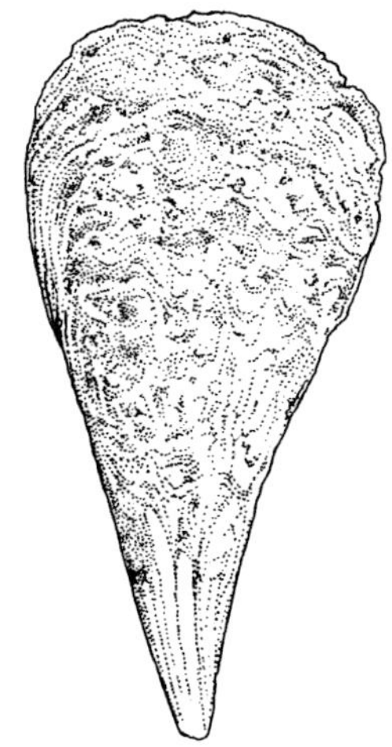

L 335 mm

This large pen shell lives only partially buried, with the upper third of its shell exposed. Its long and slender shell has fewer rows of spines than *Atrina tuberculosa*; its spines are usually worn off in older shells. The shell's pearly interior is somewhat divided lengthwise into two sections.

This pen shell is locally known in Spanish as *callo redondo* or *callo de hacha*. Its large size and elongate "male" shape give this shell its Seri name, ***seeten ctam*** 'male seeten'.

An area in the sea near the mouth of Estero Sargento is named *Seeten Ctamcö*, with the plural form of *ctam*, because of the many *P. rugosa* found there.

Said to have an extremely strong pinch that can draw blood, the commensal shrimp *seeten iixz* 'the ***seeten***'s pet' (*Pontonia* sp.) lives inside the shells of live *P. rugosa*.

Xavier Moreno, who worked harvesting pen shells, wrote an essay in Seri about the *P. rugosa*:[38]

> In earlier times, a Seri person did not eat the ***seeten ctam***. The adductor muscle of the ***seeten ctam*** is circular. Furthermore, its shell is big. One should work with it carefully. [Its broken shell] contains glochids and its shell is sharp and capable of cutting.
>
> They say that Mexicans ate them [in the past], and maybe that's so; but in the present day the Seris eat them a lot and collect them a lot [to sell]. It's true that it will be good for us to take care of the ***seeten ctam***.
>
> There are many, but it won't be good if we don't take care of them. The Infiernillo Channel is ours, it seems, but many outsiders come into the Infiernillo Channel and take away many ***seeten***. That happens to our detriment. That is happening because we don't take care of them. It will be good for us to take care of the ***seeten ctam***.

Figure 4.16. A commensal shrimp, *Pontonia* sp., in a freshly harvested *Pinna rugosa*, Isla Tiburón, 2009.

Figure 4.17. A small *Pinna rugosa* at low tide. Punta Chueca, November 2005.

Pink pen shell (unidentified)

seeten itaqueecöj caacoj '***seeten*** whose adductor muscle is big'

Apparently a new arrival in the Infiernillo Channel, this pen shell is rare.[39] It is described as having a pink-edged mantle and a large adductor muscle.

Known in Spanish by the name *callo rosa* 'pink pen shell',[40] the Seri name refers to the size of its muscle.

Ostreidae (true oysters)

Based on oral accounts and evidence in ancient camps, it is obvious that oysters were heavily consumed. Indeed, their shells are deeply layered in shell middens bordering esteros, where the oysters were plentiful. And not only an important food source for the Seris themselves, the live oysters were also harvested and sold to outsiders in the early mid-twentieth century.

María Luisa Astorga categorized the non-pearl oysters as *ziix **haxt** hant cöquiih* 'thing grouped with ***haxt***', suggesting that the large oyster was perhaps the earliest and most consumed oyster, the "basic" oyster. Then, in grouping the different oysters together, she noted the similar appearance of the internal animal.

Crassostrea columbiensis [= *Ostrea columbiensis*]

haxt
stacj caacöl 'large (pl.) ***stacj***'

Although occurring in the same estero habitat, this oyster is not as clearly identified as the larger *Crassostrea corteziensis*. While one person called it ***haxt*** (the name given to *Crassostrea corteziensis*), others described it as "***haxt** xahxaii*" '[something] like an ***haxt***', and one person suggested ***stacj caacöl***, a name more commonly given to *Myrakeena angelica*, which has a similar rounded shape. This uncertainty is hardly surprising, as the *Crassostrea* spp. oysters are not gathered in the quantities they were in the past century, so there is much less familiarity with them today.

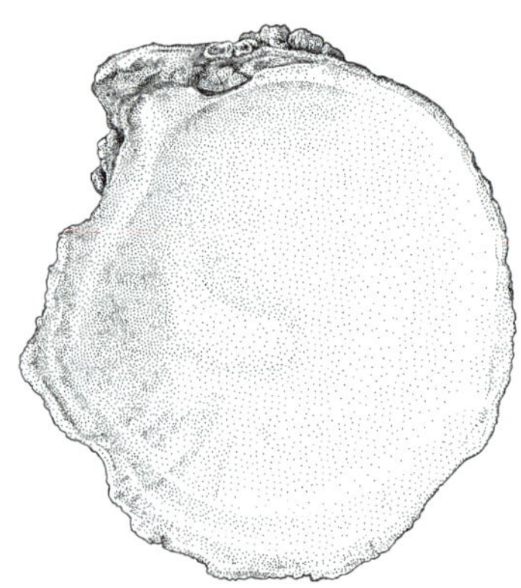

L 68 mm

Crassostrea corteziensis [= *Ostrea corteziensis*]

♦ ***haxt***
xat

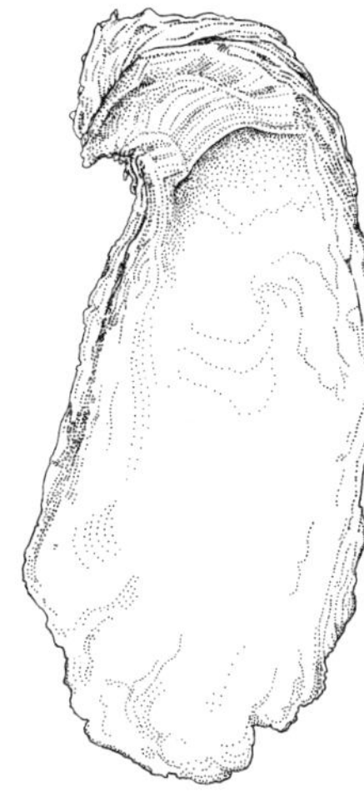

L 85 mm

This oyster was especially common in Estero Santa Cruz, most often occurring on sandy mud flats or growing attached to the roots of the red mangrove (*Rizophora mangle*). Its heavy, thick shell can reach an impressive size.

Haxt is a primary name. The phrase *pnaacoj ano* ***haxt*** '***haxt*** in mangroves' is given to those found on mangrove roots, and which are described as being clean and mud-free.

The much less common name ***xat*** is somewhat of a mystery; it is interesting to note the name's similarity to ***haxt***, the oyster's other name. ***Xat*** is the term for hail and a plant name (*Crassula erecta* [= *Tillaea erecta*]) as well.[41] The name was supplied by Victoria Astorga, who said she learned it from her aunt.[42]

These oysters were easily collected from sandy mud flats or from mangrove roots, where they were detached with a sharp rock or stick. The oysters were consumed raw or cooked by being roasted under a brush fire.

That the oysters were abundant in Estero Santa Cruz is obvious by the quantity of their shells in the surrounding middens; still recalled by name is an area in the estero where many of the oysters were found: *Haxt Án* 'area of the ***haxt***'. One woman commented that the estero used to be full of these large oysters, but that now only clams and smaller oysters are found there; she suggested that recent commercial cultivation of shellfish there had depleted the population "because of the fertilizers used."

When the Seris began selling seafood in the 1920s they harvested these oysters, transporting them by boat from Estero Santa Cruz and selling them to non-Seris in Bahía de Kino (figure 4.18).

Figure 4.18. Unloading *Crassostrea* spp. oysters from a Seri boat. Bahía de Kino, 1929. The woman on the left is identified as Ramona Blanco. Photograph by Carlos Ronstadt. Arizona Historical Society, Ronstadt Collection, Indians-Seri, #49633.

Efraín Estrella spent some of his childhood (around 1944) living at a camp bordering Estero Santa Cruz. He recounted his experiences helping to harvest the oysters during what were probably some of the last times that they were harvested there for sale by the Seris:

> *Xatj Icaiin*, *Hant Xpeecol* and *Haas Poot Quipcö* are places near Estero Santa Cruz. I will tell about when we lived there. These are the things we did.
>
> We were at *Haas Poot Quipcö*, a long time ago. My grandfather Chico Romero was a strong adult, he had good eyesight and was healthy. Lots of other [Seris] were there. Jesús Morales, Eugenio Morales, and Roberto Herrera were there. This is what happened. [One] morning a Mexican man—I don't know his name—came to us and said he wanted to buy ***haxt***. The people there were so happy and every day went into the estero to gather the oysters. [When we began] we didn't know what their value would be. When the tide was way down, we went to where our boats were anchored, in the water, and everyone who could, went. Each boat had a pole [mast], and we used blankets for sails. When the tide was way down, there was slippery mud there that was so deep it reached up to an adult's shoulders. It was so deep. We went like that, far into the estero, not in the water, but in the mud.
>
> We went deeper and deeper into the estero, and came to the place where the oysters were. It was like a shore there, with dry areas and some rocks, like where ***haxöl*** [*Leukoma grata*] are found in other places. The people put the oysters into *hapnaail hapatjc* [sacks formed from women's skirts] because there was nothing else in which to put them. We went through the mud and came to the place where the oysters were. When the people saw the oysters there, they were all so happy. The boats were really loaded with oysters. Everyone there got as many as he could. They kept them separate in the boats, so everyone knew which were his.
>
> To get to the place where more oysters were—the higher places—the people had to wade through the narrow muddy channels; the older people went first, to make sure there was nothing dangerous in the mud, like *hacat cmaam* [a stingray]. Then the children went through.
>
> The mud was so deep, sometimes it came up to one's waist. That is what it was like. All morning people worked getting oysters. Chico Romero was an adult, and he was strong, and had good eyesight, so he filled and carried a big sack of oysters.
>
> When it was time to leave, [sometimes] a *xnaai* [south wind] would be blowing, and people didn't row—the wind just sailed them along. There wasn't much water with the mud. It was easy to sail on.
>
> We got to the shore [at Bahía de Kino] and sold the oysters. We wanted to eat meat; the Mexican man brought a donkey and we bought the meat and ate it. Two pesos, maybe one peso and a half, is what meat cost back then.
>
> Every morning we went back [into the estero]. There were lots of oysters. The man put them in a truck. There were lots of people, so lots of oysters. We ate meat and we drank water [that the man brought].
>
> Then one day my grandfather Chico Romero said he was going to the ranch called Agua Zarca [a non-Seri ranch near Bahía de Kino]. We didn't know where that was. He took his belongings and walked away into the desert.
>
> The hot time of year was coming, when there were biting gnats, so my father said, "Let's go to *Tahejöc* [Isla Tiburón]." So we did. I didn't see my grandfather for a very long time. The next time I saw him was on *Tahejöc*, and he was an old man.

Figure 4.19. A dune midden bordering Estero Santa Cruz, at low tide. The abundant shells are primarily those of oysters and Venus clams.

Figure 4.20. A boy filling a metal container with *Crassostrea* spp. oysters that had been brought in by boat, Bahía de Kino, 1933 or 1934. Photograph by Gordon Gordon; courtesy of University of Arizona Libraries, Special Collections. MS316_B26_f18.

Crassostrea gigas

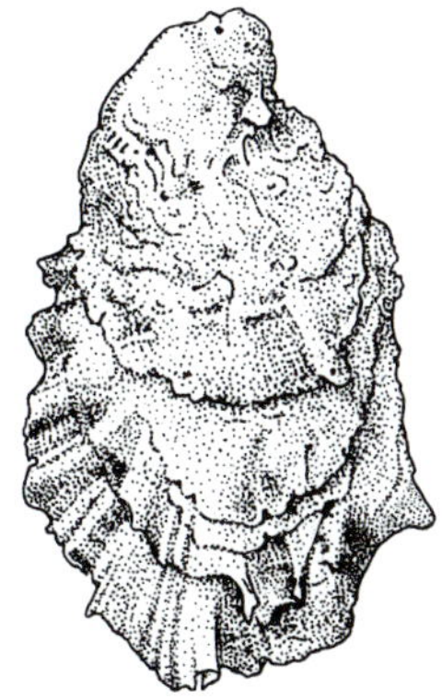

L 70 mm

This oyster has a delicate, frilled shell lightly striped with violet. A recent arrival to the region, it was introduced by non-Seri Mexicans to Estero Santa Cruz in June 1979, when a boatload was brought from Baja California.[43] The oysters were later found growing in Estero Santa Rosa as well and are eaten by the Seris today.

Myrakeena angelica [= *Ostrea angelica*]

stacj caacöl 'large (pl.) ***stacj***'

L 75 mm

This oyster is often found singly and unattached in mud, especially in Estero Santa Cruz. It grows clumped on roots of the red mangrove as well.

The oyster's resemblance to ***stacj*** (*Saccostrea palmula*) and its larger size give it the name ***stacj caacöl*** 'large ***stacj***'. Although ***stacj caacöl*** is a plural form, the name is also used when referring to a single oyster.

Cleotilde Morales recounted a time in her childhood when her mother, María Antonia Colosio, harvested and prepared the oyster near *Sacpatix*, a camp on the eastern shore of Isla Tiburón:

> My mother gathered these oysters and roasted them on the beach. After they were cooked, she removed the [top] shells and lined her basket with the [lower] shells containing the meat. She piled more and more around on top of them until her basket was piled high and brought them back to our camp in the desert.

Figure 4.21. *Myrakeena angelica* shells in a desert camp bordering Estero Santa Cruz, 2007.

In 2008, I brought some larger shells of the oyster from Bahía de los Angeles, on the east coast of Baja California and opposite the Seri region. When he saw the shells, Efraín Estrella confirmed that the oysters are found there as well as around the Gulf islands, and added that the non-Seri Mexicans collect them "by the bagfuls." He commented that the oysters from Baja California have a "fishy" odor, and he didn't like them. His wife, María Luisa Astorga, concurred, but added that the Seris who lived there in the past did eat them.

Another name given to various oysters by several older people is *cateesic*, a name related historically to *pte quiteezc* 'stacked together'. I have been unable to connect it clearly to any particular species. Efraín Estrella recalled that long ago he was shown a large oyster on Isla Tiburón by a Seri man, Pancho Contreras, who identified it as a *cateesic*. Efraín described the oyster as being similar to a ***stacj*** oyster and found on rocks, but having a much thicker and sharply fluted shell and containing only a small amount of meat.

Saccostrea palmula [= *Ostrea palmula*]

stacj
stacj hapahit '***stacj*** that are eaten'

Figure 4.22. *Saccostrea palmula* oysters growing on eroding cemented beach deposits, *Xtoozajö Cöhanípatim*, Isla Tiburón, 2009.

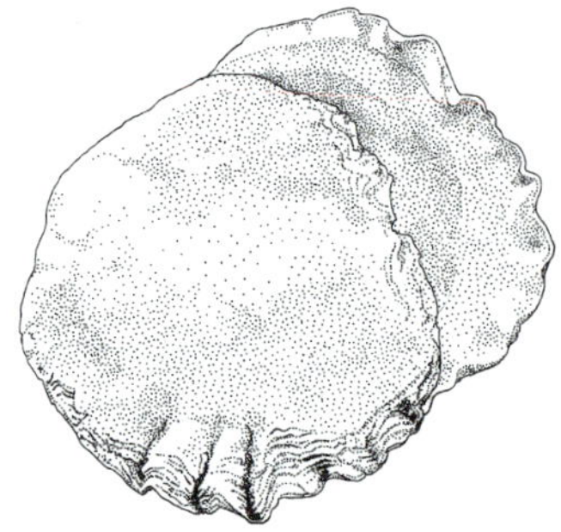

This very common oyster, found attached to rocks or mangrove roots, has a sharp-edged fluted shell with a greenish-gray interior.

Stacj is a primary name. The name ***stacj hapahit*** '***stacj*** that are eaten' refers to these medium-sized oysters, often found growing individually rather than in clumps. The oysters are good to eat, in contrast to the smaller, spiny ***stacj coiicöt***, which grow closely clumped together on rocks. Sometimes the ***stacj hapahit*** is referred to as ***stacj*** *isoj* 'real ***stacj***' to distinguish it from the smaller, less edible ***stacj coiicöt***.

The ***stacj hapahit*** that sometimes grow in clumps on rocks or mangrove roots are described as ***stacj*** *ptiiqui cosiiija* '***stacj*** lying together'.

These oysters were eaten raw or were cooked by roasting them under a brush fire. They were gathered live by removing them from rocks or roots using a sharp rock or stick.

Saccostrea palmula [= *Ostrea palmula*]

stacj coiicöt '***stacj*** that kill'
stacj quítajij (pl.) 'sharp ***stacj***'

Described as being the common oyster around Desemboque, this small oyster with sharp, fluted edges grows clumped closely together on intertidal rocks.

The name ***stacj coiicöt*** '***stacj*** that kill' suggests that these are not good to eat, while ***stacj quítajij*** 'sharp (pl.) ***stacj***' alludes to the shell's sharp edges.

These oysters, described as having a bitter taste and blackish juice with a strange odor, were not commonly eaten, as eating them was said to produce headaches, dizziness, and throat irritation. María Luisa Astorga concurred, recalling that her mother cooked up a lot of the oysters at Desemboque; the family ate them and became ill. Eva López commented that long ago people ate the oysters raw when at the beach, and became sick; she chuckled, however, dismissing the idea that eating them would kill, as the name suggests.

Anomiidae (jingle shells)

Anomia peruviana [= *Anomia adamas*]

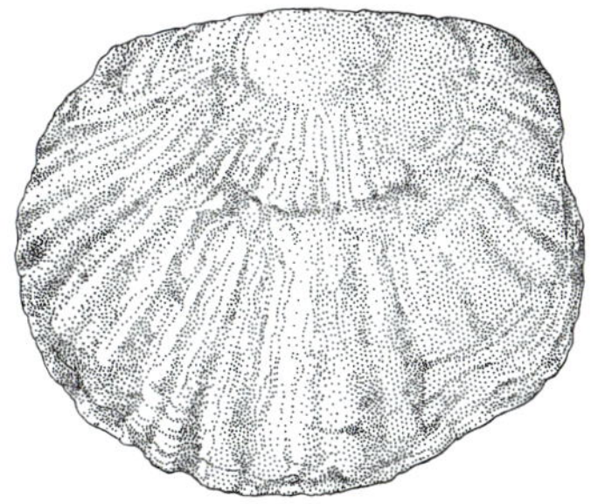

L 50 mm

haxölinaail cmasl 'yellow shells'
♦ ***haxölinaail cxatlc*** 'thin shells'
haxölinaail hamisj istj cmis 'shell like an *hamisj* (*Jatropha cinerea*) leaf'
pahotmazexe (var. ***pahotconzexe***, ***xepe hatmazexe***, ***satmazexe***)

This thin-shelled bivalve is found attached to rocks or other shells through its perforated lower valve. Its translucent and somewhat iridescent shell varies in color from white to bright orange.

The best known descriptive names, ***haxölinaail cmasl*** 'yellow shells' and ***haxölinaail cxatlc*** 'thin shells' are said to be recent. Another descriptive name, ***haxölinaail hamisj istj cmis*** 'shell like an *hamisj* (*Jatropha cinerea*) leaf', is not commonly used.

The oldest name ***pahotmazexe*** is a compound of forms of the verbs *quiho* 'to see' and *czexe* 'to cut/slice'. The name is derived from a conversation (loosely translated) between a mussel and a jingle shell:

> *Satoj quih ox tee, yoque:*
> *—¿Háquis xah tquiihtoj ma, coox hizi nsaiitaj?*
> *Pahotmazexe quih ox tee, yoque:*
> *—Tiipe, hax maa. Toc cömiihca tax, zo paaho, tmazexe ha.*

> The mussel said:
> "Where on earth were you when they were prying us loose with spears?"
> The jingle shell replied:
> "What a crying shame! If I had seen one of them, I would have cut him!"

As explained by Cleotilde Morales:

> Many people came to collect ***satoj*** [mussels]—they gathered lots of them. There was no protection by the ***pahotmazexe*** [jingle shell], who later said, "If I had seen one of them I would have cut him." This shell was a friend of the ***satoj***.

Figure 4.23. An *Anomia peruviana* shell pendant on a necklace worn by Juanita Herrera in 2007.

Alfredo López further explained that although the jingle shell was found growing with mussels it did not protect them when they were being harvested.

Part of the response by the jingle, "*zo paaho, tmazexe ha*," has been shortened and modified over time, resulting in the variations in the names ***pahotmazexe***, ***pahotconzexe***, ***xepe hatmazexe***, and ***satmazexe***.

Although not consumed today, according to some Seris the bivalve was eaten in the past. This claim is supported by the quantity of large *Anomia* shells, sometimes fire-blackened, that are common among other food shells in old seaside camps near esteros.

Today the iridescent top valves, easily collected in beach drift, are pierced and strung on necklaces together with other shells. Collections at the Arizona State Museum show the shell most often strung with *Olivella dama* and *Trivia solandri* snails. An unusual shell worn as a pendant by Juanita Herrera had been found in an old camp, and because of its uncharacteristic thickness and pronounced ribbing, neither she nor others could readily identify it (figure 4.23).[44]

Placunanomia cumingii

◊ ***Hant Ihiini stacj*** '***stacj*** from *Hant Ihiin*'
haxölinaail quihócaloj 'shell with ruffles'

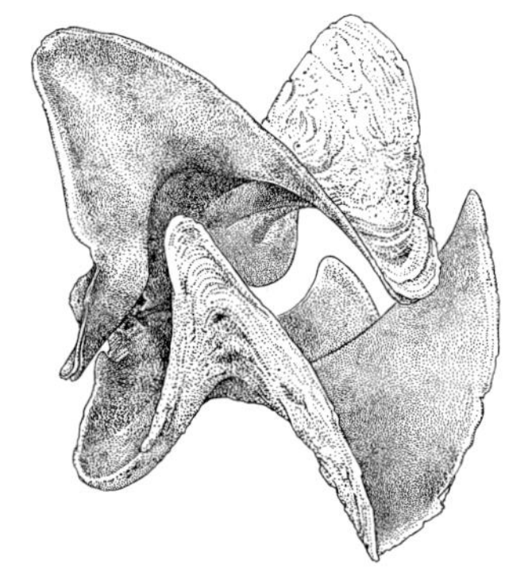

L 75 mm

The distinctive deep folds in the shell of this rarely seen bivalve make it valuable as a curiosity.

The name ***haxölinaail quihócaloj*** 'shell with ruffles' is descriptive; such deep protrusions are characterized by the verb *quihocl* 'ridged, ruffled'. The name ***Hant Ihiini stacj*** '***stacj*** (*Saccostrea palmula*) from *Hant Ihiin* (Baja California)' was given by one person because of the shell's similarity in appearance to the rock oyster, but being different enough to have come from such a distant place as Baja California.

With the increase in diving and use of nets since the 1970s, more exotic creatures such as this are being brought to the surface. The bivalve was collected in a fish net off Isla Patos. Before the shell was shown to me, it was described as being something I had probably never seen and would surprise me: it did. The conversation among the people with me involved a description of a shell baptismal font in the Hermosillo cathedral in which someone's daughter was baptized. The shell, with deeply folded margins, was that of a giant clam. María Luisa Astorga then remarked that when she first saw a *Placunanomia cumingii* long ago she dubbed it *Tear ihaxt* 'the Devil's ***haxt*** (*Crassostrea corteziensis*)' because of its strange, "fanged" appearance.

Pectinidae (scallops)

Argopecten ventricosus [= *Argopecten circularis*]

hasla 'ear'

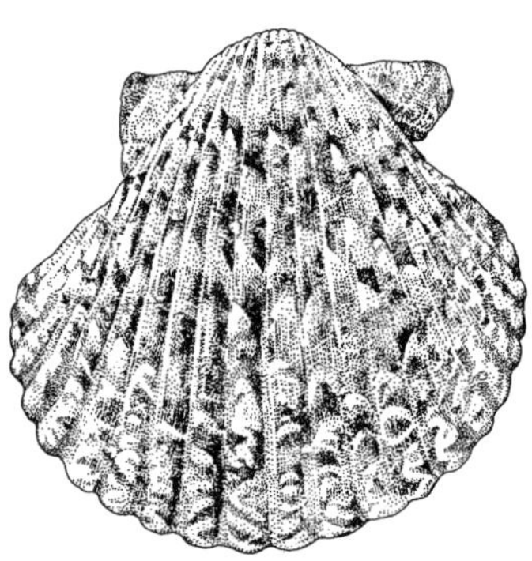

L 64 mm

The attractive shell of this scallop has variegated designs in shades varying from white, gray, pink, or purple, to a vivid orange. Although the scallop is seldom gathered live today, its shells are common in beach drift.

The shell's auricles are called *isloj* 'its ears'; however, the shell's name is given because of its ear-like appearance when turned sideways. As René Montaño recounted:

> When I was a boy I was out in a boat with my father. I had found one of these on the shore and asked my father what it was called. He held the shell up to my head, by my ear, and said the name was ***hasla***, because it looked like someone's ear.

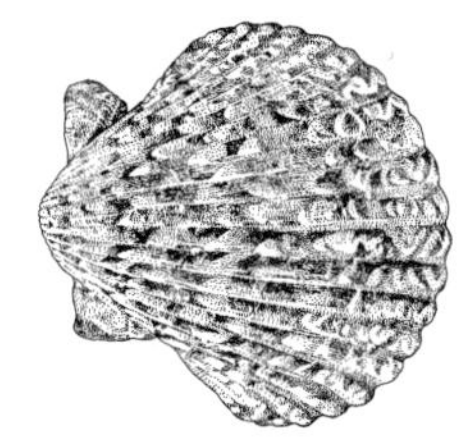

The meat is described as tasting like ***haxöl*** (*Leukoma grata*) and was considered good food, though not eaten in the same quantity as the more abundant and easily gathered shore clams. People comment that the scallop was eaten by non-Seri Mexicans in the past, although Edward H. Davis did record that it was consumed by the Seris.[45]

The shells were sometimes used in adornment and as trade items. Juanita Herrera said that, in the past, they were gathered on the southern shore of Isla Tiburón and strung with *Olivella* shells into necklaces, many of which were sold. Today the shells are sometimes strung, although this is not common.

Figure 4.24. An unusual necklace of *Argopecten ventricosus* shells, strung together with seeds of the desert shrub *Viscainoa geniculata*. Collected by Mr. and Mrs. Emmett in the 1960s. Arizona State Museum, catalog no. 87–75–7.

Sara Villalobos commented on the scallop:

> There are many at *Xpanoháx* '[fresh] water in the sea' [Puerto Libertad]. They are found at *Xapoo Yaii* 'where the sea lions are' [an area north of Puerto Libertad]. The shell was sold; when I was about five years old the *Comcaac* would sell [or trade] them there. A long time ago shells of ***xtiip*** [*Laevicardium elatum*] and ***haxöl*** [*Leukoma grata*], together with *stac imahyaa* [pumice] were taken to the ranches. [The *Comcaac*] went carrying ***haxöl*** shells—the old people took these to sell. The pumice was to rub [the ranch people's] skin. [The *Comcaac*] took ***xtiip***, ***xtapacaj*** [Turitellidae] and pumice [to sell].

Euvola vogdesi [= *Pecten vogdesi*]

cascanoohcö (convex valve)
cascanoohcö ooix (concave, almost flat, valve) 'what the ***cascanoohcö*** left behind'
cascapeetij (concave, almost flat, valve)

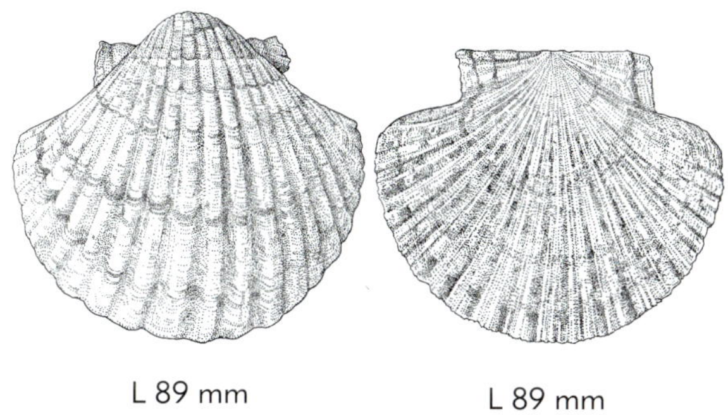

This thin-shelled scallop has two distinctive valves, one almost flat and the other deeply convex, colored in shades of buff or pink to light maroon. This is the only instance of which I am aware in which each valve of one clam has an individual name—not surprising, since each is very different in shape.

The name ***cascanoohcö*** (the convex valve) compounds *casca* with *-noohcö* 'deeply rounded' while ***cascapeetij*** (the concave, almost flat, valve) compounds *casca* with *-peetij* 'circular'.

The almost-flat valve is also called ***cascanoohcö ooix*** 'what the ***cascanoohcö*** left behind'. An archaic variation, ***cascanoohcö ooil***, is unclear.

Casca has no other meaning in Seri. Interestingly, similar words are found in two neighboring Uto-Aztecan languages unrelated to Seri, whose speakers in past centuries had contact with the Seris.[46] In Pima Bajo the word *cosca* means '*concha de nácar*' (pearl oyster shell), while in Eudeve (now extinct) the word *cóscar* was recorded as meaning '*cuenta de vidrio*' (glass bead).[47]

Accounts from folklore tell how the shells were named. In a version related by one woman, when the people were being given names by *Hant Iiha Quimx*, they sat in a large circle, each becoming the animal with that name. Because the two beings that would become this scallop were late arrivals, they missed being paired with others and had the following conversation:

> *Cascapeetij quij ox tee, yoque—Hatee hiza cöscapeetij aha—tee, yoque.*
> *Hatee hiza cöscanoohcö aha—tee, yoque.*
>
> The ***cascapeetij*** said, "I think I'll be the circular [flat] one."
> The ***cascanoohcö*** said, "I think I'll be the deeply rounded one."
>
> And that is how they got their names.

Figure 4.25. The connected valves of a *Euvola vogdesi* scallop.

Though not commonly eaten now, the meat was said to have been consumed by the Seri Ancestors, and Edward H. Davis recorded this scallop as being eaten.[48] One person commented that the meat is "strange" and is eaten by non-Seri Mexicans; however, someone else who had eaten it recently boiled in the shell and flavored with oregano, noted that it "tastes like ***haxöl*** (*Leukoma grata*)."

The convex valve was used as a container for pigments.[49] Besides the clam's most common names, Eva López also gave the name ***siml ihataxoj*** 'barrel cactus scraper', noting that it could be used to scrape a barrel cactus to obtain its juice. The name ***siml ihataxoj*** is more commonly given to the cockle *Trachycardium procerum*, and, by some, to *T. consors*. The shell was also used to scrape the air bladders of large fish, such as the *zixcám caacoj* (*Totoaba macdonaldi*). The bladders were then roasted in a fire and eaten like cracklings.

The *Hant Ihiini Comcaac* 'Baja California Seris' were a group of Seris reportedly living on Baja California in previous centuries.[50] They were said to have favored the meat of *Euvola vogdesi* scallops, which were abundant on the shores across the Gulf from the present-day Seri territory. The *Hant Ihiini Comcaac* reportedly roasted and dried the meat, transporting it in bags made from the skin of pelican throat pouches when traveling by balsa to Isla Tiburón, where they would eat it "like crackers."[51]

According to oral tradition and reported by Lorenzo Herrera, there was a large sandy bay near an estero in the area where the *Hant Ihiini Comcaac* lived. At the end of the cold time of year many of these scallops were harvested, and the people celebrated the occasion with a fiesta. The older people wore necklaces made from the strung convex valves, while the young people wore the concave, almost flat, valves strung together.

In modern necklaces the flatter valve is sometimes hung as a pendant on a necklace made of other shells, often with colorful dyed shark vertebrae glued to its surface.

Nodipecten subnodosus [= *Lyropecten subnodosus*]

hanl 'fingers'

L 135 mm

As this large scallop lives offshore, it is not often seen alive. Its attractive, thick shell, often colored with vivid orange and purple, is usually found only in beach drift. The name ***hanl*** 'fingers' alludes to the shell's distinctive knobby, ribbed sculpture that resembles knuckled fingers.[52]

According to several people, the meat was eaten in the past and cooked by roasting the scallop under a brush fire. One woman described the meat as being sweet, with a rubbery consistency. Although Edward H. Davis noted that the meat was eaten,[53] most people today say that they have not seen anyone eat it and that the scallops are not usually found alive.

The shell was used to hold sea turtle oil into which meat was dipped for flavoring or as a container for pigments (figure 4.26). Its sturdiness and large size made it useful for digging as well. One person mentioned its use in digging a pit for baking *haamjö* (*Agave subsimplex*).

The shells could be kept as *icoocmolca* (fetishes). According to María Luisa Astorga, the shells of this scallop and the *xtamaxiictoj* (*Codakia distinguenda*) were sometimes painted with stripes when used in this way.

According to several consultants, especially large ***hanl*** shells were sometimes kept by burial sponsors for digging graves (see *Laevicardium elatum*).[54] One person described fastening the shell to a stick to lengthen its reach when the grave had been dug as far as one could reach, explaining that no one would enter the pit to enlarge it.[55]

Figure 4.26. A woman, identified as Lupe Comito, painting fetishes made from *Bursera hindsiana* wood. A *Nodipecten subnodosus* shell holds the commercial *añil* blue, while the red, probably *Bursera microphylla* sap, is contained in a metal cup. Note the bundle of *Bursera microphylla* twigs (center) used in curing by a shaman. Isla Tiburón, August 1949. Photograph by William Neil Smith, courtesy of University of Arizona Libraries, Special Collections. MS316_B31_slb1_R.-182–8.

Figure 4.27. A *Nodipecten subnodosus* shell in a seaside dune midden. October 2007.

Spondylidae (thorny oysters)

Spondylus leucacanthus [= *Spondylus ursipes; S. princeps*]

oot iteexoj 'the coyote's ***teexoj***'
◊ ***teexoj ctam*** 'male ***teexoj***'

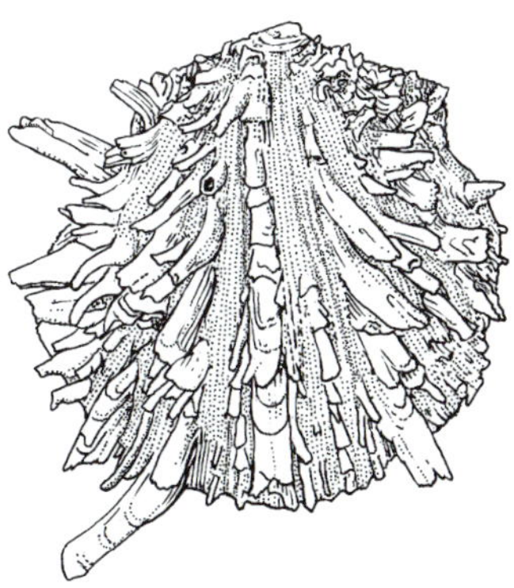

L 80 mm

The shell of this thorny oyster varies in color from white to brilliant orange. Because usually found attached to rocks in deeper water, the heavily spined shell was not often seen unless found beach-worn on shore. The shell illustrated was collected from deep water near Isla Patos by non-Seri shrimp fishermen.[56]

The name ***oot iteexoj*** reflects the shell's similarity to that of the ***teexoj*** (*Spondylus limbatus*) while its more spiny sculpture and smaller size make it different enough to be a possession of the coyote.

Teexoj ctam 'male ***teexoj*** (*S. calcifer*)' was given by one woman who explained that its similarity to the ***teexoj*** as well as its spiny 'whiskered' ("male") appearance suggested the name.

Xpeemolca are the large, double spines of the mesquite tree. René Montaño used the term *quixpeemolca* 'having *xpeemolca*' to describe the shell, noting the larger, sometimes split spines of this shell.

As such firmly attached mollusks from deeper water were not easily harvested, there is no information about its use as food, and except for undamaged specimens being kept as curiosities, the shell is not otherwise used.

In 2008, I was given a *Spondylus* sp. valve harboring an unusual crab, brought up in a net. No one had any idea what the crab was—it was attached to the shell by its posterior legs, which could be seen by tipping the shell backward, causing the crab to tip forward. María Luisa Astorga, her family, and I were joking about the crab making itself beautiful by putting the shell on its back. María Luisa said, "I'll put you on my back and be beautiful," and pretended to preen herself. We all laughed. After taking photographs of the crab and its shell, I carefully released it

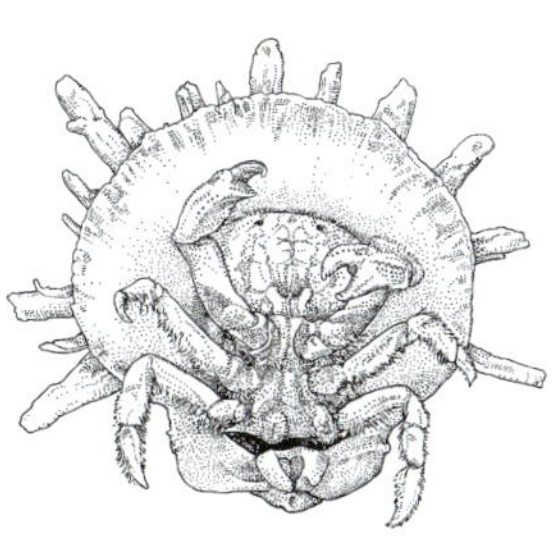

Figure 4.28. A *Hypoconcha* sp. crab in a *Spondylus* sp. shell.

at extreme low tide on an isolated beach. The crab, still brought up in conversation by María Luisa because it was so unusual, was later identified as a *Hypoconcha* sp., a creature rarely seen since the Gulf has been trawled by commercial shrimp boats.[57]

Spondylus limbatus [= *Spondylus calcifer*]

iimox (var. ***iimoj***)
♦ ***teexoj***

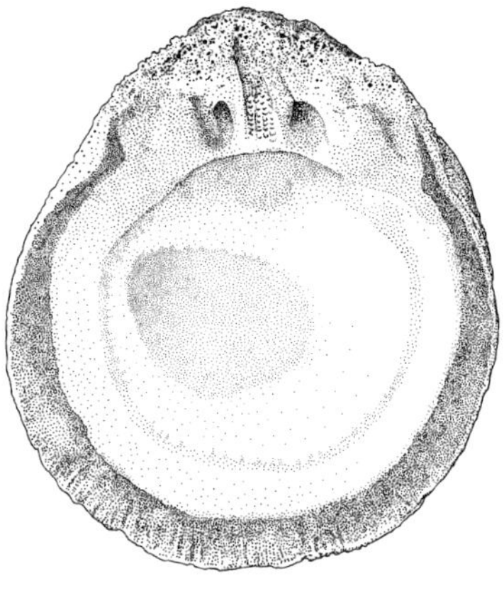

L 167 mm

This is a large, heavy-shelled thorny oyster. The shell's interior is rimmed by a deep purple. Because of its distinctive size, the shell is readily recognized and identified by name by most people.

Both the older name ***iimox*** and the most commonly used name today, ***teexoj***, are primary terms. However, according to María Luisa Astorga, these names belong to separate shells. She described the ***iimox*** as being the heavier and more massive of the two, with a more yellow-colored interior rim, while the ***teexoj*** has a deep purple rim.

Large objects were often attributed to, or said to have been used by, the Giants. Perhaps because of the mollusk's large size, it was especially noted to have been a food of the Giants. Folklore names one of the strongest Giants as *Iimox Impafc* 'Unbreakable ***Iimox***'. René Montaño describes: "The one called *Iimox Impafc* was invincible. Many lives were taken by him. He gambled using *heemot hapaxcaj* [decorated game pieces]. He bet only with his life. He wouldn't bet with things. He fed the people that he killed to his pet ravens."

Not easily accessed nor found in large quantities, the thorny oyster was not heavily eaten. When a live one was obtained, the shell was broken open and the meat removed; it could then be eaten raw or boiled, either alone or mixed with other food, and was considered to be good food.[58] One person recalled the meat being sold to non-Seris in the past.

Xavier Moreno and several other men free-dove for the thorny oyster during the 1980s in areas to the north of Desemboque. Xavier described opening its shell underwater with a knife and collecting the top (unattached) valve together with the large adductor muscle. Around four to seven specimens could be collected during one breath interval. Xavier added rather matter-of-factly that in order to dive with confidence he had to decide beforehand that if the bivalve closed on his finger, trapping him underwater, he would be prepared and willing to cut off his own finger in order to free himself.

The shell could be used to contain food. However, because of its weight the shell was not easily transported and therefore was not used as commonly as the lighter and deeper *Laevicardium elatum* shell.

Long ago non-Seri fishermen were seen eating the raw meat of the thorny oyster near the base of a seaside mountain at Puerto Libertad. A rocky point there became known as *Teexoj Tis ma Ihacanaj quih Iyat* 'point where the raw ***teexoj*** were chewed'. The use of the verb *ccanaj*, which means 'to chew strongly, with the molars' no doubt alludes to the tough consistency of the raw meat and perhaps focuses on what made the occasion memorable.

A note by Edward Moser records that the loggerhead sea turtle *xpeeyo* (*Caretta caretta*) was said to eat this mollusk.

Limidae (file clams)

Limaria pacifica [= *Lima pacifica*]

comcaai him oocatx 'what the old woman discarded'
xepenozaah 'sun in the sea'

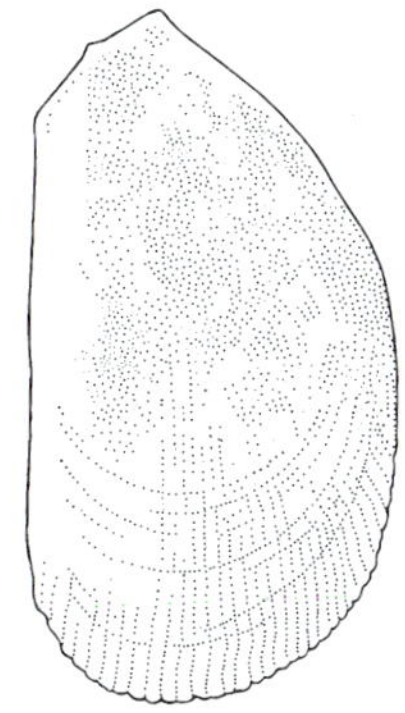

L 14 mm

Bordered by a distinctive fringed orange mantle, this small, somewhat elusive delicately shelled clam propels itself in a swimming-jumping motion by clapping its valves together. The clam lives in sandy hollows beneath submerged rocks.

The names of the live clam allude to its fringed appearance. ***Comcaai him oocatx*** 'what the old woman discarded', having a somewhat vulgar connotation, is more commonly used for the sea anemone.

The name ***xepenozaah*** 'sun in the sea' is more commonly given to the orange-colored *Berthellina ilisima* slug (see Pleurobranchidae).

I found a live clam and brought it up in a jar of seawater and showed it to two groups of Seris. Several said they had seen the clam before, and all were very interested as they watched it move about. At that time no one identified it by a Seri name; the name was given by others who later saw a color photograph of the live clam.

Figure 4.29. A live *Limaria pacifica*. Desemboque, April 2007.

Lucinidae (lucines)

Codakia distinguenda

xtamaxiictoj (var. ***tamaxiictoj***)

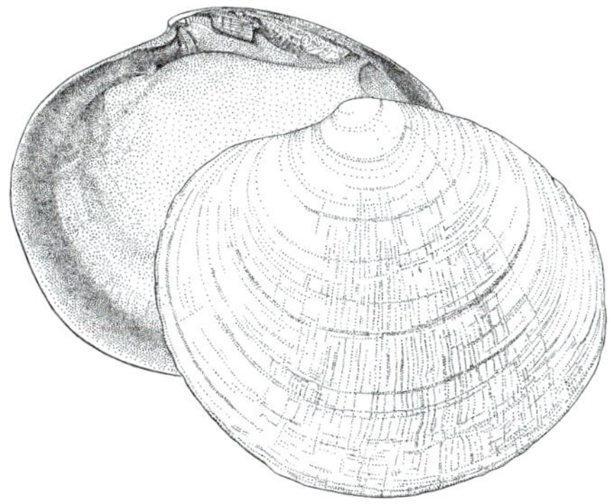

L 95 mm

This large clam has a shallow, white shell with a distinctive pinkish-red interior margin.

The name ***xtamaxiictoj*** is a compound. The sequence *tamax* is found in other mollusk names (see *Crucibulum* spp., *Acanthochitona exquisita*, *Chiton virgulatus*, and *Onchidella binneyi*). The second part, *iictoj*, connotes the color red and is related to the archaic word for red, *cactoj*.

Although not a common food clam, it was eaten raw or cooked whole by roasting it under a brush fire, and was considered a good food. Cleotilde Morales suggested that the meat "tastes like ***haxöl*** (*Leukoma grata*)." However, perhaps because of its red interior margin, the clam was considered *hapáx* (a food to be avoided) by some, and young people of childbearing age were cautioned against eating it (see Beliefs and Practices, p. 41).

The shells were used as earrings by the Ancestors. Angelita Torres recalled as a child seeing earrings made with small shells of this clam when her family was staying at *Xpanoháx* (Puerto Libertad). The earrings, strung on thread, were worn by her *amaz* (father's mother), Sara Villalobos. Angelita further noted that Sara would make jewelry out of anything: "She had rings made from sea turtle shell and even colored plastic combs. She had all different colors of rings—red, yellow, blue—and would wear them together."[59]

As with other shells of this size, that of the ***xtamaxiictoj*** was used as a container in which to mix pigments for face painting. It was also sturdy enough to be used as an implement for digging clams.

Shells were sometimes used as *icoocmolca* (fetishes); specifically mentioned were this shell and that of ***hanl*** (*Nodipecten subnodosus*). As noted earlier, such shells were sometimes painted with stripes when used for this purpose.

Carditidae

Carditamera affinis [= *Cardita affinis*]

quiit

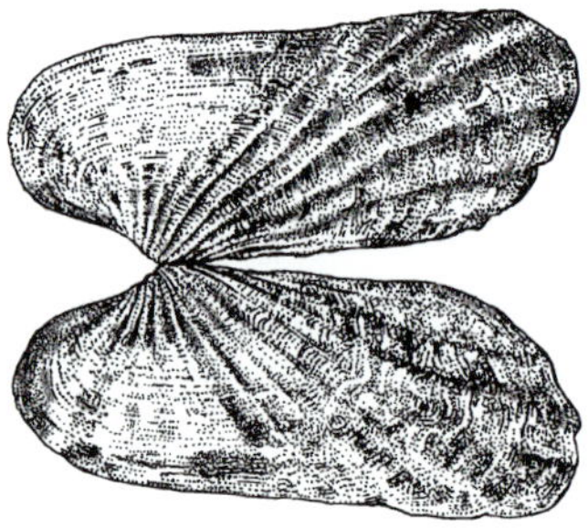

L 68 mm

This clam is common in the intertidal zone, anchored by a byssus in sand beneath rocks. Its shell is a mottled dark brown.

The word ***quiit*** may be a primary name, or it could be related to the word *quiit* 'who has head lice', derived from *hait* 'lice'.[60] A strong possibility is that it is a variant pronunciation of *quihaait* 'who has blood', since the clam has a characteristic red-colored blood.

The clam was eaten, but was considered by some as *hapáx* and to be avoided as food, because of its red blood. The meat is described as somewhat green in color. Although some Seris say that they have eaten the clam, one woman commented: "People don't eat it. It's not a good thing. It's bloody, so it isn't eaten." This said, the clam is not considered to contain real *haait* 'its blood', but rather *iix* 'its liquid'.

The clam was boiled in the shell and all of the meat consumed, as were ***haxöl*** (*Leukoma grata*) and other smaller clams. Because it is a bit larger than the ***haxöl***, it could also be roasted under a brush fire.

The shell is considered rather unattractive and is rarely used, although I have seen one necklace that included some small shells of the clam; however, it was noted as being unusual.

Carditamera affinis (an odd growth form)

oot iquiit 'the coyote's ***quiit***'

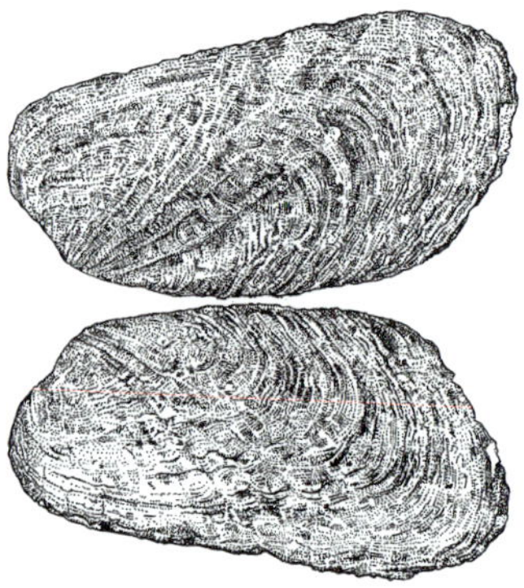

L 51 mm

The name ***oot iquiit*** alludes to the clam's strangeness as compared to the more familiar ***quiit***. Its smaller size, the indistinct radial ribbing, and prominent rough concentric ridges (lamellae) of its shell, especially evident on its posterior, are the clam's identifying features. The clams live in mussel beds visible at extreme low tide on sandy mud flats in the Infiernillo Channel, while the ***quiit*** is found elsewhere, growing together with *Leukoma grata* beneath rocks in more sandy intertidal areas.

Figure 4.30. *Carditamera affinis* clams growing in mussel beds at extreme low tide, eastern shore of Isla Tiburón, March 2009.

I was first told about the ***oot iquiit*** by Alfredo López, who described it as similar to the ***quiit***, but being smaller and having less pronounced ribbing. He later showed me part of a shell found in beach drift, but it was not until on a boat trip to Isla Tiburón that he pointed out the clams growing together with mussels at extreme low tide (figure 4.30).

There is no consensus about whether this clam was consumed; however, in shore camps bordering the mussel beds near Isla Tiburón, many of the clams' shells lie together with other food clams and oysters. María Luisa Astorga said that the clam was eaten, but noted that eating it caused an odd sensation. She described it, saying: "It knocks the air out of the mouth" (see Food).

Cardites laticostatus [= *Cardita laticostata*]

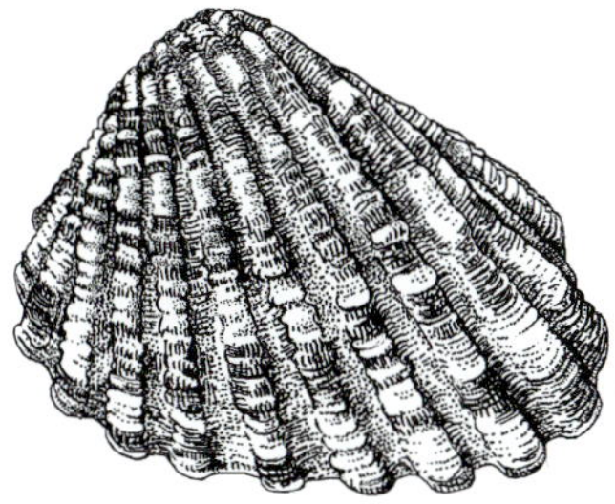

L 39 mm

haxölísotoj
mooxon iixz 'the scorpionfish's pet'
quiimosim xepenoyaafc 'what the beggar pounds in the sea'

The sturdy shell of this small clam is distinguished by its vivid brown and white bead-like markings.

The name ***haxölísotoj*** is a compound, from ***haxöl*** and the compounding stem *ísotoj* 'narrow (pl)'. ***Mooxon iixz*** 'the scorpionfish's pet' reflects the shell's similarity to the *mooxon* (*Scorpaena mystes,* scorpionfish), a vividly marked, ornate fish (figure 4.31).

The name ***quiimosim xepenoyaafc*** 'what the beggar pounds in the sea' is also given to *Anadara multicostata*, an ark clam.

Although no one recalls seeing this clam eaten, its shells are found in quantity in old camps around Estero Santa Cruz together with shells of other clams and oysters that were known to have been consumed (figure 4.32).

Figure 4.31a–b. A *Cardites laticostatus* shell and the scorpionfish showing the striking resemblance of the shell's shape, texture, and markings to the fish's fin. Photograph courtesy of Teresa Zuberbühler.

Figure 4.32. *Cardites laticostatus* and *Chione* spp. shells in a desert camp bordering Estero Santa Cruz, 2007.

Crassatellidae

Eucrassatella antillarum [= *Eucrassatella digueti*]

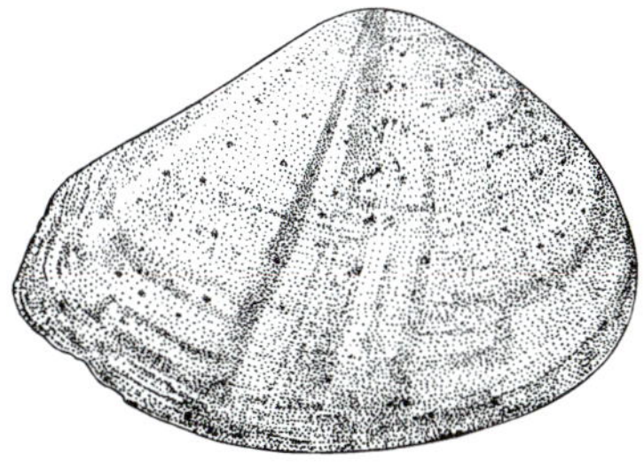

L 83 mm

The sturdy shell of this uncommon clam is white with a pinkish-brown interior. Under a brown periostracum it is marked with delicate, light yellow spots and rays. The shell illustrated was found in beach drift near Desemboque and was not identified by a Seri name.

Cardiidae (cockles)

Ctenocardia biangulata [= *Trigoniocardia biangulata*]

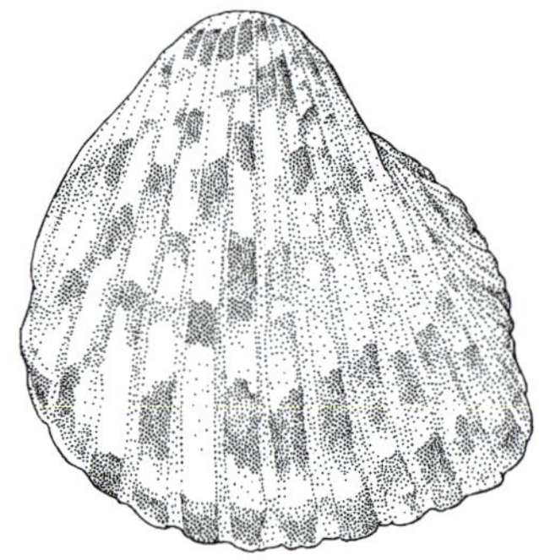

L 37 mm

The sturdy shell of this small cockle is marked with light brown and its white interior stained with a reddish brown. The clam is not recognized by a Seri name.

Laevicardium elatum

xtiip

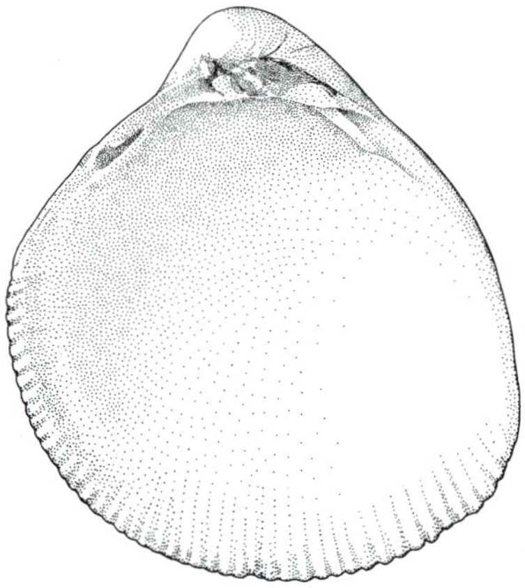

L 140 mm

This cockle's large, vivid yellow shell makes it one of the most easily identified bivalves in the area. Because of its size, sturdiness, and widespread occurrence, the shell was without a doubt the most heavily used in the Seri culture.

Xtiip is a primary name. The cockles were harvested from sand exposed at low tide. María Luisa Astorga commented that where many cockles lie buried at very low tide they provide a curious spectacle as they squirt seawater into the air.

Their meat was considered good food. The cockles could be cooked in seawater when fresh water was scarce or, more commonly because of their size, roasted under a brush fire. The whole animal was eaten, not just the meaty parts, although one woman remarked that she did not like to eat the clam because its soft parts, *itoozj* 'intestines', were too large, and she preferred smaller clams.

Figure 4.33. *Laevicardium elatum* shells on a beach near Estero Sargento, after a natural die-off. January 2009.

According to José Ángel Montaño, there are three types of ***xtiip*** shells.[61] The ***xtiip** *áa** 'true ***xtiip***' was any ***xtiip*** and was nothing special; it was used and easily discarded. The ***xtiip** *hayaa** 'owned ***xtiip***' was a shell in perfect condition, not too large and retaining its bright yellow color. This shell was a young girl's special possession. When she reached puberty and celebrated her four-day fiesta, the shell was used to hold pigments for her face paintings, which were of a traditional and different design each day. If a girl died before reaching puberty, the exterior of her shell was painted with the designs used on the first and fourth days, and the shell buried with her.[62]

The *xpacaao* ***ixtiip*** 'the mermaid's ***xtiip***' was a very large shell, used as a food container or vessel,[63] as a tool for excavating a grave or for groundwater, and as a dipper for filling a larger container.

Cleotilde Morales further described collecting water:

> They would dig for water, taste the water to see if it was good, then dip out the water [with the ***xtiip***] and fill an *hamazaj* [clay jug] and carry it on their shoulders with a *peen* [shoulder yoke]. Women would carry it on their heads. Maybe the *hamazaj* would fall and break—poor people!

Recently digitized films made on Isla Tiburón by William Neil Smith in the early 1950s give a close and personal view of the Seris' lives and show several instances of mollusk use. One short section was probably made

Figure 4.34. Julia Montaño during her four day puberty fiesta at Punta Chueca in May 2007. The design of her face painting is that worn on the first day of the festivities.

Figure 4.35. José Ángel Montaño (left) and his father, Ramón Montaño, using a *Laevicardium elatum* shell to dip drinking water from a water hole near *Xactoj* on Isla Tiburón in 1948. Photograph by William Neil Smith, courtesy of University of Arizona Libraries, Special Collections. MS316_B31_slb1_K.-226–9.

at the same time as the photograph in figure 4.35 and shows the shell used both to excavate the hole and remove the water. Another section of the film shows the shell used to break apart the root bark of the *heepol* (*Krameria grayi*) bush for use in dyeing basket-making material.

According to one woman, if such a large ***xtiip*** shell were used to dig a grave[64] it would be considered dangerous and avoided for other uses. In one account, four or five people would take turns digging the grave with the shell. After the deceased was placed in the grave, his personal belongings were buried with him. Such items were blankets, baskets, clay vessels, grinding stones, arrows, or anything else the person had used; the items were of no religious mortuary significance.[65]

The shell was used as a utensil or container for foods such as flour gruel (*atole*), eelgrass gruel (*xnoois haaztoj*), broth, sea turtle fat, meat, honey, or mesquite bean flour (*haas copxöt*). Smaller ***xtiip*** shells were used as spoons and, according to one source, were carried and kept in one's personal bags or pockets for ready use. As Eva López quipped, "The people were to be pitied—they didn't have dishes and cups, so they used ***xtiip***!" Indeed, W J McGee noted in the late nineteenth century that many Seris, including children, possessed a "shell cup," specifically identifying the *Laevicardium elatum* shell.[66] He writes that the shells were also used as eating utensils and for digging clams, a use that others confirm today.

Figure 4.36. A cache of eight *Laevicardium elatum* shells nested between two pottery vessels. Photograph by Thomas Bowen. Arizona State Museum, University of Arizona, (#14234).

Although *Trachycardium procerum* shells were those most often used to scrape the pulp of a barrel cactus to obtain liquid to drink, a medium-sized ***xtiip*** shell could also be used.

Very small ***xtiip*** shells were used to drip water sweetened with sugar or honey into a newborn's mouth; that shell often became the mother's special possession.[67] Larger shells were used to soothe the skin or calm inflammation. The shell was placed near a fire and heated, then applied to one's chest area to soothe painful breasts.

Nested shells have been found in dune camps, perhaps left for future visits (figure 4.36).[68] María Luisa Astorga noted that such items were often buried beneath large tree roots or the sagging arm of a large

Figure 4.37. A dipper made with a *Laevicardium elatum* shell, collected on Isla Tiburón by Edward H. Davis in 1924. National Museum of the American Indian, Smithsonian Institution, catalog no. 112184.000. Photograph by the author.

columnar cactus to mark the location of the cache or protect it from scavengers.[69]

The shell was used to dip cactus fruit wine from a container. Wine made from *ziix is ccapxl* 'whose immature fruit is sour' (*Stenocereus gummosus*) was stronger than that made from other fruit. According to Lorenzo Herrera, there was a saying that "drinking four ***xtiip***-fulls" would make a person really drunk.

Lorenzo then described another use for the shell: long ago the people would rub their skin with animal fat as a skin protector and softener. When fat was rendered from chuckwallas (*Sauromalus obesus*) by roasting, the drippings were collected in the shell. A note of Edward Moser's mentions that children were sometimes frightened by being told that a malevolent spirit, the *slootxöla*, might kill and roast them on a spit, with a ***xtiip*** shell under their feet to collect the fat.

A mixture of sea turtle oil and blood was cooked in clay vessels, and the shell was used to dip out the hot mixture. Armando Torres remarked that the food was "delicious, just like *xnoois haaztoj* (eelgrass gruel or atole)." An unusual object in the National Museum of the American Indian collections is a *Laevicardium elatum* shell attached by wire to a wooden handle and still containing a dark residue (figure 4.37). In 1924, Edward H. Davis recorded collecting a *cucharón* (dipper) that was being used for a pot of sea turtle blood.[70] Perhaps an indication of the object's uniqueness, when I questioned Victoria Astorga about it and showed her its photograph, she commented with much surprise that her husband Miguel Barnett had made such a shell dipper and had sold it to *Maricaana Coojöquim* 'Barking American' (Davis's Seri nickname) on Isla Tiburón.[71] Another person who recognized the utensil noted that *Ziix Anxö Coohit* 'thing that eats a lot', a legendary Giant, was said to use such a dipper to stir pots of food.

The shell was used as a container for pigments used for face painting, most often that made from the sap of the elephant tree (figure 4.38). The bark of the tree was removed and the wood scraped. A small amount of water was added to the scrapings and the mixture squeezed through a cloth to press out the blood-red sap (figures 4.39 and 4.40).[72] The shell containing the sap was known as ***xtiip*** *hahiixat* '***xtiip*** that was caused to have liquid'.

Figure 4.38. A *Laevicardium elatum* shell containing processed *Bursera microphylla* sap. Prepared by Elvira Valenzuela in the late 1970s.

Figure 4.39. A woman preparing *Bursera microphylla* sap in a basket. Note the *Laevicardium elatum* shell containing the pigment (center left). Isla Tiburón, 1947. Photograph by William Neil Smith, courtesy of University of Arizona Libraries, Special Collections. MS316 B29a f10 25–13.

Figure 4.40. The sap, mixed with water, is squeezed through a cloth into the shell. Note the balls of processed inner bark or wood, lower left. Isla Tiburón, 1947. Photograph by William Neil Smith, courtesy of University of Arizona Libraries, Special Collections. MS316_B29a_f10_25–13.

Other pigments, such as pulverized red-ochre (*xpaahjö*) or the store-bought *añil* blue were mixed in this shell as well. Faces of men, women, and children were painted with designs using such pigments. A common use for the pigments was as a protection for one's face from the sun; a slip made from a special pink clay from Isla Tiburón, *casiime*, was used for this purpose. According to María Luisa Astorga, such a mask was called a *samiipni*. The clay was also used to soothe one's skin; Victoria Astorga kept some of the clay in a shell for this purpose (figure 4.41).

Figure 4.41. Victoria Astorga holding two *Laevicardium elatum* shells, one of which contains *casiime* clay. Punta Chueca, 2011.

When talking about the ***xtiip*** shell, Angelita Torres recalled learning how to make face paintings as a young girl by practicing on the shell's exterior. She added that bleached bones found on the beach, such as those of a porpoise or sea turtle, were also used.

In one of twelve personal wooden storage boxes collected by William Neil Smith in the mid-twentieth century is a cloth bag containing a supply of sea salt and a *Laevicardium elatum* shell used for removing the salt (figure 4.42). Another box contained the shell holding what appears to be the dried processed blue pigment (popularly called "Seri blue") made from a mixture of the sap the *Guaiacum coulteri* tree, the root bark of *Franceria ambrosioides,* and a clay, combined in a secretive process (figure 4.43).[73] A text recorded around 1968 describes the process and notes that the mixture is put into a ***xtiip*** shell; and, "if it isn't put into a ***xtiip*** shell, [the mixture] won't turn blue."[74]

Edward H. Davis recorded the use of a "large shell used for eating" (most likely *Laevicardium elatum*) in making cocoon rattles.[75] The shell was pounded into small pieces and several were inserted into an empty cocoon of the *Rothschildia cincta* moth. After the end was stitched closed, the cocoon was sewn together with others into strands, called *ziix quiinla* 'thing that rings' and worn by a dancer around his lower leg and ankles (see figure 3.56 , p. 66).[76]

Figure 4.42. A cloth bag containing sea salt, with a *Laevicardium elatum* shell used to remove the salt. Collected by William Neil Smith. Private collection.

Figure 4.43. A *Laevicardium elatum* shell containing dried, prepared natural blue pigment. Collected by William Neil Smith. Private collection.

Laevicardium substriatum [= *Laevicardium elenense*]

◊ ***xtiip heeque*** 'small ***xtiip***'

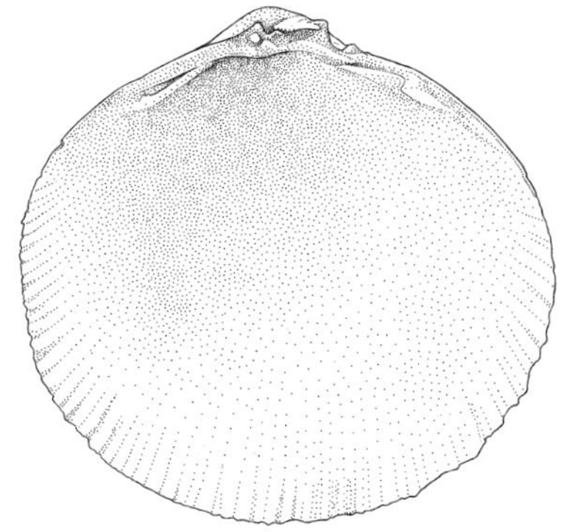

L 25 mm

This small cockle has a very fragile, light yellow shell faintly blotched with reddish brown.

The shell was recognized as being different from that of *Laevicardium elatum*, and the name ***xtiip heeque*** 'small ***xtiip***' was supplied by one person.

Papyridea aspersa

snoomtxö 'one-pelican-skin covering'

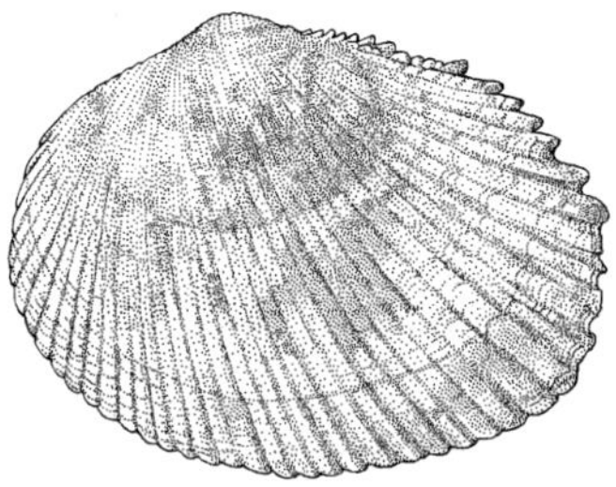

L 44 mm

This small, somewhat uncommon cockle has a fragile, finely ribbed and delicately patterned pinkish shell.

Snoomtxö is a personal covering made from a single pelican skin. Larger coverings made by stitching together two or six skins had names as well, and were used into the early twentieth century. The shell's similarity in its ribbing and designs to a feathered wing suggests the source for its name.

Trachycardium consors

oot izacz 'the coyote's ***zacz***'
siml ihataxoj 'barrel cactus scraper'
♦ ***zacz*** (var. ***xcacz***, ***zcaczj***)
zacz ctam 'male ***zacz***'
zcaczj quictamo 'fierce ***zcaczj***'

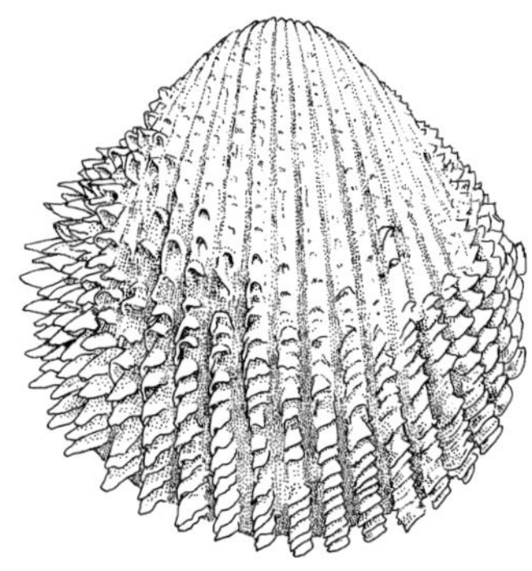

L 55 mm

This medium-sized cockle has a sturdy shell characterized by a sharp, spiny sculpture, described as *quicös* 'spiny'. It is found buried on intertidal sand flats.

The name ***oot izacz*** 'the coyote's ***zacz***' suggests that another cockle, *Trachycardium procerum*, may be the original bearer of the name ***zacz***. However, that the names are shared by both suggests a loss of the distinction today.

A few people gave the name ***siml ihataxoj*** 'barrel cactus scraper' to the cockle. However, that name is more commonly associated with *T. procerum*, which has a larger shell and would be more easily used for the purpose suggested by the name—scraping a barrel cactus in order to collect its juice for drinking.

The most commonly used (and primary) name, ***zacz***, has several slight variations.[77]

The names ***zacz ctam*** 'male ***zacz***' and ***zcaczj quictamo*** 'fierce ***zcaczj***' reflect the shell's spiny, "whiskered" appearance.

The meat was eaten. The whole cockle was either boiled or roasted under a brush fire. However, someone remarked that eating the meat irritated the throat.

Trachycardium procerum [= *Trachycardium panamense*]

L 67 mm

♦ ***siml ihataxoj*** 'barrel cactus scraper'
zacz (var. ***xcacz, zcaczj***)

This good-sized cockle has a sturdy shell with a smooth, deep radial sculpture and light brownish-red markings.

The name ***siml ihataxoj*** describes the shell's function as a scraper (from *ctaxoj* 'to scrape') for use with the barrel cactus. The shell's prominent ribbing creates a deep-toothed margin ideal for such use.

Figure 4.44. A *Trachycardium procerum* shell held as a scraper. Desemboque, 2008.

Figure 4.45. A *Trachycardium procerum* shell in a desert camp, 2007.

According to some, the original name of this shell is the primary name ***zacz***, while the descriptive name ***siml ihataxoj*** is more recent. The name ***zacz*** is also given to *Trachycardium consors*.[78]

The meat was eaten, cooked in the shell by boiling, or roasted under brush.

The shell was a common tool. María Luisa Astorga told of using it to scrape off the root bark of the *xpaxoocsim* (*Batis maritima*) plant; the inner roots were then washed, mashed, and boiled to make a liquid that was used as a sweetener. Perhaps the shell's best-known use was for scraping the interior of the barrel cactus *siml áa* (*Ferocactus tiburonensis* and related species) into a pulp from which to squeeze liquid to drink when water was scarce. One barrel cactus found in the area, *F. emoryi* (= *F. covillei*), was not used, as it is regarded as toxic.[79]

María Luisa Astorga recalled living as a child with her family in the desert north of Desemboque and having to survive on the cactus juice for a long time. She described how the cactus's spines were first burned off, the cactus then cut in half and scraped. Her parents took turns scraping the pulp, which was then squeezed by hand into a one-liter can. One good-sized cactus could yield five or six liters of juice; if available, honey was used as a sweetener. María Luisa related that when people who were unaccustomed to it drank the juice, they suffered adverse effects. However, she and her family did not get sick because they were used to it. She further recalled that when her family found fresh water and she drank some of it, she got a stomachache and went back to drinking the cactus juice until she slowly adjusted to the water.

One short section of film made by William Neil Smith in the 1950s shows two children scraping the flesh of a barrel cactus with the shell and one child drinking the liquid using the same shell.

Chamidae (jewel boxes)

Chama buddiana

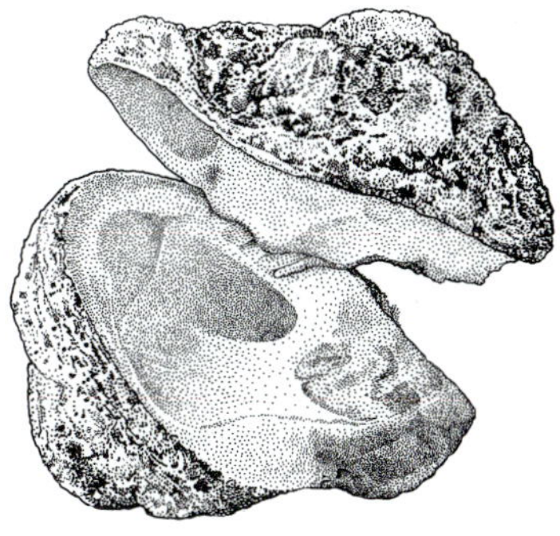

L 50 mm

♦ ***hapos cahaaxat*** 'what irritates the throat'
zeel

This irregularly shaped and thick-shelled bivalve is found attached to rocks or other shells. The shell's interior is tinged with violet and its exterior marked with violet and orange.

The name ***hapos cahaaxat*** 'what irritates the throat' describes what some people say about the meat—that eating it irritated the throat, making one cough.[80] The primary name ***zeel***, known by far fewer people, seems to be an older name for the clam.

The clam was eaten and could be easily collected at very low tide. Its shell is common in middens. The meat is not often eaten today, although one woman did say that her son ate it, implying that this was rather strange.

Mactridae (surf clams)

Mactrotoma nasuta [= *Mactra nasuta*]

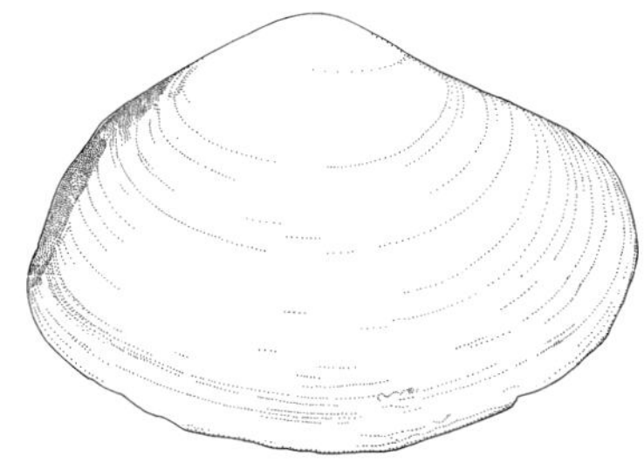

L 50 mm

This surf clam has a polished white shell covered by a tan periostracum.

There is no Seri name for this clam. Two valves were found in beach drift near Desemboque in 2010 by a woman who said she had never seen the clam before.

Raeta undulata

Juan Tomás ihaxöl 'Juan Tomás's ***haxöl***'

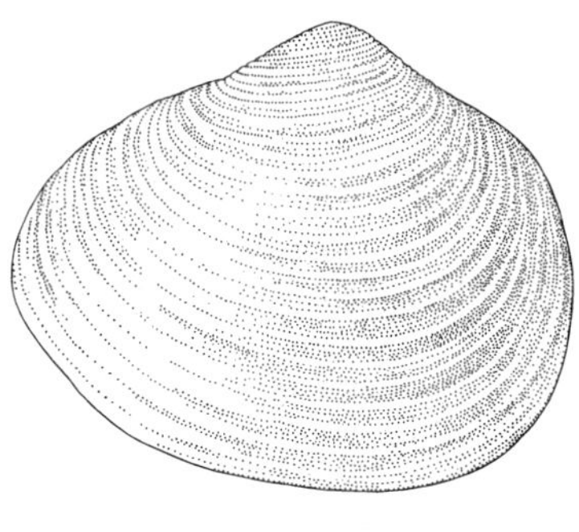

L 93 mm

This uncommon, thin-shelled clam has deep, fragile white valves.

Several people provided the name ***Juan Tomás ihaxöl*** 'Juan Tomás's ***haxöl***', explaining that a Seri man, Juan Tomás (d. 1932), was said to have used its shell as a spoon. It is interesting that the use of this shell as a utensil was unusual enough to merit the inclusion of the user in its name.

When I asked José Juan Moreno about the shell, he chuckled as he recounted being present on Isla Tiburón when Edward H. Davis jokingly pointed out the many shells on shore, saying that all of them were "Juan Tomás's spoons." Perhaps Davis's comment was the origin of this shell's name.

This clam is uncommon; no information was provided about it other than the description mentioned above.

W J McGee collected a shell he labeled a "mortuary cup" (see Beliefs and Practices, p. 41), a use uncorroborated by the Seris (figure 4.47).

Figure 4.46. Juan Tomás with Edward H. Davis. Isla Tiburón, 1922. Photograph by Edward H. Davis, courtesy of the National Museum of the American Indian, Smithsonian Institution (N23733).

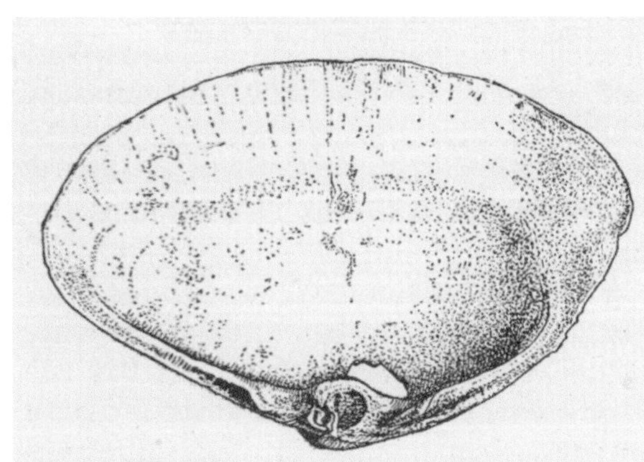

Figure 4.47. Illustration from McGee (1898:291, figure 42) of a shell, possibly of the family Mactridae, that McGee claimed was used as a "mortuary cup."

Simomactra dolabriformis [= *Mactra dolabriformis*]

haxöl icaai 'clamshell for making pottery'

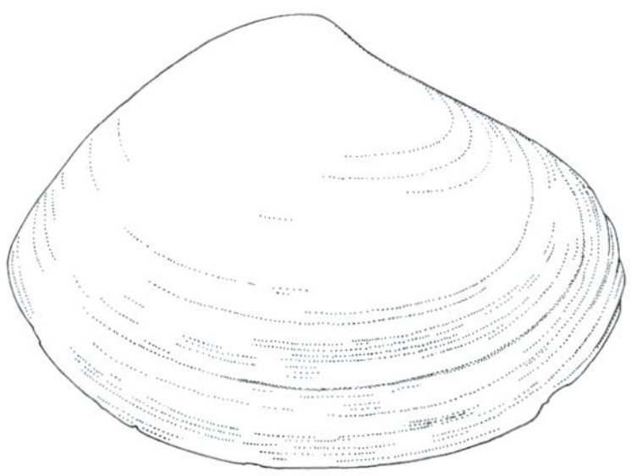

L 92 mm

The ***haxöl icaai*** is a surf clam with a sturdy, smooth white shell. Not gathered live today, the clam's empty shells are most often found in beach drift on sand beaches.

The shell is named for its use by the potter. A minor variation is ***haxöl icaa***. The name ***haxöl icaai ctam*** 'male ***haxöl icaai***' is given to the longer, more narrow clams *Tagelus peruvianus* and *Pholas chiloensis*.

This clam is edible, although it was not common enough to be gathered in quantity in most areas. When talking about the clam, Eva López explained: "The things in the ocean, on the shore—all were eaten. The Ancestors who went ahead of us ate them first, and they didn't die, so we eat them now."

The shell was used by a potter to smooth *hamazaj* (clay vessels). Sherds of very thin-walled clay vessels found in the region, popularly referred to as eggshell pottery, have a thickness of two to five millimeters; the ratio of wall thickness to volume makes these vessels remarkable.[81] More recent pottery is thicker, softer, and not as finely made. Clay vessels have not been made for actual use since the mid-1900s, when some of the last ones were made on Isla Tiburón.[82]

When the *Simomactra dolabriformis* shell was used in forming the vessel, the coils (*hanteezj hamaanoj* 'rolled clay') were pressed together using the exterior of the shell (*hapx iicp* 'next to the outside') held by its dorsal (hinge) area. After this, while still damp, the inside of the pot was shaped and thinned by scraping the wall using the sharp ventral edge (*iteel*) of the shell. Finally, the shell's exterior was repeatedly dipped in water and used to smooth and burnish both the inside and outside of the thinned pot. A special smooth stone called *iqueemt* (also called *hast cooil* 'blue/green stone', as some have natural blue markings) was also used to finish the prefired pot.

The shell was also used in forming clay figurines (figure 4.48). In 2010, Victoria Astorga was using the shell to shape figurines and small pots; she was probably the last person to use the shell for this purpose. Today only a few women make small clay objects for sale to tourists.

Several women remarked that the shell was a potter's special possession, since it was somewhat uncommon and not easily replaced. Victoria showed me several she stored carefully wrapped in tissue paper in her bag of personal belongings. When I asked Juanita Herrera about several perforated *Simomactra dolabriformis* shells in the collections at the National Museum of Natural History, she replied that the shells were sometimes kept on a string so that they would not be lost.[83] Other ***haxöl icaai*** shells in the collection were found in a pottery bowl contained in the grave of a woman and were identified by McGee as "mortuary food" (see figure 3.13, p. 43).[84] The Seris have no knowledge of such a custom and say that it is most likely that these were belongings of the deceased that, in a common practice, were buried with the woman—possibly a potter.

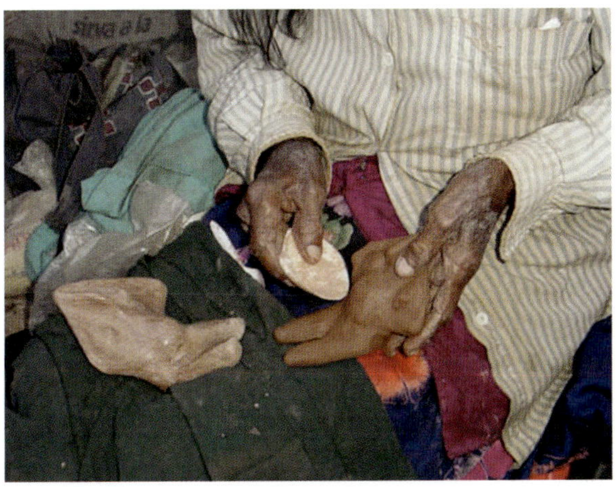

Figure 4.48. Victoria Astorga using a *Simomactra dolabriformis* shell to form clay figurines. Punta Chueca, 2003.

Besides being a potter's tool, the shell was also used as a utensil for eating; specifically mentioned by one person was its use for eating *haaizj* (mashed cooked cactus fruit).

Tellinidae (tellins)

The names of clams found on sandy shores and estero mud flats sometimes reference shore birds, as such shells are very common where herons and other birds fish at low tide, leaving behind their clearly visible tracks. Two such clams have shells with a posterior fold and have names alluding to their appearance of having been stepped on by such birds.

Psammotreta cognata [= *Florimetis cognata*]

haxöl icaai sleecoj iti yazám 'haxöl icaai stepped on by a heron'
sleecoj cooxp ihaxöl 'the Snowy Egret's ***haxöl***'

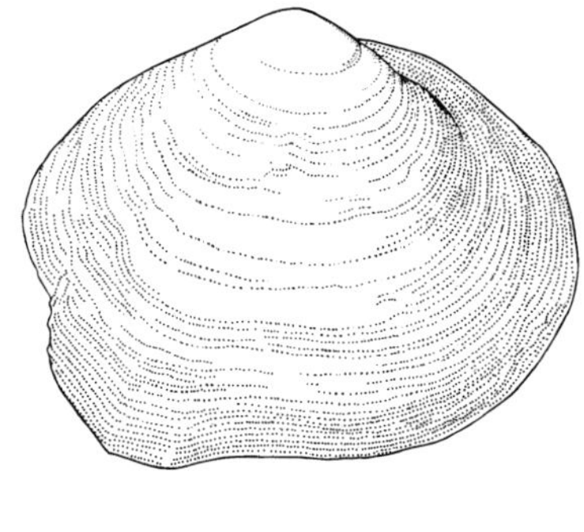

L 92 mm

The distinctive posterior fold on this clam's white shell makes it look somewhat misshapen, as suggested in the name ***haxöl icaai sleecoj iti yazám*** '***haxöl icaai*** stepped on by a heron'. The name is shared by the *Tagelus* sp. clams, which display a similar feature.

The clam's other name, provided by María Luisa Astorga, and one that references another shore bird (*Egretta thula*), is ***sleecoj cooxp ihaxöl*** 'the Snowy Egret's ***haxöl***'.

Psammotreta pura

hatj ihaxöl cooxp 'white ***hatj ihaxöl***'

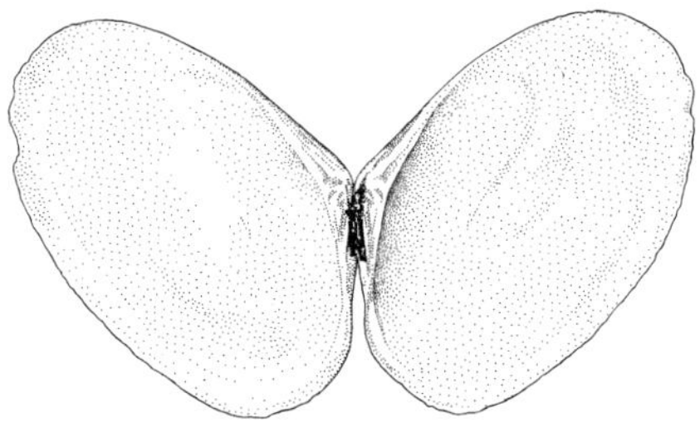

L 38 mm

This somewhat uncommon, white-shelled clam is found on sandy shores.

The name ***hatj ihaxöl cooxp*** 'white ***hatj ihaxöl***' alludes to the clam's similarity to the ***hatj ihaxöl*** (*Tellina regia*).

The clam was not eaten, perhaps because the clams were not gathered alive in quantity. The shell is sometimes strung in shell jewelry.

Tellina cumingii

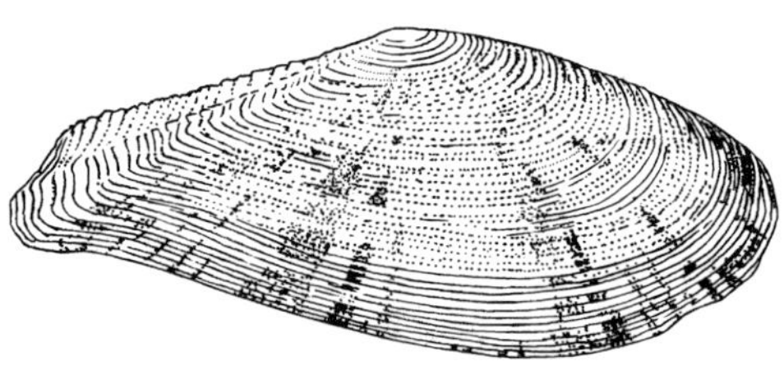

L 62 mm

The shell of this clam has raised concentric ridges marked with light brown broken rays. As the clam is uncommon, it is usually only seen as a shell in beach drift. The clam is not recognized by a Seri name.

Tellina regia [= *Tellina simulans*]

hatj ihaxöl 'the Tern's ***haxöl***'
slenapzj ihaxöl 'the Little Blue Heron's ***haxöl***'

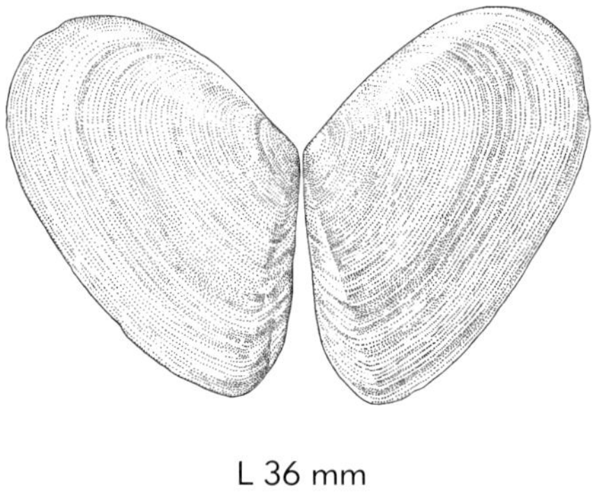

L 36 mm

The clam's attractive rose pink and finely ridged shell make it very distinctive.

The name ***hatj ihaxöl*** alludes to the ownership by the *hatj* bird, the Tern (*Sterna* sp.), which is said to eat the clam. As was described, the bird flies with the clam in its beak and drops it on rocks in order to break open its shell. A similar name, ***slenapzj ihaxöl*** 'the Little Blue Heron's ***haxöl***', refers to that heron (*Egretta caerulea*).

The clam was not eaten; its shell is sometimes used in shell jewelry.

Donacidae (beach clams)

Donax punctatostriatus

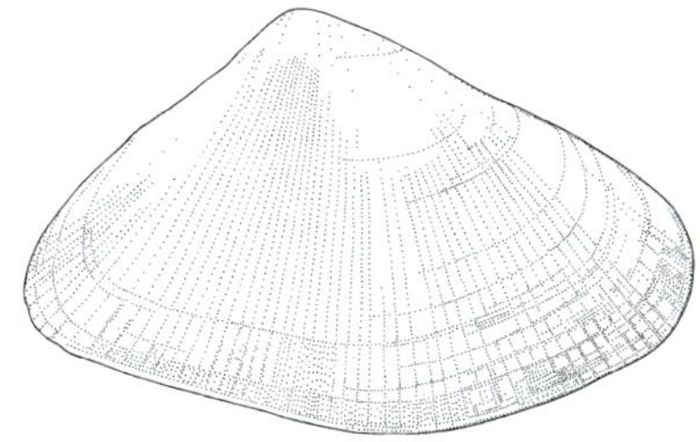

L 32 mm

This small clam has a sturdy, polished shell and is found on sand beaches; it is especially abundant at Bahía de Kino. Although it is common, no one provided a Seri name.

The clam was said to have been eaten in the past; in fact, quantities of the shells are found together with those of other known food clams and oysters in old desert camps around the Bahía de Kino area.

Psammobiidae

Gari helenae

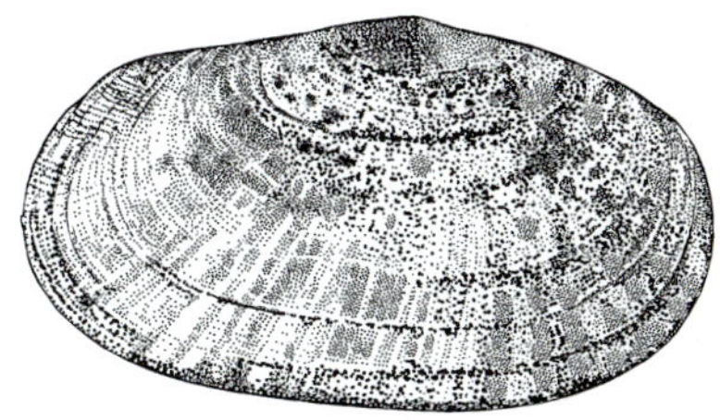

L 48 mm

This rather uncommon clam has a fragile, grayish-violet shell marked with spotted rays. Its shells usually only found in beach drift, the clam is not recognized by a Seri name.

Heterodonax pacificus

taijitiquiixaz

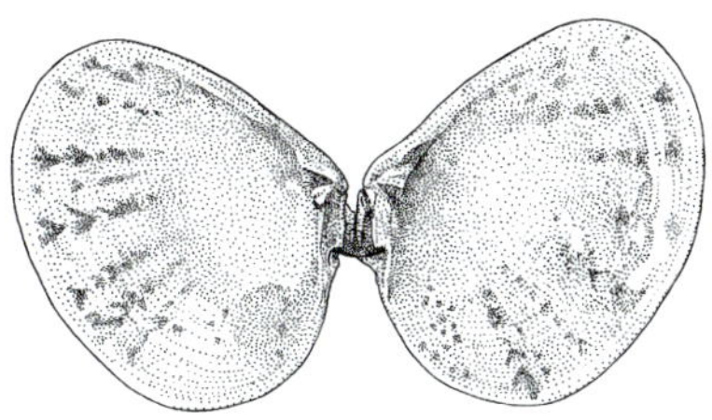

L 12 mm

The delicate shell of this very common, attractive little clam is lightly rayed and variously colored, sometimes in bright pinks and yellows.

The name ***taijitiquiixaz*** is derived from *itaaij iti quiixaz* 'what tinkle in the surf (swash zone)' and reflects the sound made when the empty shells are swept back and forth together by the lapping waves. The name is aptly given, as the clam is found on sand beaches shallowly buried in moist sand above the low intertidal area, where it is easily collected.

The clam was not a food, although Roberto Herrera commented that children sometimes ate the clams when playing. He added that there was a saying that if one ate a lot of these he would go blind, a comment made by others about the *Olivella dama* snail.

Figure 4.49. A child holds *Heterodonax pacificus* clams. Desemboque, 2012.

Attractive and its shell easily pierced, this is the most common bivalve used in shell jewelry today. An interesting past use came to light when I asked whether shells were ever used as decorations on clothing. An older woman, Chavela Torres, said that very long ago the people would sew this or other shells, specifically mentioning the ***xepe quilp*** (*Theodoxus luteofasciatus*) snail, onto their clothing—often made of deerskin.

In the few songs that refer to mollusks, sometimes the creature's name is merely inferred, and might be named only in the song's explanation. One of several that do mention a name references the ***taijitiquiixaz*** and another creature, the mole crab *iicj ano moosni* (*Hippa pacifica*).[85] Armando Torres presented the song:

Taijitiquiixaz cöicoos

Xe pe té li xe pe yaó za que ya
Xepe iteel, xepe yahoozaj
Tai ji ti quí xaz ya táo co quí coj íc ja no mó sa ni hax pó a tom
Taijitiquiixaz taax cohpiij xoc, iicj ano moosni iihax pooitom
Ic ja no mó sa ni hax pó a tom i ta ya
Iicj ano moosni iihax pooitom
Xe pe ha tá sim ya hai ya cá pat ya
Xepe hataasim, hai yacapat
Tix i mí max te
Tiix imiimx tee.

Song about the ***taijitiquiixaz***

On the sea shore, at the very edge of the wave,
The ***taijitiquiixaz*** says, "I wish I were there, and I would say to the mole crab,
I would say to the mole crab,
'The wind is blowing away the sea foam'."
So he said.

Sanguinolaria tellinoides

hatj ihaxöl 'the Tern's ***haxöl***'
◊ ***haxöl icaai*** 'clamshell for making pottery'

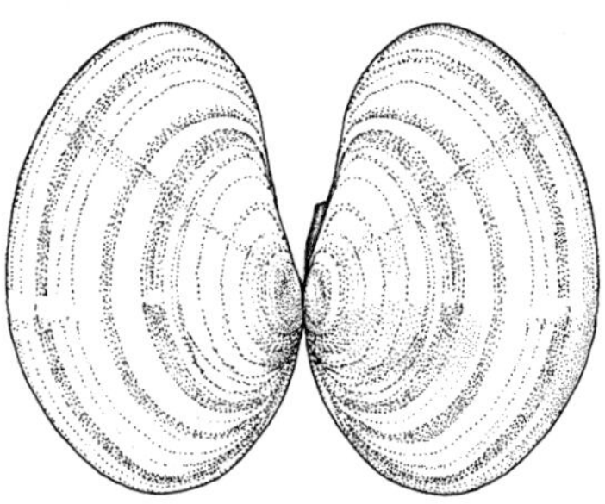

L 53 mm

The dark salmon pink color of this clam's shell makes it very conspicuous on sandy beaches, where it is most often found empty in beach drift with both valves connected.

The name ***hatj ihaxöl*** 'the Tern's ***haxöl***' is usually given to *Tellina regia*; however, the name was recently shared with this clam because of its similar shape and color.

The shell's form is similar to that of *Simomactra dolabriformis* and that clam's name, ***haxöl icaai*** 'clam shell for making pottery', has occasionally been shared with this clam.[86]

The clam has been rediscovered in Puerto Libertad, a non-Seri fishing town about thirty-five miles north of Desemboque. Not since the mid-twentieth century have the Seris regularly lived in that area, but with several people recently living there, the clam's shells are more frequently collected for use in necklaces. Because of their unusual color, the shells are clearly a special item shared among the women; one often hears the name of the person who collected and shared them.

Solecurtidae

Tagelus peruvianus

ctam yazexe 'with what the man incises'
haxöl icaai ctam 'male ***haxöl icaai***'
haxöl icaai sleecoj iti yazám '***haxöl icaai*** stepped on by a heron'
ipac casa ihaxöl 'the outcast's ***haxöl***'
xlehaxöl
xpajísotoj (var. ***pajísotoj***)
♦ ***xtipisot***

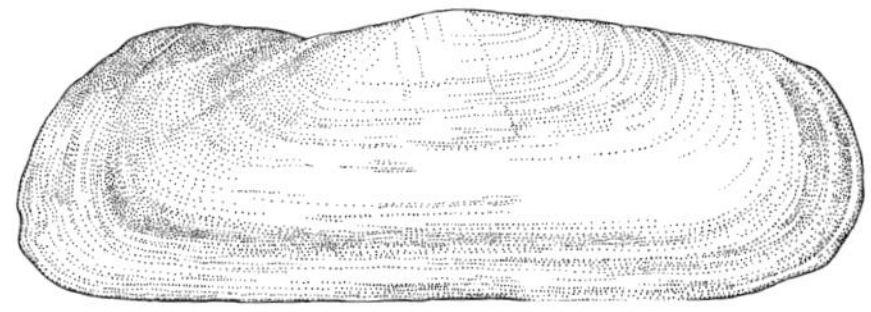

L 81 mm

The clam has a narrow, fragile white shell. Several species occur in the area and the variety of names (the most of any mollusk) appears to apply to any of the clams. The clams are common in esteros on muddy sand flats.

Ctam yazexe 'with what the man incises' is unclear in exact meaning, but obviously references its use as a knife.

Haxöl icaai ctam 'male ***haxöl icaai*** (*Simomactra dolabriformis*)', a name shared by the *Pholas chiloensis* clam, refers to its similarity to *S. dolabriformis*, while its elongate shape connotes the "male" form (see Names and Naming, p. 25).

According to María Luisa Astorga, the name ***haxöl icaai sleecoj iti yazám*** '***haxöl icaai*** stepped on by a heron', suggests the similarity between the shell's long, thin shape and the footprint of a heron on the mud flats where the bird commonly fishes. A posterior fold on the shell also suggests an appearance of having been stepped on by the bird, a characteristic shared by another shell with that name (see *Psammotreta cognata*).

The name ***ipac casa ihaxöl*** 'the outcast's ***haxöl***' is somewhat unclear. It either refers to its being eaten by, or its shell used as a spoon by a vagrant, *ipac casa*—described as someone without a home or family to care for him.

Xlehaxöl is possibly a compound derived from *xeele* 'fog, dew' and ***haxöl***. No Seri has suggested such an etymology, however.

Xpajísotoj and its variant ***pajísotoj*** derive from *xepe* 'sea' and *-ísotoj* 'narrow (pl.)'. The name is given as well to *Pholas chiloensis*, which is similar in shape.

Xtipisot is a compound derived from ***xtiip*** (*Laevicardium elatum*) and *-isot* 'narrow'.

The clams were not recalled as being a food. However, in referring to them, Eva López made a somewhat sweeping statement: "These were eaten by the Ancestors. All [mollusks] were eaten, except for ***xiica cooxp*** (*Olivella dama*)!" And indeed, the shells of this clam are common in middens bordering esteros.

The shell was used in forming clay vessels in the same way as the more commonly used ***haxöl icaai*** (*Simomactra dolabriformis*). One woman described the shell's narrow shape as being suited for shaping a section of a vessel that was more constricted, such as the rim. In fact, it is quite obvious when comparing the shell's shape to the interior of a narrow clay vessel (such as a drinking goblet) that this shell would be a perfect fit for shaping the clay.

María Luisa Astorga described the sharp shell being used to slice open the skin of the fruit of the *ziix is ccapxl* (*Stenocereus gummosus*), pitahaya agria cactus.[87] And in noting the shell's use as a spoon by the older people, one woman recalled seeing her father, Jesús Félix (d. ca. 1940), eating with this shell.

Veneridae (Venus clams)

The Venus clams are included under the phrase *ziix* ***haxöl*** *hant cöquiih* 'thing grouped with ***haxöl***', the "prototypical" clam being ***haxöl*** (*Leukoma grata*) (see Classification and Naming).

These sturdy-shelled clams were a major food; especially favored were the abundant and easily accessed *Leukoma grata*, *Chionista fluctifraga*, and the *Chione* species. Clam beds were often spotted at low tide because of the water being spurted up by the many clams. Such sand-dwelling clams were harvested by digging with a wedge-shaped prying stick, *hehe hapxeezc* 'wedge-shaped stick', or a large shell. Today a metal implement might be used. Often clams were found and turned up by foot when people walked through the wet sand. Juanita Herrera described digging for the clams, saying that when she was young her family once found an *hantx mocat* 'old fashioned' rake (with metal teeth) near Bahía de Kino and used it to harvest the clams. She added that other Seris would keep bothering the family to borrow it for that purpose, because it worked so well.

Besides consuming the clams themselves, during the early to mid-twentieth century the Seris harvested clams in the esteros bordering the Infiernillo Channel for sale to commercial buyers.

The Venus clams were eaten raw, boiled in the shell, or roasted under a brush fire. Today, they are usually boiled and the meat might be reheated with other foods such as tomatoes and onions.

Amiantis callosa

The polished, ivory-white shell of this uncommon clam has prominent concentric ridges. The shell was not recognized by those to whom I showed it.

L 64 mm

Chione californiensis

spiitquim (var. ***spiitcam***)

L 62 mm

This clam is very common on estero sand flats. Its sturdy white shell is marked by a prominent cancellate sculpture.

Spiitquim is a primary name. The shells were inserted into rock crevices, trees, or columnar cactuses such as a cardón as an offering to solicit good luck from the spirit of the area or plant.

Boys played games of skill with the empty shells. In one game, the boys would pile the shells on the shore behind a line in the sand and would throw them into the ocean to see who could throw the farthest or highest—whoever's shell hit the water last, won. Another game, called *iti icamaptx* roughly translated 'head injury game', involved throwing the shells high into a cardón cactus to embed the shells in its flesh. Xavier Moreno explained that the shell's sharp edge could easily slice into the cactus. Lorenzo Herrera gave details of the game as he played it as a boy: each player had five shells, and the boy who had all five shells remain in the cactus was the winner. The shells were thrown very hard and could ricochet off the cactus at high speed, at times back at the boys playing the game. Lorenzo remembered a childhood friend who suffered a head injury, adding dryly that the boys' parents did not like the game.

The shells were said to have been briefly used as tokens for payment by an early non-Seri commercial fish buyer (see Trade and Commerce).

Figure 4.50. A shell lodged high in a cardón cactus.

Chione tumens

haxöl aapa 'huge or true ***haxöl***'
oot ihaxöl 'the coyote's ***haxöl***'

L 62 mm

The shell of this smaller *Chione* species is distinctive because of its markedly raised, rounded ridges and striking pattern of fine brown V-shaped markings.

The origin of the name ***haxöl aapa*** 'huge or true ***haxöl***' is somewhat confusing, as the *Leukoma grata* clam is considered to be the true ***haxöl.***

The name ***oot ihaxöl*** reflects the clam's somewhat unusual shell sculpture, as well as its striking markings that are similar to those of the *Leukoma grata* clam.

This clam was not a common food, as it was not found alive in quantity.

Chione undatella

spiitquim coospoj 'spotted ***spiitquim***'
xepe an hapafc 'what is pounded in the sea'

L 55 mm

The shell of this medium-sized clam has raised concentric ridges marked with delicate brown V-shaped designs.

The name ***spiitquim coospoj*** 'spotted ***spiitquim***' is clearly descriptive. ***Xepe an hapafc*** 'what is pounded in the sea' is probably a variant of ***quiimosim xepe-noyaafc*** 'what the beggar pounds in the sea', a name given to *Anadara multicostata* and *Cardites laticostatus.*

This clam was eaten, although in lesser quantities than other more common *Chione* clams.

Chionista fluctifraga [= *Chione fluctifraga*]

haan

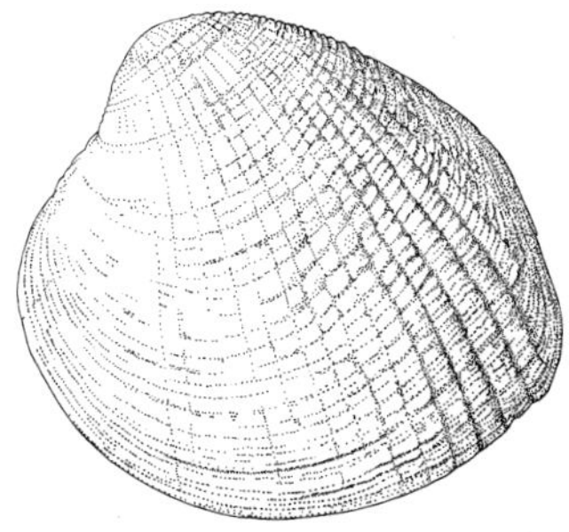

L 46 mm

This clam, common on estero sand flats, has a somewhat smooth, white shell with a deep purple interior margin.

Haan is a primary name.

The clam was an important food. It was especially abundant on shores along the Infiernillo Channel and Estero Santa Cruz, where it was sometimes harvested and sold to commercial fish buyers in the mid-twentieth century.

A number of place names reference the name ***haan***; all are areas close to *Haanc,* a point on the eastern shore of Isla Tiburón known as Punta San Miguel, where the clam was abundant (see Place Names).

Chionopsis gnidia [= *Chione gnidia*]

queeex 'who grunts'
spiitquim cpaxz 'rough ***spiitquim***'
♦ ***spiitquim ctamoo*** '***spiitquim***-shaman power'
spiitquim quictamo 'fierce ***spiitquim***'

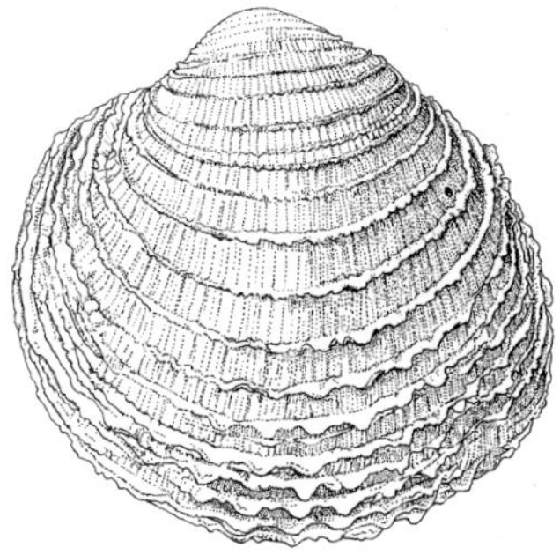

L 80 mm

Easily identified by its large size and ornate, sharply ridged white shell, this clam was considered to be good food, although it was not as abundant or widespread as some of the other Venus clams.

Queeex 'who grunts' is an archaic name. When someone purposefully groaned or grunted in front of another person, it was understood that he was putting a curse on that person.[88]

Spiitquim cpaxz 'rough ***spiitquim***', somewhat descriptive of the shell's physical characteristics, is not a well-known name. The clam's most common name, ***spiitquim ctamoo*** '***spiitquim***-shaman power' is also given to *Periglypta multicostata*, another Venus clam. The term *ctamoo* is archaic.[89] Less known is ***spiitquim quictamo*** 'fierce ***spiitquim***' a name possibly alluding to the prominent, sharp ridges on the clam's shell.

Dosinia ponderosa

halít cahooxp 'what causes white hair'

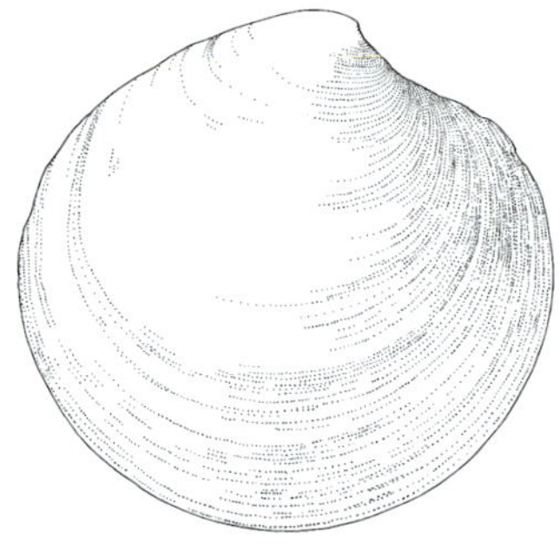

L 100 mm

The white shell of this large clam is sturdy, heavy, and very smooth. The clam is found on intertidal sand flats.

The name ***halít cahooxp*** 'what causes white hair' comes from the idea that eating too many of the clams will turn one's hair white. However, Francisco Morales laughed as he related how the clam's name and corresponding belief began: an old man eating the clams informed some young people sitting nearby that eating the meat would turn their hair white—because he wanted to eat the clams himself.

Although some say that the meat was *hapáx* and was something to be avoided by young people, others said that the clam was good to eat.[90] One person said that the clam had to be boiled rather than roasted under a brush fire since the shells were similar to those of ***camazjij*** (*Megapitaria squalida*) and would easily shatter in the fire. Others, however, did not concur and described how the clams were placed hinge-down, close together in rows in the sand, and roasted under a brush fire. When the brush had burned down and the shells were open, the clams were cooked. The meat was often flavored with oregano, although one woman said that she liked sweetening the clams with sugar.

Because the shell was strong it was used as a digging tool (figure 4.51). In relating another use, María Luisa Astorga said that in the past the shell's ventral edge was chipped away to create a rough, sharp edge, and the shell was used to scrape sea lion skin or deerskin or the air bladders of large fish (see *Arca pacifica*, *Modiolus capax*, and *Euvola vogdesi*).[91] Several people reported that the shells were used to harvest agave plants as well.

A rounded core left after the shell's edges were chipped away and smoothed was used as a tool to burnish clay vessels during their manufacture. A small smooth stone was also used.

Figure 4.51. José Manuel Romero (left) and Roberto Herrera using shells to dig for fresh water. José Manuel sports a strand of shells around the crown of his hat. A *Dosinia ponderosa* shell lies in the foreground. Tecomate, Isla Tiburón, 1946. Photograph by William Neil Smith, courtesy of University of Arizona Libraries, Special Collections. MS316_B31_slb1_R.-114–12.

Hysteroconcha lupanaria [= *Pitar lupanaria*]

L 48 mm

The white porcelain-like shell of this clam is distinguished by a row of long, slender spines, marked at their base by a deep violet.

Usually found south of the present day Seri territory, the clam was not recognized by anyone to whom I showed it. I have only recently seen the shells in beach drift on sand beaches in the Bahía de Kino area.

Lamelliconcha concinnus [= *Pitar concinnus*]

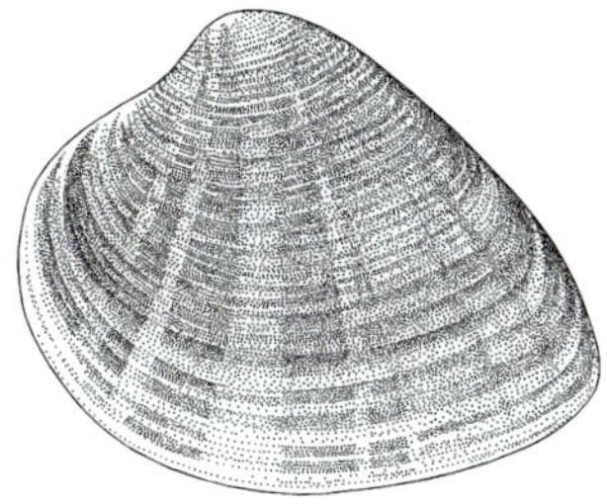

L 77 mm

The very attractive white shell of this clam is marked with brownish-red rays.

Although the Seris today have not provided a name or information about this clam having been eaten, Edward H. Davis did collect the shell and recorded that the clam was eaten by the people.[92]

Leukoma grata [= *Protothaca grata*]

haxöl
haxöl áa 'true ***haxöl***'

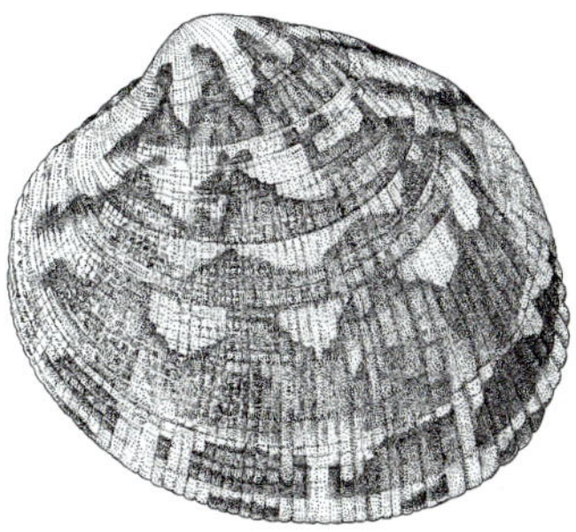

L 35 mm

This very common clam occurs throughout the region, found buried in sand beneath intertidal rocks. The clams' attractive shells are marked with distinctive and variable designs in shades of brown and tan.

Haxöl is a primary name. Since other bivalve names incorporate the word, the name ***haxöl áa*** 'true ***haxöl***' perhaps implies that this clam is the original bearer of the name—the simple or real ***haxöl***.

The people harvested the clams by digging in the sandy substrate beneath intertidal rocks with a wedge-shaped stick, a larger shell, or spoon. If the clams were to be consumed promptly they were rinsed in seawater at the beach; this helped to separate any dead, dirt-filled shells called *hant ccam* 'full of dirt', so they could be discarded.

Heavily consumed, the clams could be eaten raw or cooked by boiling. When eaten raw, a clam was broken open by smashing it against a rock so that its meat could be easily removed. When cooked, small clams such as these were not roasted under a brush fire—as one person observed, their shells might break apart in the fire. When boiled, the whole clams were placed in either seawater or freshwater; if cooked in freshwater, the broth could be consumed. The cooked meat was sometimes removed from the shell and reheated with other ingredients.

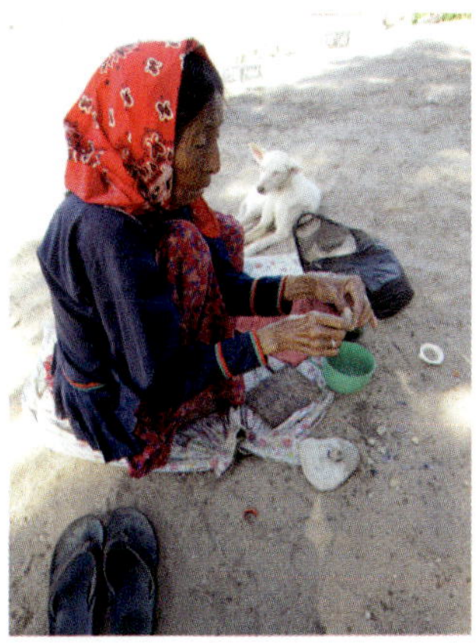

Figure 4.52. Angelita Torres breaking open raw *Leukoma grata* clams to eat. Desemboque, 2013.

When talking about this clam, Angelita Torres recalled living as a child on Isla Tiburón, and her parents collecting a large quantity of the clams. The clams were cooked, their meat removed from the shells, and spread out to dry in the sun. When dried, the meat was carried in a cloth bag to an inland camp, where it was later rehydrated and cooked with sea turtle oil and seasoned with oregano. Angelita commented that the clams were delicious when eaten that way.

An early gesture of Seri hospitality is recorded in the voyage account of a vessel exploring the northern Gulf, made by a Jesuit missionary, Juan de Ugarte, in the summer of 1721. He was accompanied by William Stratford, an English shipwright. Their return journey southward brought them a second time to Tecomate, on the north shore of Isla Tiburón. With their crew suffering from scurvy and Ugarte from severe rheumatism, they found themselves welcomed by the Seris and given food and water. Ugarte wrote that "todas las mujeres y niñas fueron a la playa y vinieron cargadas de unas almejillas delicadas" [all the women and girls went to the beach and brought back some delicious little clams]—no doubt the abundant ***haxöl*** found on that shore.[93]

Figure 4.53. *Leukoma grata* shells strung on a necklace with dyed *Asclepias subulata* stems. Collected by John Houser in 1968. Arizona State Museum, catalog no. 78-25-3.

Edward H. Davis recorded that the clam was eaten by the people. A note accompanying Davis's shell collection reads: "Small clams found on stony flats at low tide. Found in sand below the rocks. Boiled in the shell, the tiny clams and broth are delicious."[94]

María Luisa Astorga recalls her grandmother carrying the live clams to *Hatajc* (Pozo Coyote), a day's walk inland from Desemboque, and cooking them there.

Because they are not easily pierced without breaking, the shells were rarely used for adornment; however, one unusual necklace in the Arizona State Museum collections is composed of the shells strung together with sections of plant stems (figure 4.53).

The shells were used as tweezers for plucking whiskers; Juanita Herrera recalled by name a man she saw using large ***haxöl*** shells for this purpose.[95]

The varied and striking markings on the shell have been sources of inspiration for basket designs.[96] When talking about this, one woman also recalled using the V-shaped patterns she found on one shell for her face-painting designs. Especially noteworthy were shells having human-like figures on them, and certain women who liked incorporating such designs on their baskets were singled out as having done this. According to one woman, young girls were encouraged to use a simple design taken from a clam when making their first baskets.

Cleotilde Morales related how the clams received their designs:

> The clams' markings are from the time of the great flood Catastrophe, it is said. At that time they were cows, called *hant csii* 'what smells the earth'. They went toward the sea and entered it. The flood covered everything, and [the cows] had the markings, and were called *hant csii*. It is said that the cows went into the sea when the Catastrophe happened, and the cows' markings remained.[97]

Figure 4.54. *Leukoma grata* clams with their varied designs.

Angelita Torres described a shell having a light violet–colored beak as being *haxöl inaail hayeen yahihjö* '***haxöl*** shell having a reddened face'. She added that such clams were not *hapáx* and to be avoided as food; however, another woman did comment that such a clam would not be eaten by a woman of childbearing age.

Certain fish are known to eat the clams. Efraín Estrella noted that the *tjámoja* fish (Gulf opaleye, *Girella simplicidens*) eats ***haxöl***, and Francisco Morales described how the *ziix hant cpatj* (triggerfish) searches for clams head-down on the ocean floor. Xavier Moreno recalled a ditty his grandmother sang while cleaning some triggerfish he had caught:

Figure 4.55. A *Leukoma grata* clam with a violet-colored beak.

Ziix hant cpatj cöicoos

Hant coaan ano hptiti
Hiza hptiti
Hast ihtcaxaj íi!

Song of the triggerfish

Here I am, head-down in the swirling sand
Here I am, head down
[Oops!] I just bit down on a rock!

There are a few accounts of individuals having adverse reactions involving marine creatures or organisms, two of which relate to *xepe án iheel* 'red (state) of the sea area', the toxic red tide. Cleotilde Morales recounted an incident that took place when a group of people died from eating contaminated clams:

> A group of *Comcaac* gathered ***haxöl*** at *Hajháx* [Tecomate], not knowing that they were going to die. They were hungry, so they ate [the clams]. They vomited, their heads hurt, their eyes popped out, they died there. Their heads just broke, that is why red tide is so dangerous. They didn't know. Not knowing they would die, they ate [the clams]. At *Hajháx* they all died. Other *Comcaac* came from the *Xepe Coosot* [Infiernillo Channel] and saw the people. They didn't know what was wrong. Only one person, an old woman, still had a little bit of life in her. She knew what had happened. The people who came still didn't know what the others had eaten that had killed them. The old woman said: "They ate these things, and they all died." The old woman said this. Then she just died. She could only say one or two words; she said them, and then died. So those [other] people that came there knew that they shouldn't eat this bad food. Not one person was left alive who had eaten it. They ate whatever food they liked. The poor people didn't know. They all died.

A second account of probably the same incident was given by Eva López in which her great grandfather, Antonio, is figured:

> *Cmaacoj Piipa Ano Coopis* 'Old Man Pipe-Smoker' had a Mexican name, Antonio. His mother planted a mesquite seed from a [non Seri] ranch, and it grew into the huge mesquite at *Hajháx*. He and his family were inland Island people. They lived at *Xeescl Cmoiij*. They were tired of eating deer, and were temporarily at *Hajháx*. At that time there was red tide in the area. The people ate clams and all were poisoned except for a little girl who was still nursing. Other *Comcaac* came from the *Xepe Coosot* area and saw the dead, and Antonio dying. All he could say was "*hao<u>l</u>a, hao<u>l</u>a*" [***haxöl***]. He pointed to a pot—they could see what had been cooked. This happened long before Sara Villalobos was born [ca. 1900].

Figure 4.56. The extensive rocky intertidal area that gives Desemboque the name *Haxöl Iihom* 'place of the ***haxöl***'. 2013.

Figure 4.57. The base of *Haxöl Hast* (foreground). October 2007.

The clam's name occurs in two recorded place names. The name for Desemboque, *Haxöl Iihom* 'place of the ***haxöl***', aptly describes it as a place where many of the clams are found in the rocky intertidal areas. *Haxöl Hast* '***haxöl*** mountain' is a seaside mountain north of Bahía de Kino. According to Lorenzo Herrera, the mountain is so named because the clams were abundant in the area. In stormy seas they would be cast onshore into hollows in the rocky cliff base, where they could be harvested without digging.[98]

Megapitaria aurantiaca

camazjij caacoj 'large ***camazjij***'
zaah 'sun'

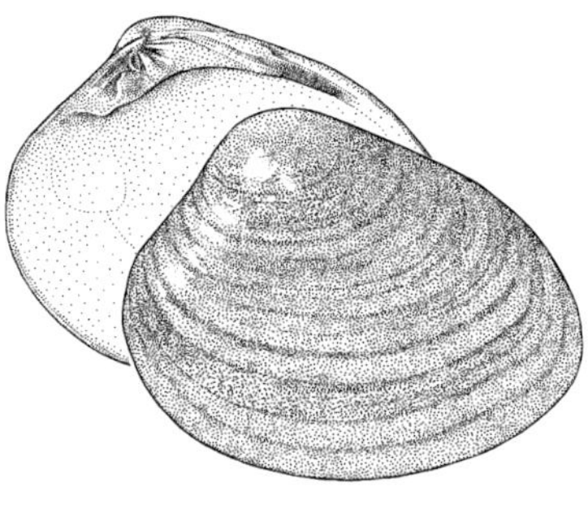

L 111 mm

This large clam is found in submerged sandy areas and is rather uncommon. Its sturdy shell has a violet-tinged white interior and a pinkish-orange exterior, covered by a shiny periostracum.

Many people call the clam ***camazjij caacoj*** 'large ***camazjij***' because of its similarity to the smaller ***camazjij*** (*Megapitaria squalida*).

Zaah 'sun', an older name unrecognized by most people, was said to have been given because of the sun's bright reflection on the shell's varnished, brownish-orange surface when fresh from the sea. Victoria Astorga provided the name, having saved some of the shells after the clams' meat had been eaten. She thought that the shell and its name would be new to me, so she seemed rather delighted to be able to surprise me. As she talked with me about the shells, she kept using one of them to scrape off the other's varnished brown periostracum, in order to "clean" it. I'm glad I refrained from asking her to stop. Those scratched-up shells are a much more poignant reminder of the day she introduced them to me than they would otherwise be, lying pristine in a collection.

This clam was eaten, although probably not in such quantities as *M. squalida*, which is much more common.

Megapitaria squalida

camazjij

This medium-sized clam is fairly common on some intertidal sand flats. Its shell is distinctive because of a varnished brown periostracum.

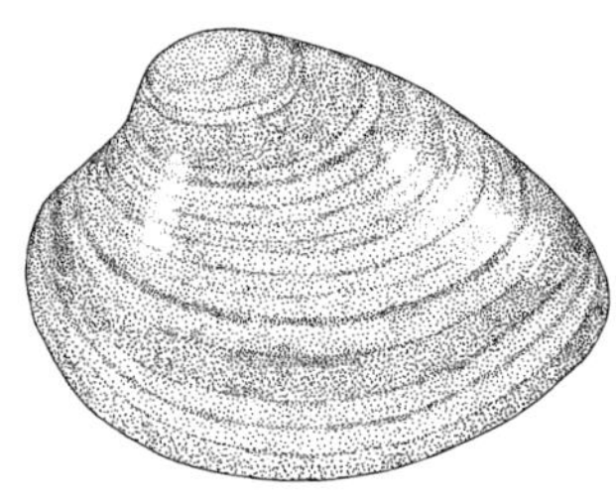

L 77 mm

The name ***camazjij*** is related indirectly to the intransitive verb *cmazjij* 'what cracks open', alluding to how the shell cracks and shatters in a fire.

This clam was considered good food, and some people told me that they had eaten it recently, describing its meat as sweet-tasting. The clam was usually boiled in the shell, since its shell was said to break apart in a fire, although one person did say that in the past it was roasted under a brush fire.

Periglypta multicostata

spiitquim ctamoo '***spiitquim***-shaman power'

The very robust shell of this rather uncommon large clam has a strong cancellate sculpture.

L 125 mm

The name ***spiitquim ctamoo*** '***spiitquim***-shaman power' is shared with *Chionopsis gnidia*, a more common clam. Although the difference is obvious, the large size and ornate sculpture of both clams make them somewhat similar.

This clam was eaten, and was cooked by roasting under a brush fire.

Tivela byronensis

oot icamazjij 'the coyote's ***camazjij***'

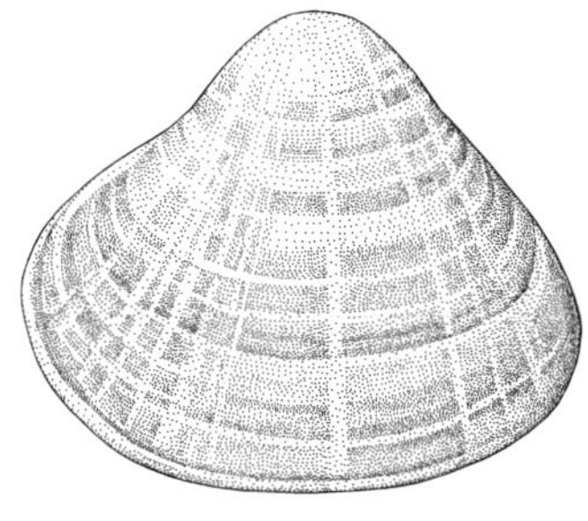

L 54 mm

The polished shell of this clam is finely marked with muted brown rays.

The name reflects the shell's similarity to that of the ***camazjij*** (*Megapitaria squalida*), with its similar varnished periostracum, while being different enough to be called a possession of the coyote.

Although there is little information about the clam having being eaten, Edward H. Davis collected the shell and recorded that the clam was eaten by the people.[99]

Raquel Moreno pointed out the shell's attractive striping and recalled that a woman she knew used such a design in a basket (figure 4.58). According to Raquel, a basket incorporating designs taken from such clams was called *haxöl hasaj* 'clam (design of) basket', and she called the design on the clam *haxöl yap* 'clam basket design'.

Figure 4.58. María Elena Romero sewing a basket with a design inspired by the *Tivela byronensis* clam. Desemboque, ca. 1970. Photographer unknown.

Petricolidae

Petricolaria cognata [= *Petricola cognata*]

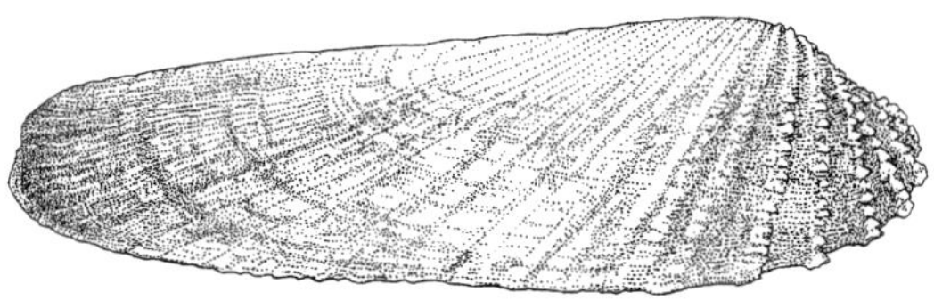

L 53 mm

This "nestling clam" (so named because the clam sometimes lives nestled in rock crevices) has a long, narrow, and fragile shell. Mostly white, it is tinged with light brown on its posterior margin; anteriorly strong radial ribs intersect with the comarginal ridges, forming sharp protrusions. The clam is not well known, as it occurs at the southern extent of the Seri territory today.

Pholadidae (piddocks)

Pholas chiloensis

haxöl icaai ctam 'male ***haxöl icaai***'
heexoj
xpajísotoj (var. ***pajísotoj***)
ziic ccap cooxp ihaxöl 'the Tern's ***haxöl***'

This boring clam has a distinctive rough-sculptured, fragile white shell. Since the clam occurs around Bahía de Kino and to the south, and not in the present-day Seri region, it was not recognized by many people.

The name ***haxöl icaai ctam*** 'male ***haxöl icaai***' notes the shell's similarity in shape to that of the ***haxöl icaai*** (*Simomactra dolabriformis*), while its elongate shape and spiny surface texture suggest "male" characteristics. Although it is rather fragile and not smooth-surfaced like the ***haxöl icaai***, one person said that the shell was used as a tool in shaping clay vessels. The name ***haxöl icaai ctam*** is shared by the *Tagelus peruvianus* clam.

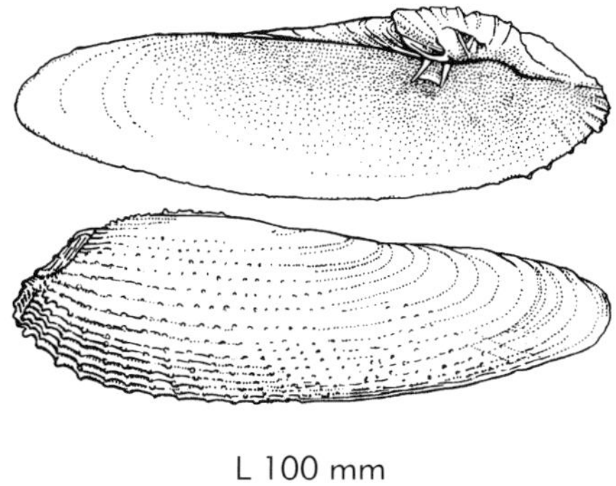

L 100 mm

The primary name ***heexoj*** also refers to a torch made from the dried ribs of the organ pipe cactus (*Stenocereus thurberi*), although no relationship has been suggested. ***Xpajísotoj***, as well as its variant ***pajísotoj***, is a compound of *xepe* 'sea' and *-ísotoj* 'narrow (pl)'. The name is also given to the similarly shaped *Tagelus peruvianus.* According to Cleotilde Morales, the descriptive name ***ziic ccap cooxp ihaxöl*** 'the Tern's ***haxöl***' was given to the shell by her mother, María Antonia Colosio, and was apparently a name only she used. Cleotilde explained that *ziic ccap cooxp* is a name given to the Tern (*Sterna* sp.) to replace the name *hatj*, when a man named *Hatj Quiho* 'he who sees the Tern' died, and his name was no longer spoken. Other clams, however, do contain the name *hatj* (see Tellinidae).

In what I believe to be the only previously published reference to the use of this shell by the Seris, M. R. Harrington notes that Edward H. Davis recorded that it was sharpened on a stone and used as a knife to cut the umbilical cord of a newborn.[100]

Lyonsiidae

Entodesma pictum

oot isatoj 'the coyote's ***satoj***'

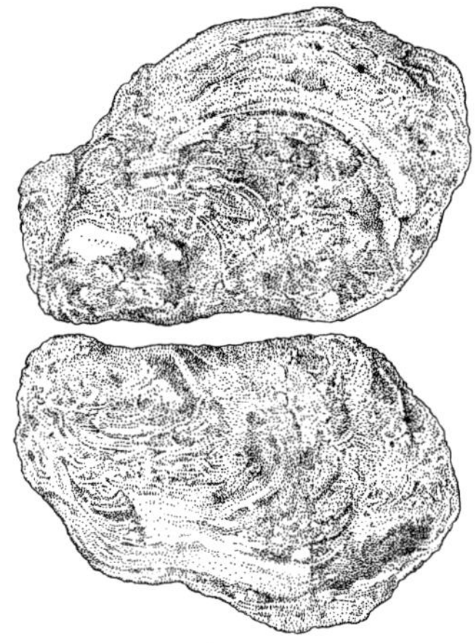
L 43 mm

This small bivalve is unusual and uncommon in the area. Its very fragile shell has a pearly interior and is covered with a thin periostracum.

The name ***oot isatoj*** 'the coyote's ***satoj***' notes the shell's similarity to that of a mussel, ***satoj***, while being odd enough to be owned by the coyote.

A shell of this clam was given to me in 2009 by Alfredo López, who had collected it live from a mussel bed in the Infiernillo Channel. Alfredo informed me that this is the real ***oot isatoj***, rather than the *Mytilus californianus* or *Modiolus tumbezensis* (mussels for which he had earlier provided the name). However, since *Mytilus californianus* and *Entodesma pictum* are little known today, most people identify *Modiolus tumbezensis* with the name ***oot isatoj***.

GASTROPODA

Like the bivalves, the gastropods were eaten, although probably not in the same quantities since many are small and the amount of meat they contained was apparently not worth the work involved in removing it. Their larger shells were used, although, again, not as much as those of the bivalves, nor in the same way. Their smaller shells were heavily used, and continue to be, in shell jewelry.

There is no Seri term for "gastropod." More than eighty species are recognized by Seri names, some with multiple names. A few that are unnamed are included here as well, for the record.

Fissurellidae (keyhole limpets)

Diodora inaequalis

hast iti queemij 'what moves slowly on rocks'
mosniil inoosj 'the *mosniil* sea turtle's claw'
mosniscl inoosj 'the *mosniscl* sea turtle's claw'

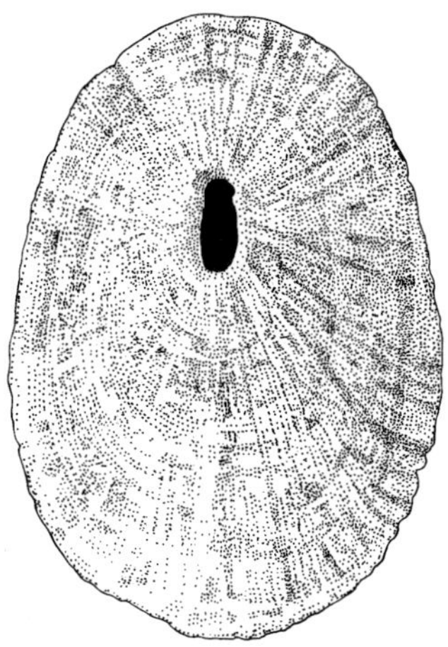

L 22 mm

This small keyhole limpet is common beneath intertidal rocks. The interior of its shell is white, while its exterior, with its characteristic notched apex, has strong radial ribbing lightly rayed with grayish-brown.

The name ***mosniil inoosj*** 'the *mosniil* sea turtle's claw' alludes to the shell's similarity in shape to the claw of a kind of black sea turtle that is no longer seen. ***Mosniscl inoosj*** incorporates another name for the turtle.

According to Alfredo López, the name ***hast iti queemij*** 'what moves slowly on rocks' describes the limpet's movement.

While shells of the keyhole limpets are common in middens on Isla Tiburón, indicating that they were probably eaten,[1] no one has confirmed that the limpet was considered food.

Lottiidae (limpets)

Lottia dalliana [= *Collisella dalliana*]

cmacjij (var. ***cmahjij***, ***cmatjij***)
hast iti cöcooyam 'what passes over [high places of] rocks'
♦ ***haxölinaail cooil*** 'blue/green shell'
haxölinaail cooil caacoj 'large ***haxölinaail cooil***'
◊ ***haxölinaail cooil ctopis*** '***haxölinaail cooil*** that attaches by suction'

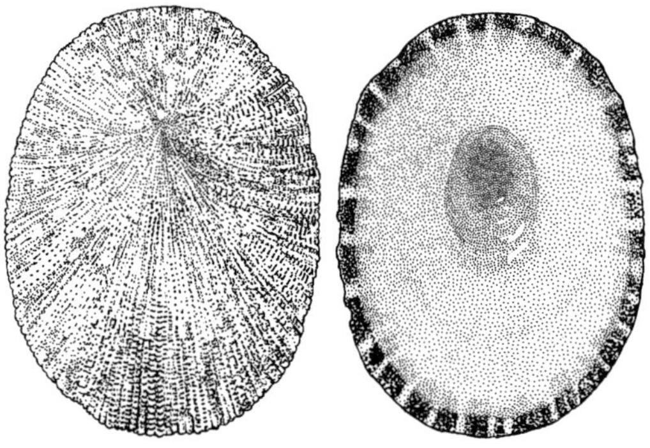

L 44 mm

This limpet, the largest of its genus in the area, has a shell with an attractive light blue interior.

The primary name ***cmacjij*** (or its variant pronunciations) is said to be the oldest name for this limpet. ***Hast iti cöcooyam*** 'what passes over [high places of] rocks' is another old name, alluding to the limpet's movement over rocks exposed at low tide.

This limpet, along with *Lottia stanfordiana* (both of which have bluish interiors) is most commonly called ***haxölinaail cooil*** 'blue/green shell'.[2] To distinguish between them, the name ***haxölinaail cooil caacoj*** 'large ***haxölinaail cooil***' is sometimes used for the larger *L. dalliana*.

The name ***haxölinaail cooil ctopis*** '***haxölinaail cooil*** that attaches by suction' was given by one person and seemed somewhat improvised.

These limpets were eaten in the past, something evident in dune middens where the shells are found together with those of other commonly eaten shellfish. After the limpets were boiled in saltwater, their meat was easily removed and was said to taste like ***haxöl*** (*Leukoma grata*).

This shell is a favorite for stringing on necklaces. A recent use is to cut the shell into small shapes which, when put together with a complete shell, take the form of a tiny sea turtle. The shell and fragments are glued onto small cloth bags filled with the fragrant leaves of desert lavender, which are then attached to necklaces.

Lottia stanfordiana [= *Collisella stanfordiana*]

hast iti cöcooyam 'what passes over [high places of] rocks'
♦ ***haxölinaail cooil*** 'blue/green shell'
◊ ***haxölinaail cooil ctopis*** '***haxölinaail cooil*** that attaches by suction'
naj

This small limpet is common on intertidal rocks. Its shell has a striking teal-green interior.

According to Juanita Herrera, the name ***hast iti cöcooyam*** 'what passes over [high places of] rocks' is an old name for this shell, and alludes to the shell's slow movement over rocks exposed at low tide.

This limpet and the *L. dalliana* (both with bluish interiors) are most commonly known as ***haxölinaail cooil*** 'blue/green shell'. A similar name ***haxölinaail cooil ctopis*** '***haxölinaail cooil*** that attaches by suction' is shared by *L. dalliana*.

According to some, the primary term ***naj*** is the oldest name for this limpet.

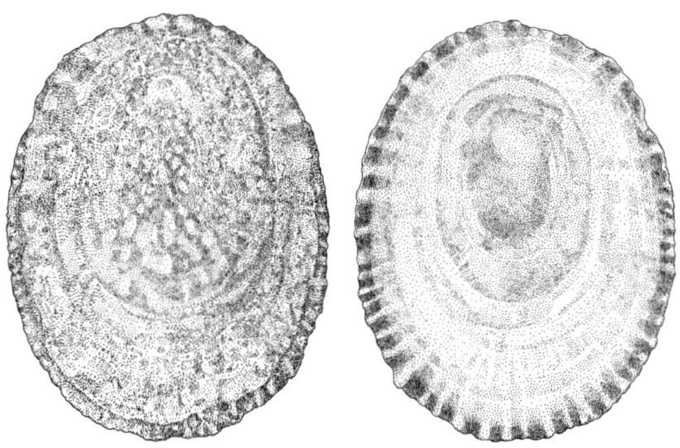

L 31 mm

Although not eaten now, the limpet was said to have been eaten by the Ancestors.

The limpet is easily collected from rock surfaces at low tide and is a favorite for use on necklaces because of its striking and distinctive color. In preparing the shells for stringing, hot water is poured over the limpets, and the meat is then removed.

Lottia strigatella [= *Collisella strigatella*]

◊ ***haxölinaail cooil cooxi*** 'faded ***haxölinaail cooil***'

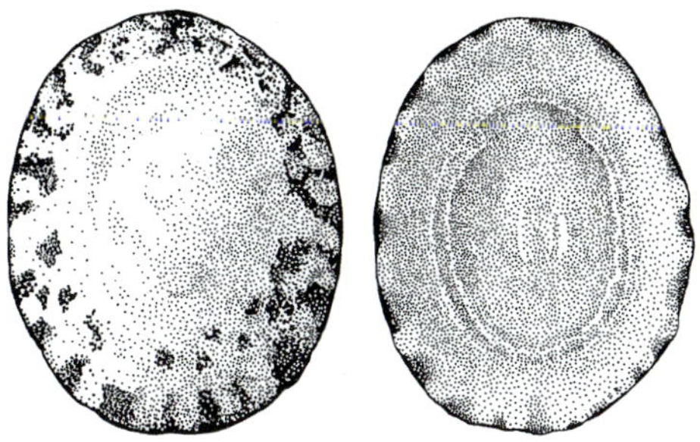

L 18 mm

This small limpet has a rather pale shell lacking the colorful interior of the other better-known blue-green limpets.

The name ***haxölinaail cooil cooxi*** 'faded ***haxölinaail cooil***', given by one person, seemed somewhat improvised, rather than being a well-used name.

The shell is not strung on necklaces as are the other more colorful members of this family.

Figure 5.1. María Magdalena Estrella preparing *Lottia* spp. shells for stringing. Desemboque, 2013.

Trochidae (top shells)

Tegula rubroflammulata

cotopis quislitx '***cotopis*** with an ear canal'

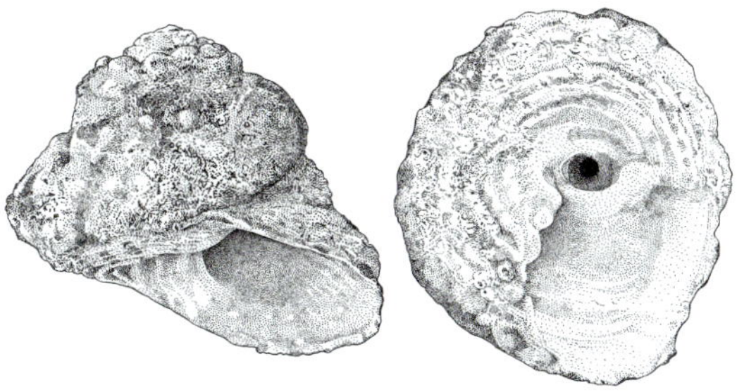

L 21 mm

The shell of this snail has ridged whorls and a pearly, light green interior. The heavily encrusted snail was found live on a pen shell from the Infiernillo Channel in 2009, and Alfredo López provided its name.

The name ***cotopis quislitx*** '***cotopis*** with an ear canal' references the shell's umbilicus. The ***cotopis*** (*Turbo fluctuosus*) snail has a similar pearly green interior, but without such an umbilicus.

Tegula rugosa

haapx

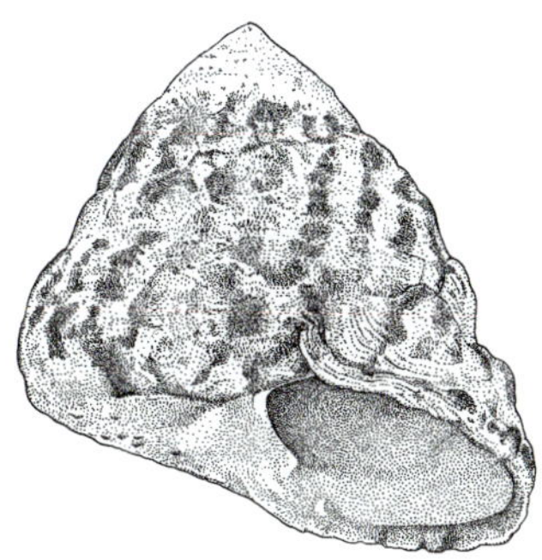

L 25 mm

This snail is very common on intertidal rocks. Its grayish shell, marked with a darker striped pattern, is often weathered and covered with encrustations, giving it a somewhat nondescript appearance.

Haapx is a primary name. The name occurs elsewhere: a similar but smaller unidentified snail is ***pohaapx camoz*** 'what thinks it is an ***haapx***'.[3] An unidentified beetle is called *haapx quiimosim* 'what begs from the ***haapx***'.

This snail was eaten. Large ones are said to occur around Isla Tiburón, and the shell is common in middens throughout the area. The meat could be eaten raw, often by children playing at the beach. When cooked, the snail was boiled or roasted under a brush fire; some people said the whole animal was eaten, while others preferred the solid meaty parts.

As Sara Villalobos commented, "These were good food. The Ancestors ate them when there wasn't any other food, and they were hungry. [They would say] 'We don't have blankets, and what will we eat?'"

Lorenzo Herrera told about finding snails "like ***haapx***" at a beach area north of Desemboque. He described them as being similar in size to the larger *Turbo fluctuosus* snail and provided the name ***hapxaacoj*** 'large ***haapx***'; others confirmed the name.

The *Tegula rugosa* snail, along with similar intertidal snails, was used for bait. The shell was either broken open and the animal put on a small hook, or the snails were crushed and thrown into the sea to attract fish, which were then speared.

Turbinidae (turban snails)

Astraea unguis

Hast Án ano haapx '***haapx*** from *Hast Án*'
◊ ***xepenyaaan***

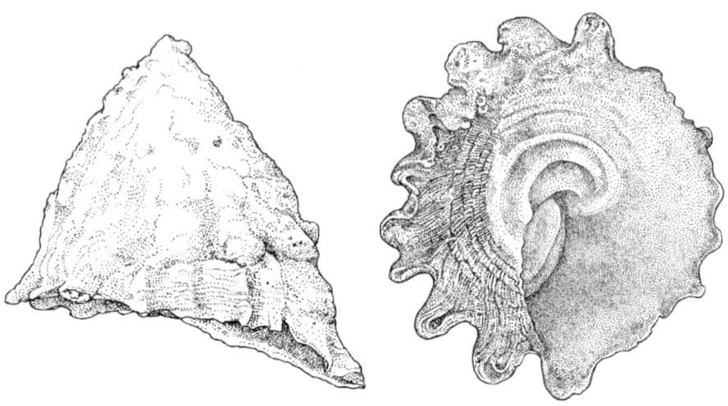

L 32 mm

The shell of this snail is characterized by a pearly, greenish-white interior and a base with rounded projections. Its operculum is a porcelaneous white.

The name ***Hast Án ano haapx*** '***haapx*** from *Hast Án*' relates the snail both to where it is found and to the ***haapx*** snail (*Tegula rugosa*), found in similar rocky habitats. According to Alfredo López, this uncommon snail is found among large intertidal rocks at *Hast Án* 'rocky area', a location on the western shore of Isla Tiburón. Occurring only into the southernmost limits of the Seri territory, the shell was not recognized by most people to whom I showed it.

The name ***xepenyaaan***, from *xepe* 'sea' and the verb *coaan* 'stirred up' was given by Lorenzo Herrera. He suggested the name reflects that when large numbers of this snail were seen offshore through calm, clear water, they gave the impression of a stirred up, murky area.

Turbo fluctuosus

cotopis 'what attaches by suction'

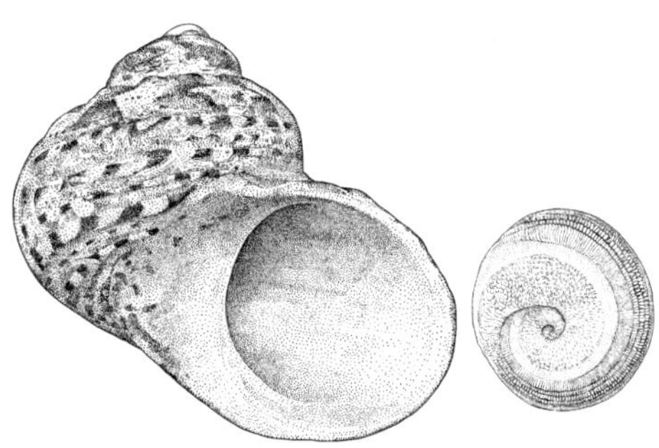

L 43 mm

This common snail is found strongly fastened to rocks in the low intertidal or shallow submerged area. The shell's exterior is a patterned brown and the interior a pearly, pale green. The sturdy operculum has a distinctive spiraled exterior.

The name ***cotopis*** 'what attaches by suction' is highly descriptive, as this snail has a noticeably strong attachment to intertidal rocks; perhaps its good size makes this characteristic more obvious. The snail's hard calcareous operculum, and those of similar snails, is called the *iti yahimoz* '[a hard thing] with which it covers itself'.

Easily collected at low tide, this snail was considered good food. According to some, both men and women could be involved in its harvest, since men would sometimes help with gathering and carrying shellfish because of the weight and distance involved. The snail was boiled or roasted under a brush fire and the meat removed from the shell after cooking. One woman said she liked eating the meat flavored with a little sugar, while another commented that the meat was good cooked with rice and tomatoes.

Cleotilde Morales said that the Ancestors often ate the snails and described how they were kept and carried: "They could be kept alive for four or five days by keeping them moist. That is why going up [into the desert, away from the sea] you will see these shells. They could be wrapped in cloth or put in a [clay] jug and could be taken far inland. They stay alive because they close themselves up with seawater inside."

According to one consultant, the meat of this snail was *hapáx*, a food to be avoided. Boys were cautioned against eating too many of these snails, because their foreskins would retract, exposing their penis heads.

Sara Villalobos remarked that the shells were "just played with by children." Angelita Torres recalled seeing the opercula used as *hapéxeca* (game pieces) for the game of *raama* (from Spanish *damas*, checkers), saying that her father had a glass jar in which he kept the opercula. They were painted with commercial paint used for Seri boats; half of them were blue and the others red. Other people have described the game where one player used the (unpainted) opercula placed face-up, while his opponent placed his face-down (see Recreation).

The opercula were also used as *haheet* (game counters). Curiously, a clutch of opercula collected by Edward H. Davis in the National Museum of the American Indian is labeled as "money." When I asked Victoria Astorga about this, she chuckled as she recalled that someone traded Davis a bag of opercula on Isla Tiburón. She told me they were not used as money, since "money wasn't known [at that time]"; rather, they were used as counters in gambling games.[4]

The shell was used medicinally. It was toasted until ashy and ground into a fine powder. The powder was applied to a baby's umbilical stump if it were slow in healing, helping it to dry and drop off in a short time.

The live snail was broken open and the meat used as bait, either put on a hook or thrown into the sea to attract fish for spearing. According to Efraín Estrella, during a short time in the early 1950s a group of Seris were trapping lobsters off the southern shore of Isla Tiburón,[5] and the ***cotopis*** meat was found to be the best bait for attracting the lobsters (see Bait).

Cotopis Inaail quih Yeen 'face of the ***cotopis*** shell' is an area by *Hasteemla* (a mountain south of Desemboque) where Seris ate the ***cotopis*** snails. The "face" here is said to refer to the area, not the shell.

A song composed and sung by Ramón López in 2008 refers to the snail:

Cotopis Cmaacoj Cöicoos

Có to pis cma coj ya
Cotopis cmaacoj
Xé pe cyail a no mo ta ya
Xepe cyaail ano mota
Xé pe té li nái ya
Xepe iteel inyaai
Xé pe té li nái ya
Xepe iteel inyaai
Có to pis cmá coj ya
Cotopis cmaacoj
Có to pis cmá coj ya
Cotopis cmaacoj
Xé pe cyail a no mó ta
Xepe cyaail ano mota
Xe pe té li nái i ta
Xepe iteel inyaai
Hái com a nox há mo sá xoe ya
Hai comaanoj haa mosa xoee
Có to pis cma coj zix i sax cöi ma naj xoé
Cotopis cmaacoj ziix iisax cöimaanaj xoee.

Song of Old Man ***Cotopis***

Old Man ***Cotopis***,
Came from the deep sea,
He went to the shore,
He went to the shore,
Old Man ***Cotopis***,
Old Man ***Cotopis***,
He came from the deep sea,
He went to the shore,
He said, "The turbulent wind is coming."
"Old Man ***Cotopis*** wasn't knocked about," he said.

As Ramón explained, Old Man ***Cotopis*** came from the deep sea, toward the shore at *Socaaix* (Punta Chueca). He arrived at the shore and sat there. A turbulent wind blew, but he was secure and nothing happened to him, because he was firmly anchored.

Neritidae (nerites)

Nerita funiculata

comcaii inoosj heeque 'small ***comcaii inoosj***'

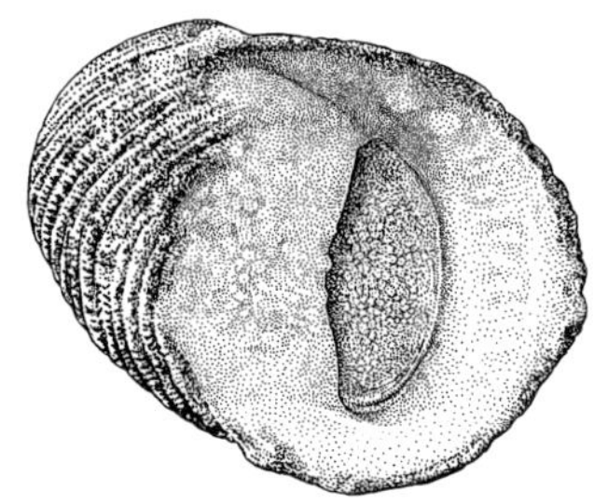

L 15 mm

This small snail is very common on rocks in the intertidal area. Its shell is black with a white interior and a pustular surface around its aperture.

The snail shares its name with the larger *Nerita scabricosta*, but was not eaten because of its small size.

The shells are strung on necklaces; sometimes they are bleached in hot sand and a commercial bleach solution, turning them a pale yellow.

Nerita scabricosta

comcaii inoosj 'old woman's [toe]nail'

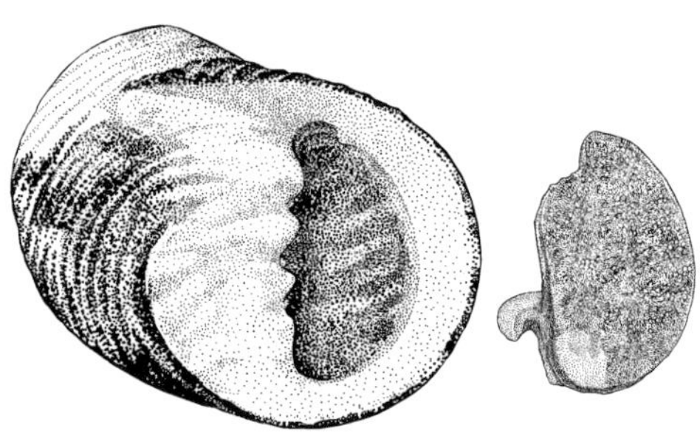

L 28 mm

Colonies of this snail are common in exposed rock crevices in the splash zone. The snail's shell is black with a white interior and a toothed aperture. The apparent similarity of the snail's distinctive, rough-surfaced operculum to an old woman's toenail gives the snail its name.[6] As Eva López explained:

> An old woman was there, saw this shell and said, "How interesting! The operculum of this snail looks just like my [toe]nail." The one there naming things [*Hant Iiha Quimx*] said, "Then that will be the name for this snail."

The snail was eaten; Cleotilde Morales noted that it tastes like ***cotopis*** (*Turbo fluctuosus*), and Juanita Herrera added that the snails were good because they did not contain sand. The snails were either boiled in the shell or baked under hot coals, and all of the animal was eaten, not only the muscle. One woman, however, remarked that she did not like the meat because it "stinks."

The larger ***comcaii inoosj*** snails, as well as others, were sometimes used as bait. The meat was put on a hook, or the snails were crushed and thrown into the sea to attract fish for spearing.

A past custom involved placing the live snails in a wooden fishing boat for good luck. Angelita Torres said that she remembered her father doing this when she was young. She described that in groups of three or four, the snails were placed in the boat's prow, stern, and by its ribs, where they would reproduce and colonize. Soon there would be many small snails spread around inside the boat. This was said to bring good luck—bad weather or high waves wouldn't affect the boat, and the fishing and sea turtle hunting would be good. As another woman commented: "The boat that didn't have these snails in it didn't exist!"[7] One man explained that the wooden boats were not watertight, and the snails would receive moisture.

Figure 5.2. *Nerita scabricosta* snails in rock crevices in the splash zone, where they are moistened by waves and sea spray. A shore north of Bahía de Kino, 2009.

Victoria Astorga remembers that her father carefully placed these snails in his boat: "They were alive. When you would pick one up it would be stuck on [with suction], and would have water in it." She added that an orange-colored starfish (possibly *Phataria unifascialis*) could be used as well.

Alfredo López recalled that live sea turtle barnacles (*Chelonibia testudinaria*) were also placed in a boat; these would attract the turtles, which could then be easily harpooned.

A camp located near Bahía de Kino, named *Comcaii Inoosj An Ihatootoj* 'where the ***comcaii inoosj*** were baked', was described by Roberto Herrera as an overnight camp where one might cook a meal, rather than being a place of long-term residence. Although the occasion providing the name is unknown, the name concords with the fact that nearby beaches have rocky outcroppings where large *Nerita scabricosta* snails are found (figure 5.2).

Theodoxus luteofasciatus

Hant Iiha Quimx itaajc ootp 'saliva that *Hant Iiha Quimx* spit out'
♦ ***xepe quilp*** 'seawater squirter'
xepe quis

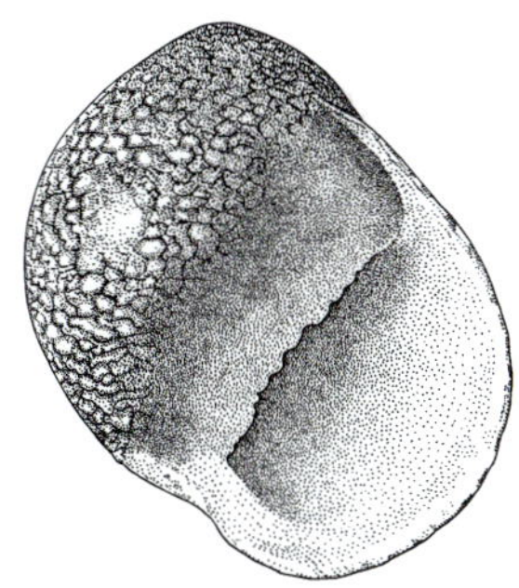

L 12 mm

This small snail is common in esteros and on mud flats. Its attractive polished shell is marked with intricate designs that vary on each shell.

Hant Iiha Quimx ***itaajc ootp*** 'saliva that *Hant Iiha Quimx* spit out' is an old name; the verb ***cootp*** specifically refers to spitting out small objects. *Hant Iiha Quimx* was a principal figure in the world's creation (see Mythology and Folklore). As explained by María Luisa Astorga, *Hant Iiha Quimx* spoke so energetically that drops of spittle flew to the ground and became the snails.

The name ***xepe quilp*** 'seawater squirter' is said to describe the way the snail spurts out a bit of liquid when it retracts. The word *quilp* has both a meaning of squirting something from one's mouth or of taking small sips or sniffs. ***Xepe quis***, not a well-used name and only given by one person, is perhaps a variant of the name. The word *quis* means 'raw'; the connection is unclear.

As with others of this size, this snail was not eaten.

Because of its attractive designs, the shell is a favorite for stringing on necklaces, either in its natural state or whitened by toasting in hot sand followed by a soak in a commercial bleach solution. The shell is often colored with commercial dyes.

Figure 5.3. A strand of *Theodoxus luteofasciatus* snail shells.

According to an older woman, the shell of this snail and other shells were sewn onto clothing for decoration long ago (see *Heterodonax pacificus*).

Littorinidae (periwinkles)

Echinolittorina aspera [= *Littorina aspera*]

haxölinaail cooscl 'speckled shell'

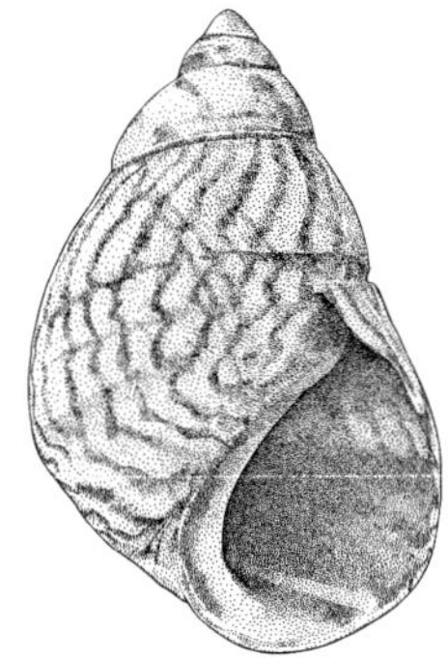

L 13 mm

This snail is found in crevices of large rocks in the high intertidal zone. Its attractive gray shell is marked with fine brown zigzag designs.

The name ***haxölinaail cooscl*** 'speckled shell' describes the shell's finely marked appearance. The word *cooscl* also connotes a muted color, which fits this somewhat gray-colored shell. The name is also given to the more familiar *Costoanachis coronata* snail.

The use of this shell in jewelry is relatively recent, as the snail is not common in the mainland Seri region; I first saw the shell strung in Desemboque in 2009.[8]

Turritellidae (turret snails)

The turret snails are very common throughout the area, but were not eaten. The primary name ***xtapacaj*** defines a generic group described as *ziix* ***xtapacaj*** *hant cöquiih* 'thing grouped with ***xtapacaj***'. The group includes similar high-spired snails of the Cerithiidae and Terebridae families that include ***xtapacaj*** along with other descriptive terms in their names.

Turritella banksi

xtapacaj heeque

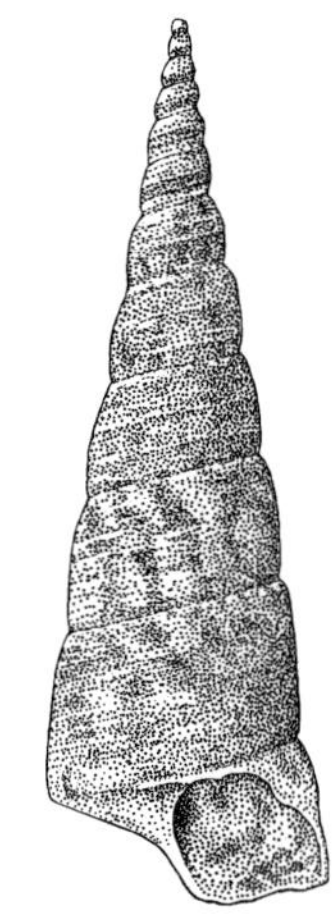

L 43 mm

The fragile shell of this small turret snail has somewhat flat, reddish-brown whorls patterned with darker brown designs.

The shells of such small snails, pierced and hung by their anterior ends, are sometimes strung on necklaces together with other shells.

Turritella gonostoma

xtapacaj

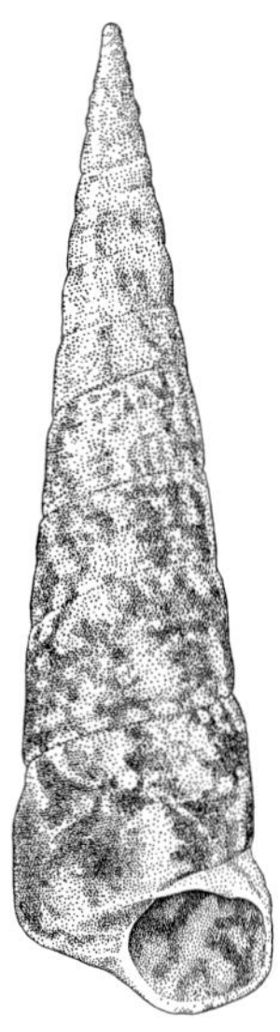

L 120 mm

The sturdy shell of this turret snail is patterned in shades of tan and brown.

The shells of such large turret snails were used as game pieces in a gambling game, a throwing game, and as toys and augers.

One version of a game involved several players taking turns flipping the shell into sand to see who could make it enter point-down. The game was also played with knives, pieces of metal, or sharp sticks.

The game *icacoml* 'what one pierces with' was played by men or boys.[9] Eight pieces of the skin of cactus fruit or watermelon, cut into very narrow strips about three to four inches long, were placed in rows between two players. Each took turns throwing a large mesquite thorn (*xpeemoja*) or sharp stick at the pieces, to see who could pierce them. When one missed, it was the other's turn; the one who pierced the most pieces won. Some were said to be really good at the game. Today a length of wire about five inches long is used. A similar game of skill played by men in the past was described by Cleotilde Morales and María Luisa Astorga. In this game, a ***xtapacaj*** shell was flipped between two fingers at the astragalus bone of a deer or cow—the goal was to have the shell's tip hit the bone's center hollow area. The winner was the one who scored the most hits.

Figure 5.4. A *Turritella leucostoma* shell and cane tube, used in a game of skill. Collected by Edward H. Davis in 1929. National Museum of the American Indian, Smithsonian Institution, catalog no. 164342.000. Photograph by the author.

Figure 5.5. Jesús Morales holding four cane tubes. From the tube in his right hand projects the shell of a turret snail. Bahía de Kino, 1929. Photograph by Edward H. Davis, courtesy of the National Museum of the American Indian, Smithsonian Institution (N24014).

A gambling game, described by Edward H. Davis,[10] was played by two men. A ***xtapacaj*** shell was filled with sand and held point-up in a player's right hand, his left hand holding a cane tube on the ground, open end up. The shell was quickly thrown point down into the cane tube; if it entered the tube, a point was scored.

In the National Museum of the American Indian, collected by Davis, are five *Turritella* shells are labeled as "punches or reamers." Such a use was confirmed by a consultant who noted that shells of large turret snails were used to punch holes in a pliable material such as deerskin. And, according to one person, the shell was inserted base down into the open end of a cane pole and used for spearing.

Juanita Herrera said that when playing, girls would sometimes push large *Turritella* shells point-down into sand, to outline an area in which to place their dolls.

Sara Villalobos included the shell among those sold or traded long ago to non-Seris living on ranches in the area.

Turritella leucostoma

xtapacaj

The shell of this turret snail is finely marked with reddish-brown designs. On some sandy beaches the shell is very common.

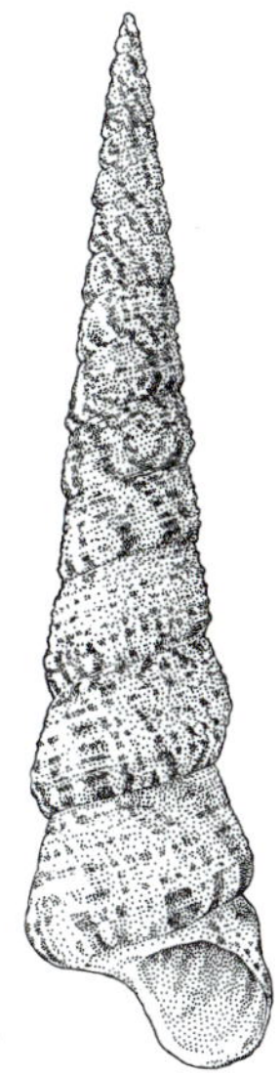

L 105 mm

Vermetidae (worm snails)

The worm snails comprise a generic group of *ziix* ***xtoozajö*** *hant cöquiih* 'thing grouped with ***xtoozajö***'. The primary name ***xtoozajö*** is given to any member of the family.

Tripsycha tripsycha

xtoozajö (var. ***toozaxö***)

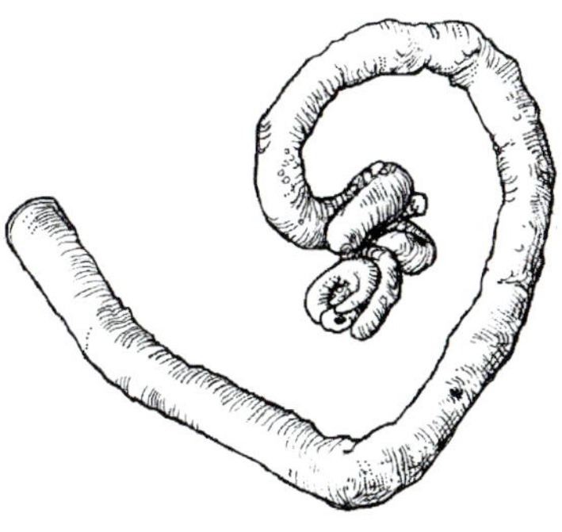

L 300 mm Dia 8 mm

Worm snail shells are irregularly coiled tubes. Large ones, from deeper water, are often found in beach drift left from dredging by commercial shrimp boats.

A short straight section of a large shell, around four inches in length, was used for smoking tobacco and called *hapis ano cöihiip* 'where the *hapis* (tobacco) is inserted'. The dried leaves were stuffed into the tube using a large bird feather quill or stick. The tobacco especially favored was *hapis áa* (*Nicotiana obtusifolia* [= *N. trigonophylla*]) from an area north of Desemboque called *Hapis Iihom* 'where the tobacco lies'. Also smoked was the *haaxat ihixcoocoj*, the leaf gall on creosotebush (*Larrea divaricata*). Pipes for smoking were also made of clay,[11] reedgrass (***xapij***, *Phragmites australis*), or even a section of pelican bone.

Figure 5.6. Eroding worm snail shells at *Xtoozajö Cöhanípatim*, Isla Tiburón, March 2009.

Sections of reedgrass were used like straws for drinking cactus wine from narrow-mouthed clay vessels made especially for holding the wine. Not commonly known today, but mentioned by Edward H. Davis as being used in the same way, were long pieces of a worm snail shell.

Prehistoric tubular beads formed from short sections of the smoothed shell are found in the Seri region, but the people today have no knowledge of their use or manufacture. Such items are attributed to the Giants.

A seaside camp on the eastern shore of Isla Tiburón called *Xtoozajö Cöhanípatim* 'where the ***xtoozajö*** struck it' has that aptly descriptive name because eroding cemented beach deposits containing masses of horizontal worm snail shells give the appearance of having been deliberately thrown (figure 5.6).[12]

The meat of the worm snails was not eaten. As aptly observed by Sara Villalobos: "Is the animal [even] seen?!"[13]

Small intertidal worm snails

cozaaij 'what makes a circular cut'
xtoozajö

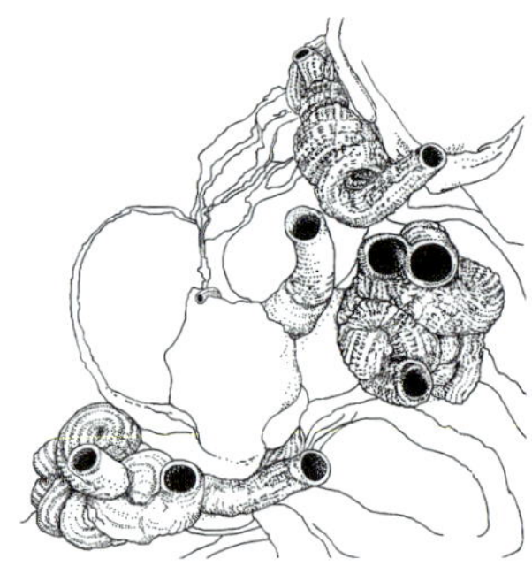

approx. Dia 4 mm

Although all worm snails have the generic name ***xtoozajö***, those with small diameters are also called ***cozaaij*** 'what makes a circular cut'. Since such worm snails often grow firmly clumped on other shells or hard substrate, stepping on a shell with a bare foot can produce a neat, round cut in the sole, an attribute reflected in the shell's name. The name is also given to tusk shells (*Dentalium* spp.).

Cerithiidae (ceriths)

Cerithium maculosum

napxeec 'the Turkey Vulture's daughter'

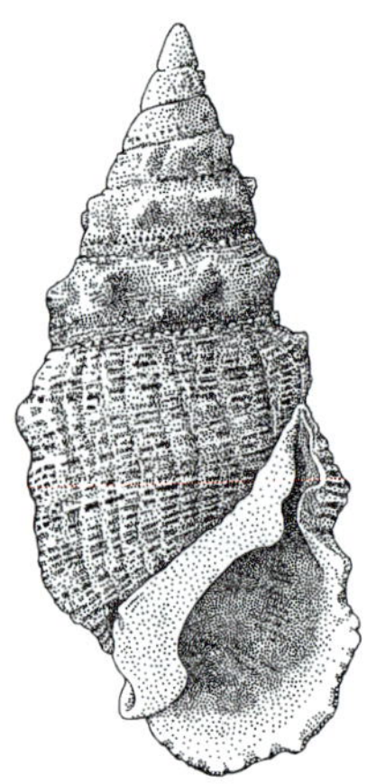

L 39 mm

This very common snail is abundant on muddy sand flats; in the intertidal zone its shell is often found inhabited by hermit crabs.

The name ***napxeec*** is a compound of *naapxa* (*Cathartes aura*, Turkey Vulture) and *eec* 'his daughter'. The Turkey Vulture (who was once a "poor, luckless man") is considered *captasa* 'unlucky', since it does not hunt on its own, but eats what others have left behind. Although the snails of this genus are considered rather drab and "not pretty," the connection to the bird is not clear. The name is given to other snails as well (see *Cerithium stercusmuscarum*, *Mancinella tuberculata*, *Melongena patula*, and *Strombus galeatus*).

The shell is occasionally strung on necklaces.

Cerithium stercusmuscarum

napxeec 'the Turkey Vulture's daughter'

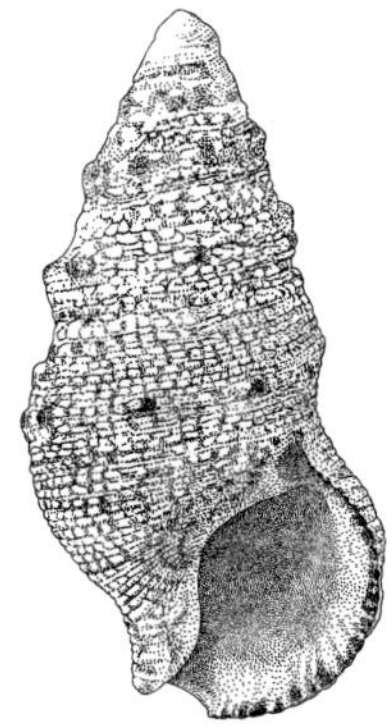

L 17 mm

This small, dark-shelled snail is found in great numbers on muddy sand flats and is especially abundant in esteros.

The name ***napxeec*** is a compound of *naapxa* (*Cathartes aura*, Turkey Vulture) and *eec* 'his daughter' (see *Cerithium maculosum*).

The names ***hast iti cöcooyaj*** 'what travel on rocks', ***hast iti cöquiipax*** 'what climbs up rocks', and ***hast iti cpaainalca*** 'what tumble off rocks', are given to this snail and others of similar size that are inhabited by hermit crabs and found perched on intertidal rocks.

The shell is occasionally strung on necklaces.

Liocerithium judithae

haxöl inaail quítajij 'pointed shells'
◊ ***xtapacaj heeque*** 'small ***xtapacaj***'

L 20 mm

The fragile, gray shell of this snail has delicately patterned whorls with fine spiral ridges.

The name ***haxöl inaail quítajij*** 'pointed shells' is most often associated with the *Cerithidea mazatlanica* snail, while the name ***xtapacaj heeque*** alludes to the shell's similarity to that of the turret snail, ***xtapacaj***.

The shells are not common in shell jewelry. An unusual item in the Arizona State Museum collections, possibly a hatband, has three *Liocerithium judithae* shells strung together with *Trivia solandri*, *Melampus* sp., and a single *Zonaria annettae* shell (see figure 5.23, p. 172).

Potamididae (horn shells)

Cerithidea mazatlanica

♦ ***haxölinaail quítajij*** 'pointed shells'
◊ ***xtapacaj coopol*** 'black ***xtapacaj***'
◊ ***xtapacaj heeque*** 'small ***xtapacaj***'

L 27 mm

The sturdy shell of this small snail has a sharply pointed apex. These snails are especially abundant on tidally wet muddy sand flats bordering esteros, together with *Melampus mousleyi* snails. The surfaces of inland dry lakes around Cerro Tepopa, part of an ancient shoreline, are littered with this fossilized shell.

The name ***haxölinaail quítajij*** 'pointed shells' is descriptive, as the shells are quite sharp, especially when stepped on. The name is shared with *Liocerithium judithae*.

The names ***xtapacaj coopol*** 'black ***xtapacaj*** (Turritellidae)' and ***xtapacaj heeque*** 'small ***xtapacaj***' were supplied by a couple of people and do not appear to be well-used names.

The (non-fossilized) shells are sometimes strung on necklaces.

Strombidae (conchs)

Strombus galeatus

haxölinaail an icoopxoj 'shell that is blown into'
napxeec 'the Turkey Vulture's daughter'
♦ ***xepe yamaasij*** 'what is rolled by the sea'
xpa̲leemelc caacoj 'large ***xpa̲leemelc***'

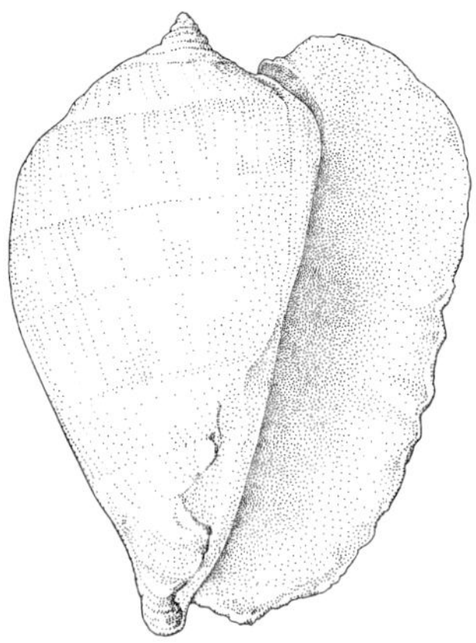

L 220 mm

This is the heaviest-shelled gastropod in the area. Its shell is a creamy white under a thin brown periostracum; the interior is a polished white, with a rose or light orange color near its outer lip.

The name ***haxölinaail an icoopxoj*** 'shell that is blown into' reflects the shell's use as a "trumpet" (see the discussion below).

The name ***napxeec*** 'the Turkey Vulture's daughter' is a compound of *naapxa* (*Cathartes aura*, Turkey Vulture) and *eec* 'his daughter'. Other snails sharing the name are *Cerithium* spp., *Mancinella tuberculata,* and *Melongena patula*. The connection of the snails' name to the bird is unclear.

The snail's most common name, ***xepe yamaasij*** 'what is rolled by the sea', is given to other snails as well (see *Cypraecassis coarctata*, *Semicassis centiquadrata*, and *Malea ringens*). The name, alluding to the shell's smooth, rounded shape, suggests it being easily rolled by the surf. Although ***xepe yamaasij*** refers to the whole animal, only the empty shell is said to be rolled.

The name ***xpa̲leemelc caacoj*** 'large ***xpa̲leemelc***' refers to the snail's larger size as well as its having a typical ***xpa̲leemelc*** shape.

Because it was from deeper water, the snail was not often harvested alive, so was not a common food. When the live snails are found today, sometimes brought up by divers, the meat is removed raw and then cooked by frying, sometimes with other ingredients.

In the past, girls used the shell as a doll. Its somewhat round form was considered a "female" shape.[14]

Chico Romero said that the shell was used to summon Seris together,[15] and another person suggested an event such as when a deer had been killed.[16] The shell is mentioned by Dane and Mary Coolidge as being used to "summon home the children, when they are out hunting cactus-pears in the hot season" and to summon people together for feasts or in war.[17] More recently, during the 1960s through the mid-1980s, the Seri Apostolic church in Desemboque used a *Strombus galeatus* shell "trumpet" to announce the church services. A hole made in the apex of the shell was blown into, creating a resonant, easily heard sound.

Cleotilde Morales sang a song about the ***xepe yamaasij*** shell:

Xepe Yamaasij Cöicoos

Xe pe té lim i mi tax i
Xepe iteel iti miitax
Xe pe tél com ha ti ya
Xepe iteel com haa ntiya
Xe pe tél i ti mi catx
Xepe iteel iti imiicatx
I ti ma sij te
Iti mmaasij tee
Xe pe téel i ti ma sij
Xepe iteel iti mmaasij.

Song of the ***Xepe Yamaasij***

It was going on the edge of the sea,
It was going there, on the edge of the sea,
It was put there, on the edge of the sea,
The waves rolled it,
On the edge of the sea it was rolled.

As Cleotilde explained:

> The ***xepe yamaasij*** shell is rolling at the edge of the sea. The waves roll it along. That's it. [The snail] is singing and that's what it said. It's singing about its arrival to the shore and that's what it said. It arrives there, it rolls on the shore, it travels, the waves roll it along, it goes, it wants to stand up, and that is what it said.

Strombus gracilior

haxölinaail cotitla cmaam 'female ***haxölinaail cotitla***'
haxölinaail cotázita 'shell that pinches (small pinches)'
♦ ***haxölinaail cotítzilca*** 'shells that pinch'

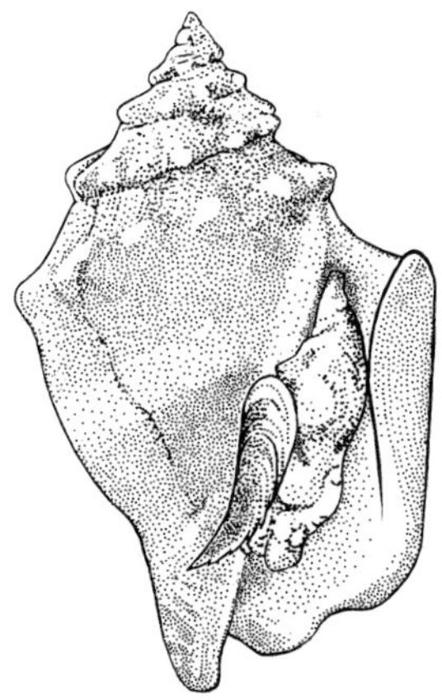
L 77 mm

This small conch has an attractive, banded creamy-orange shell. It is rather common on some sand flats and in especially notable numbers when breeding.

The name ***haxölinaail cotítzilca*** 'shells that pinch', derives from the claw-like operculum, *inoosj* 'its claw' that is said to pinch when closing.[18] Variations of the name include a shortened form, ***cotitla*** 'what pinches', use of the singular, ***ziix cotitla*** 'thing that pinches', or ***xiica cotítzilca*** 'things that pinch'.

Haxölinaail cotázita 'shell that pinches (small pinches)' refers to small shells of the species.

In distinguishing this shell from *Strombus granulatus* (a knobby conch) the name ***haxölinaail cotitla cmaam*** 'female ***haxölinaail cotitla***' is used, as it is the smoother of the two.

Eva López commented that eating the meat of this small conch caused dizziness. However, many people say that the meat was eaten, some further commenting that only the foot muscle was eaten. Angelita Torres recalled a Seri man bringing large quantities of the snails from an estuary on Isla Tiburón and selling the cooked meat to non-Seri Mexicans in the mid-twentieth century.

While the opercula are occasionally strung on necklaces, sometimes the smallest shells are strung as well.

Strombus granulatus

haxölinaail cotitla ctam 'male ***haxölinaail cotitla***'
cotitla oot quih yaa 'cotitla '***cotitla*** belonging to the coyote'
♦ ***haxölinaail cotítzilca*** 'shells that pinch'

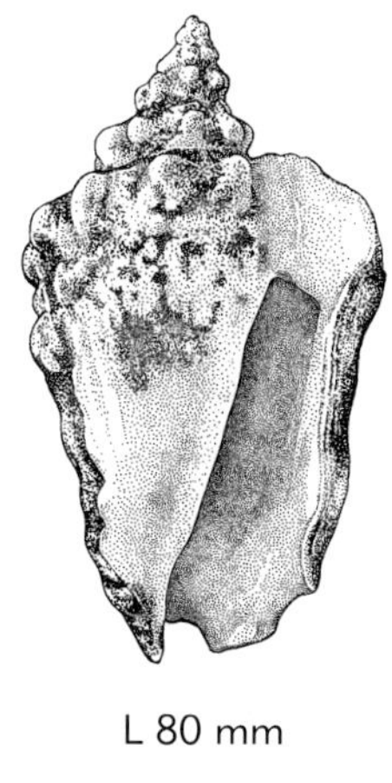
L 80 mm

The knobby shell of this conch is a creamy-white, splotched with light brownish-purple. The conch is not usually found alive.

Besides the snail's most common name, ***haxölinaail cotítzilca*** 'shells that pinch' and its variations (see *Strombus gracilior*), the more knobby "male" appearance of this snail's shell is alluded to in the names ***haxölinaail cotitla ctam*** 'male ***haxölinaail cotitla***' and ***cotitla oot quih yaa 'cotitla*** belonging to the coyote'. These names are used when differentiating this shell from *S. gracilior*, the smooth conch.

The live snail is not as commonly found as the *S. gracilior* (which was eaten), and it is unclear whether its meat was eaten.

The shells are not used. Unlike the opercula of the *S. gracilior*, those of this snail are not used for stringing on necklaces, perhaps because the snail is not collected alive.

Janthinidae (purple sea snails)

Janthina prolongata

◊ ***hant haaco iima ano haxölinaail*** 'shell of the Catastrophe'
◊ ***xepe ano xojmás*** '***xojmás*** in the sea'

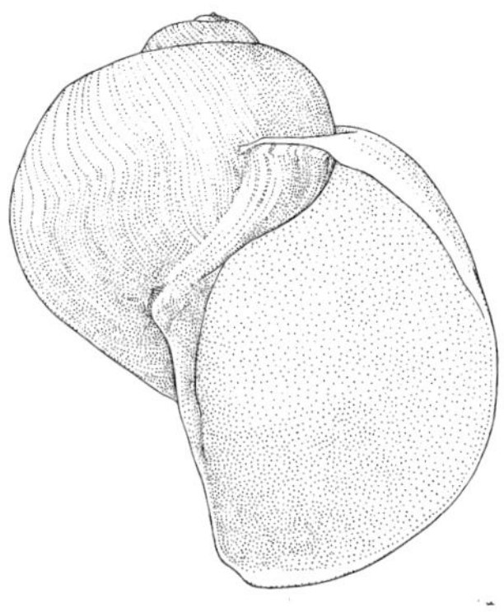

L 21 mm

These snails live floating on the ocean's surface on a bubble raft of their own making. Their fragile, light violet shells are found only rarely, in beach drift.

The name ***hant haaco iima ano haxölinaail*** 'shell of the Catastrophe' was provided by a woman who had once seen a great many of the shells together in beach drift. The name is probably a descriptive, personally given name noting the shell's strangeness and manner of occurrence. The phrase *hant haaco iima* denotes a catastrophic event, pertaining both to the flood Catastrophe at the world's beginning and the world's end and is sometimes used in an exclamation regarding something extremely strange. The woman who supplied the name for this snail has personally named other shells, even though in one case she knew a commonly used name.

The name ***xepe ano xojmás*** '***xojmás*** in the sea' (perhaps literally 'marine land snail') suggests the snail's similarity to ***xojmás***, equally fragile-shelled terrestrial or freshwater snails.

Eulimidae

Niso splendidula

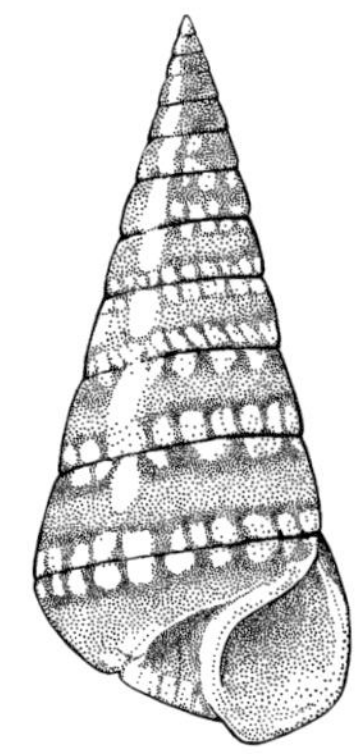

L 28 mm

The shell of this small, rare snail is highly polished and beautifully marked. It was not recognized by anyone who saw it.

María Luisa gave me two shells, found by her son-in-law in a baited crab trap. When she handed them to me she quipped, "Here is gold, I think!" indicating how unique they were.

Hipponicidae (hoof snails)

Hipponix antiquatus

cmatjij oot quih yaa '***cmatjij*** belonging to the coyote'
hax cöoaal 'water servant'
mox

L 28 mm

The cream-colored shell of this common snail has pronounced scale-like concentric ridges.

The name ***cmatjij oot quih yaa*** '***cmatjij*** (*Lottia dalliana*) belonging to the coyote' was given by one person and did not seem to be a well-known name. The name alludes to the roughly ridged surface of the shell, as compared to the smooth surfaced *Lottia dalliana* limpet.

The name ***hax cöoaal*** 'water servant' was supplied more recently by one person. The ***hax cöoaal*** were normal people who worked as slaves for the Giants, typically fetching water for them. The connection to the shell is unclear.

The name ***mox*** was recorded in early notes and is not used today. The name differs slightly from those given to the similarly shaped barnacles: goose barnacles are called *moocj*, rock barnacles are *mooc* or *cmooct*.

Calyptraeidae (slipper snails and cup-and-saucer snails)

Both the slipper snails and the cup-and-saucer snails are very common throughout the area.

Because of their appearance, the slipper snails are most commonly known as ***miizj hant quixquim quih iti ihiij*** 'where the navigator sits'. The name describes the shells' similarity to a boat, with the shelf-like septum being the stern seat where the man guiding the boat sits. The name is recognized as having been given by Loreta Marcos, considered to be one of the *hant iiha cöhacomxoj* 'those who have been informed about the ancient things', who pass along knowledge.[19]

According to Cleotilde Morales, a name similar to ***miizj hant quixquim quih iti ihiij***, although apparently older, is ***casquim quih iti ihiij*** 'where the paddler sits'.

The name ***oot iscám*** 'the coyote's balsa' notes the shell's form as well as its small size, and therefore lack of value as a balsa. It is like a copy of the real thing, and hence is a possession of the coyote.[20]

Crepidula onyx

casquim quih iti ihiij 'where the paddler sits'
matcz ital coofizlc 'who accompanies your younger sister to gather firewood'
♦ ***miizj hant quixquim quih iti ihiij*** 'where the navigator sits'
oot iscám 'the coyote's balsa'

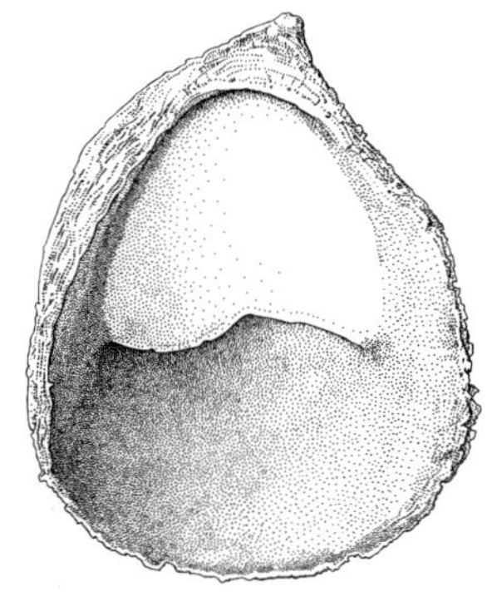

L 37 mm

The snail has a dark brown shell with a distinctive contrasting white shelf-like septum.

While its most common names refer to the shell's boat-like appearance, another name, provided by only a few people, is ***matcz ital coofizlc*** 'who accompanies your younger sister to gather firewood'—a rather intriguing name similar to ***hitcz ihizcoofizlc*** 'the firewood my younger sister gathered' given to the *Vokesimurex* spp. snails.

Crepidula striolata

casquim quih iti ihiij 'where the paddler sits'
♦ ***miizj hant quixquim quih iti ihiij*** 'where the navigator sits'
oot iscám 'the coyote's balsa'

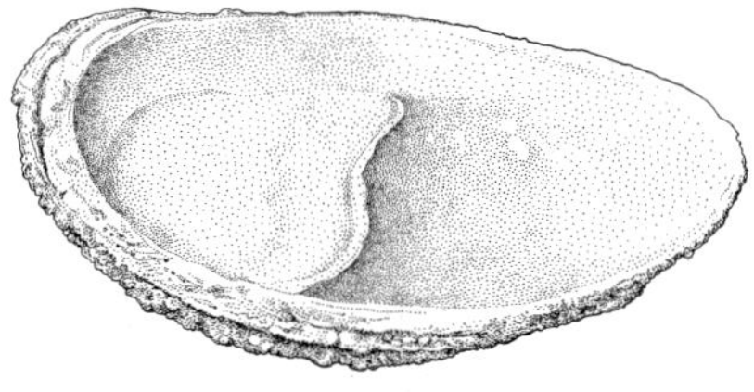

L 47 mm

Both the shell and septum of this snail are white.

Although the snail is not eaten today, its shell is common with other food shells in inland camps.

Although someone once supplied the name ***cahiixona*** for this snail, the name is most often associated today with the cup-and-saucer snails.[21]

Crucibulum scutellatum

♦ ***cahiixona***
casooca
catamax
hasajitij
tásona

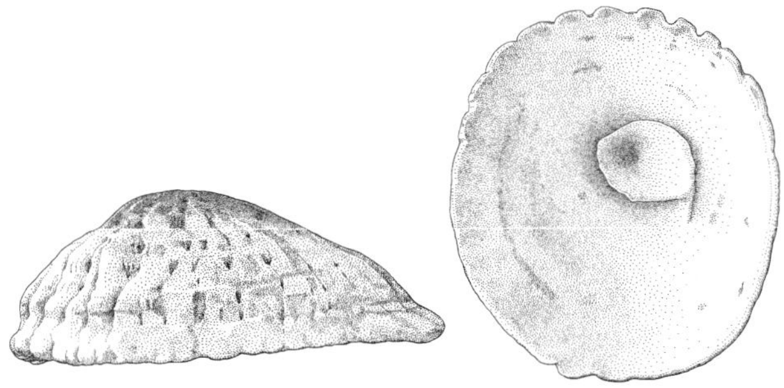

L 60 mm

This snail is very common in the intertidal zone. Its sturdy shell has rough, deep radial grooves and contains a distinctive cup-like septum.

The etymology of the name ***cahiixona*** is unclear. Eating the meat was said to affect (irritate) the throat. In a recorded interview, Roberto Herrera related the name with that characteristic.[22]

Several people mentioned that when one sees someone wearing a conical hat, one might comment: "*Hoox cahiixona quiionam*!"—"[look] how [that person] has a hat like a ***cahiixona***!"

The name ***casooca*** and ***catamax*** are primary names.[23] ***Hasajitij*** 'basket's base . . . ' is obviously a compound, but is not completely clear. Eva López named the shell's internal septum *hasajitij iti yahimoz* (a snail's operculum is the *iti yahimoz* '[a hard thing] with which it covers itself'). When talking about the shell, Eva compared the septum to a hat that was misshapen because of being smashed down on one's head, and noted that the similarity could be pointed out, and the person wearing the hat would be urged to discard it because it had lost its shape and was no good. ***Tásona*** is a primary name recorded by Gwyneth Harrington in 1941.[24] Unknown by most people asked, the name was confirmed by Raquel Moreno.

The meat of this snail was said to have been eaten in the past, cooked or raw "when one was in the sea [at the beach]," although some agreed that eating the meat irritated one's throat, causing a "rough voice."[25]

As with others of this size and shape, this shell was sometimes used as a container for pigments.

Crucibulum spinosum

cahiixona
catamax
♦ ***tamax cosi*** 'spiny ***tamax***' (var. ***catamax cosi***, ***xtamax cosi***)

The shell of the spiny cup-and-saucer snail is covered with rows of fragile tubular spines that are usually broken and worn when the shell is found in beach drift.

The name ***tamax cosi*** 'spiny ***tamax***' is somewhat confusing and may involve a name transfer. I have been told that the "real" creature with the name ***tamax cosi*** is the glass-hair chiton *Acanthochitona exquisita*. Since the chiton is no longer eaten, has a rather narrow distribution, and is not found with its spines well preserved in beach drift, few people seem to recognize it by name or sight. Some early clues suggest that the spines of the shell of the *Crucibulum spinosum*, as well as its similar habitat, are what provoked the name transfer.[26]

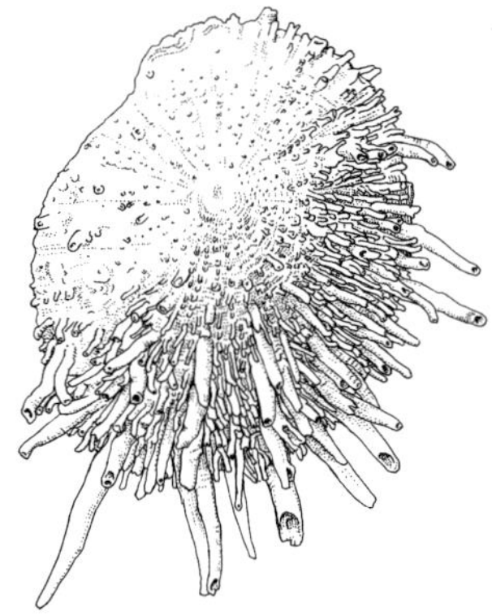

L 40 mm

Naticidae (moon snails)

With the exception of the unnamed *Sinum debile*, moon snails are grouped as *ziix* ***xepeiinoj*** *hant cöquiih*, where the prototypical snail is the large ***xepeiinoj*** 'the sea's roar' (*Neverita reclusiana*). The moon snails were not eaten.

Natica chemnitzii

xepeiinoj mayoocö quih yaa '***xepeiinoj*** belonging to the Yellow-crowned Night Heron'

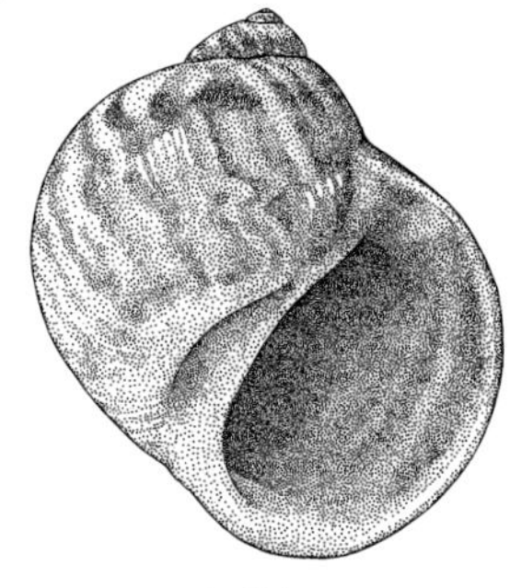

L 27 mm

The shell of this moon snail is grayish-blue and buff, patterned with darker designs. A somewhat uncommon snail, it is found on intertidal sand flats.

The name ***xepeiinoj mayoocö quih yaa*** '***xepeiinoj*** belonging to the Yellow-crowned Night Heron' suggests a similarity in color and markings to the plumage of the *Nyctanassa violacea* bird.

Neverita reclusiana [= *Polinices reclusiana*]

xepeiinoj 'the sea's roar'

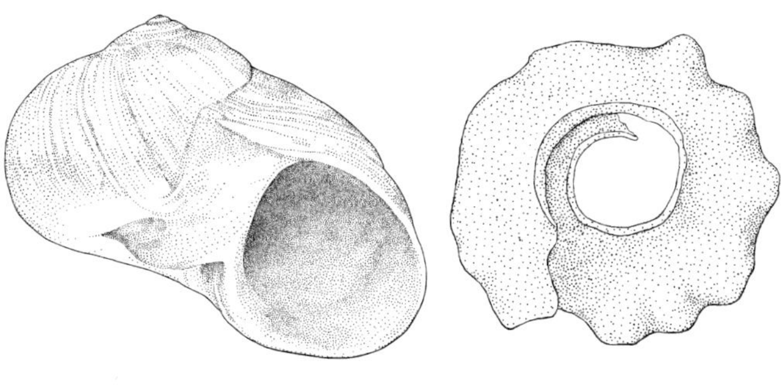

L 50 mm

This large moon snail has a banded cream and buff-colored shell. The snail is beautiful to watch as it slowly glides with its wide foot on submerged sandy areas.

The name ***xepeiinoj*** 'the sea's roar' alludes to the sound heard when the shell is held up to one's ear. Since other moon snails in the area are smaller, sometimes the inclusion of *caacoj* 'large' in its name distinguishes this snail.

As Eva López described, the shell was used to "listen to the ocean" when people were inland:

> The Ancestors used these as "telephones"—they kept them in their bags. If they were away in the desert and wanted to hear the sound of the ocean, they held them to their ears. If there were a lot of noise, the tide was incoming or high; if there were little sound, it was low tide.

Others have confirmed this use for the shell. Angelita Torres said that one could hear whether the seas were rough when at a distant inland camp.

The sandy collar-shaped structure containing the snail's eggs is called *toerc hamcanoiin* 'pot of the Willet (*Catoptrophorus semipalmatus*)' or *xatoscö imazaj* 'clay pot of the Eared Grebe (*Podiceps nigricollis*)'. Such names allude to its resemblance to the rim of a round clay vessel.

Polinices bifasciatus

xepeiinoj heecto 'small ***xepeiinoj***'

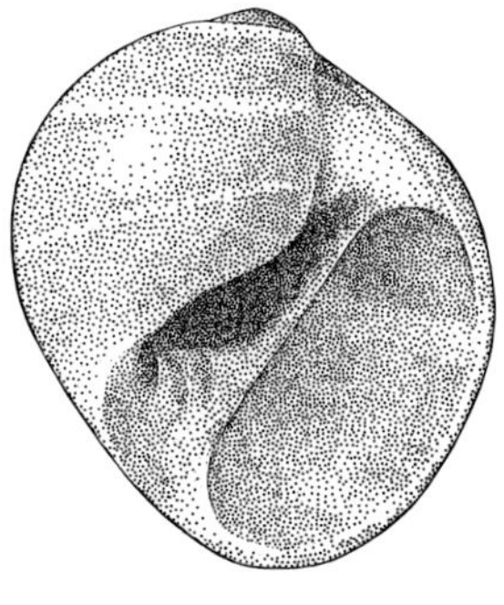

L 12 mm

This small moon snail has a buff-colored, lightly banded shell.

The name ***xepeiinoj heecto*** 'small ***xepeiinoj*** (pl.)' alludes to its small size as compared to other moon snails. The shell's coloring is similar to that of the larger ***xepeiinoj*** (*Neverita reclusiana*) snail.

The shell is sometimes strung as a pendant on shell necklaces.

Polinices uber

xepe quilp cooxp 'white ***xepe quilp***'
♦ ***xepeiinoj cooxp*** 'white ***xepeiinoj***'

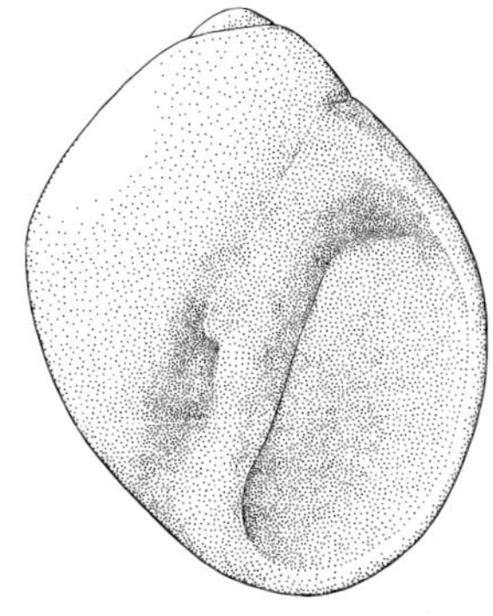

L 20 mm

This somewhat uncommon moon snail is distinctive because of its striking white shell.

Although most commonly known as ***xepeiinoj cooxp*** 'white ***xepeiinoj***', some people call it ***xepe quilp cooxp*** 'white ***xepe quilp*** (*Theodoxus luteofasciatus*)' relating it to that snail.

The live snail is sometimes gathered, though in low numbers, on sandy submerged intertidal areas. Because of its pure white color the shell is a favorite for stringing, especially with the contrasting black seeds of the *xneeejam is hayaa* (*Viscainoa geniculata*) shrub, used as common fillers in necklaces.

Sinum debile

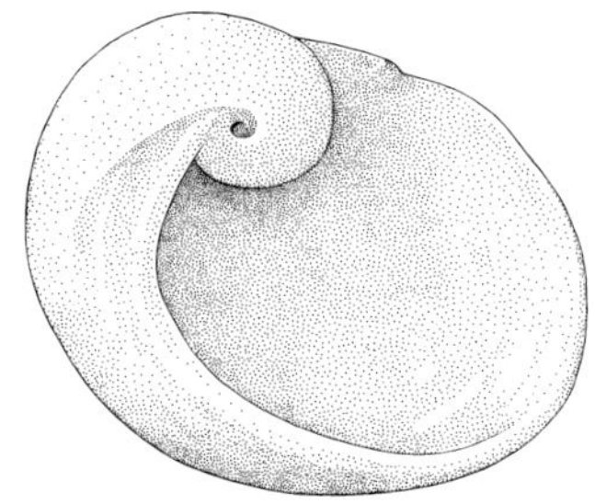

L 25 mm

The white shell of this uncommon, beautiful moon snail has a distinctive, flattened shape. When talking about it, someone compared its shape to that of the ***itx imonaaaij*** (*Architectonica nobilis*) snail.

The snail has no name recalled today.

Shells of such uncommon snails are sometimes hung as pendants on shell necklaces.

Triviidae

Trivia solandri

camonlc 'puckered [shells]'
hasit camós 'puckered vulva' (var. ***halít camós*** 'puckered head or hair')
♦ ***haxölinaail cheel*** 'red shell' (var. ***haxölinaail cquihjö*** 'red shell', ***haxölinaail cöquihmeel*** 'pink shell')

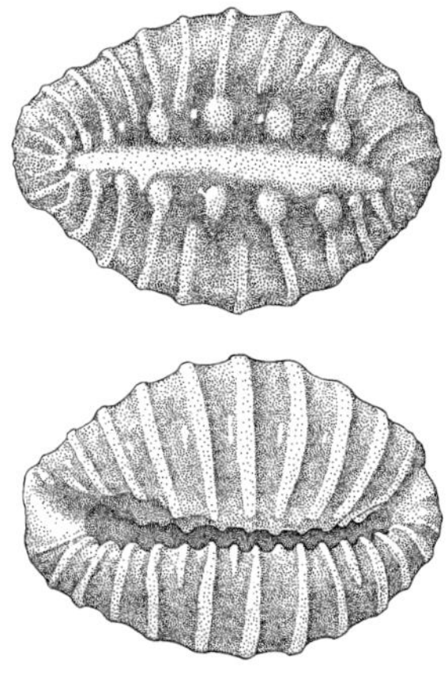

L 16 mm

This small, pink-shelled cowrie-like snail is quite common. When alive, its spotted mantle can cover most of the shell, and the creature is fascinating to watch moving slowly along a surface in the submerged low tide zone.

The name ***hasit camós*** 'puckered vulva' is considered vulgar and is little used. The name ***halít camós*** was given by one person, and is probably a euphemism. The snail's name is sometimes shortened to ***camonlc*** 'puckered [shells]'.

The most commonly used name today is ***haxölinaail cheel*** 'red shell', with some variation in the color term.

Easily gathered in beach drift, the shell was common in necklaces, hatbands, and belts in the mid-twentieth century. It remains a favorite for stringing, often used together with *Olivella dama* shells and the small black seeds of the *xneeejam is hayaa* (*Viscainoa geniculata*) shrub.

This and other small shells were often used as earrings for a child's doll made from the cranium of a pelican, called *toohit imcám* 'who eats but doesn't get full', since sand poured into an opening in the bone (its "mouth") would pass right through (figure 3.40, p. 60).

Cypraeidae (cowries)

The cowries, because of their different form, were considered special, and were often used as amulets. They were not eaten.

Macrocypraea cervinetta [= *Cypraea cervinetta*]

camós caacoj 'large ***camós***'
haxölinaail cheel caacoj 'large ***haxölinaail cheel***'

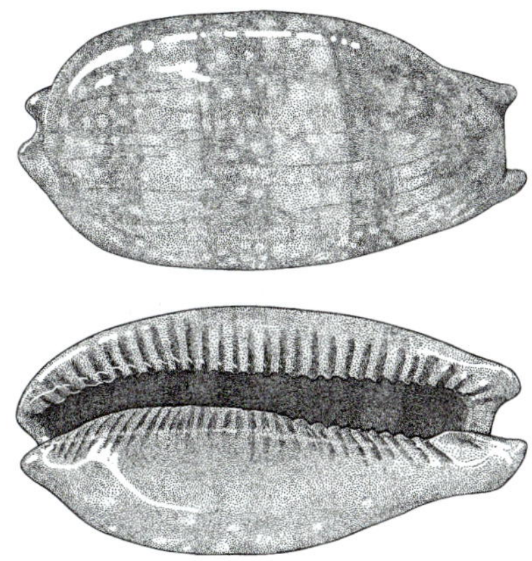

L 100 mm

Figure 5.7. A child's rattle or toy made with the shells of *Macrocypraea cervinetta, Zonaria annettae*, and *Euvola vogdesi*. Collected by Gwyneth Harrington, 1941. Arizona State Museum, catalog no. E-1029.

This uncommon, large cowrie is said to occur around the shores of the Gulf's islands. Its striking and highly polished shell is a variegated brown spotted with tan, giving it a somewhat fawn-like appearance.

The name ***haxölinaail cheel caacoj*** 'large ***haxölinaail cheel*** (*Trivia solandri*)' reflects its similarity in form to that shell.

The name ***camós caacoj*** 'large ***camós*** (*T. solandri*)' incorporates a shortened version of ***hatajeen camós*** 'puckered genitalia'.

Perhaps because of its size, striking appearance, and relative scarcity, this shell was used as an amulet or in babies' toys (figure 5.7).[27]

Zonaria annettae [= *Cypraea annettae*]

hatajeen camós 'puckered genitalia'

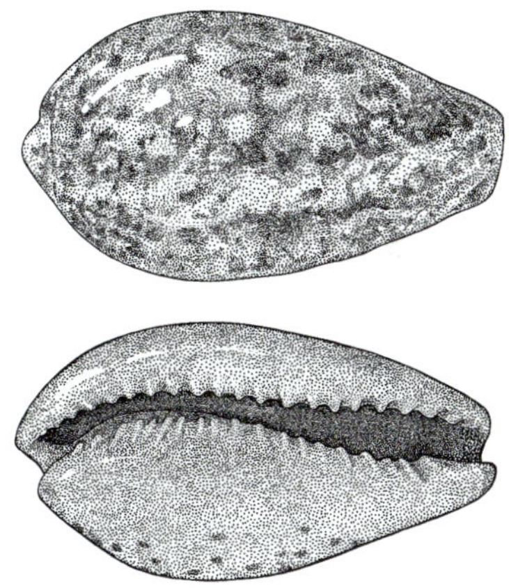

L 48 mm

Figure 5.8. A *Zonaria annettae* shell and the flat valve of a *Euvola vogdesi* scallop strung together with sinew. Collected by Edward H. Davis in 1924. National Museum of the American Indian, Smithsonian Institution, catalog no. 129697.000. Photograph by the author.

The shell of this somewhat common cowrie is a pinkish-cream color, spotted with brown. When not beach-worn, it is highly polished, having been protected by its enveloping mantle when alive.

The name ***hatajeen camós*** 'puckered genitalia' is suggested by the shell's appearance.[28]

This cowrie was often used in adornment, and as evidenced by its use, was obviously considered special. Early collections show it strung with other shells in children's amulets or rattles (figures 5.7 and 5.8). It was often hung as a protection around the neck of the *caal*, the youngest child of a pregnant mother (see Beliefs and Practices, p. 41).

In the past, cowries were strung together with *Trivia solandri* in hatbands or belts. Somewhat worn specimens of *Z. annettae* can be gathered in beach drift; however, highly polished shells have recently been collected live in Puerto Libertad and are valued for stringing in shell jewelry.

An unusual object in the Arizona State Museum collections is a dried plastron of a juvenile hawksbill sea turtle adorned with two *Z. annettae* shells

Figure 5.9. The dried plastron of a juvenile hawksbill turtle on which are hung two *Zonaria annettae* shells. Collected by Gwyneth Harrington in 1944. Arizona State Museum, catalog no. E-2054.

(figure 5.9). Seris who have seen its photograph identify it as an amulet or fetish, perhaps placed in a boat for protection on the sea. One person suggested that it was used as a doll.[29]

Ovulidae

Cyphoma emarginatum

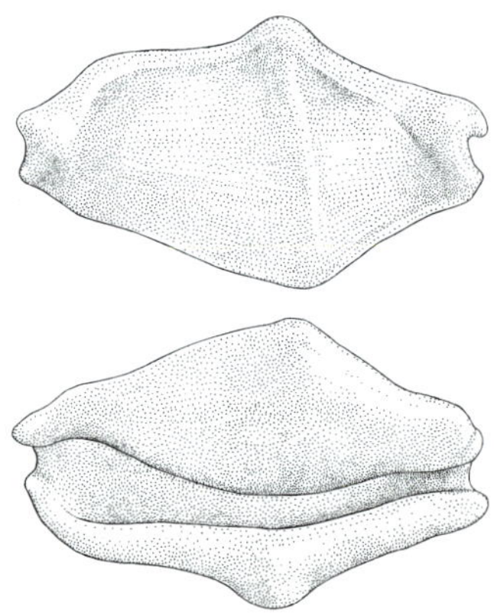

L 25 mm

This very uncommon, porcelain-like white shell was found in the stomach of a flounder. The woman who gave it to me, and others who saw it, commented that they had never seen the shell before.

Jenneria pustulata

♦ ***hatajeen camós cahtxima*** 'rich ***hatajeen camós***'
hatajeen camós coopol 'black ***hatajeen camós***'
xepe quih iti cöiixi iiqui izooxolcam 'eighth tide line'

L 22 mm

This uncommon snail is distinguished by a shell ornately decorated with colorful orange nodes on its dorsal surface.

The snail is one of several that have been recently rediscovered at Puerto Libertad (see Names and Naming, p. 25). It is also said to be found at *Xapoo Yaii* 'where the sea lions are' (an area north of Puerto Libertad), as well as on certain rocky shores of Isla Tiburón.

The name ***hatajeen camós cahtxima*** 'rich ***hatajeen camós*** (*Zonaria annettae*)' notes the shell's similarity in form to the cowrie, while also being more ornate and colorful. The modifier *cahtxima* is elsewhere applied to colorful species such as the *cooiz cahtxima* (*Argiope* sp., a garden spider) or *caatc cahtxima*, a multicolored grasshopper.

Hatajeen camós coopol 'black ***hatajeen camós***' is a recently given name that refers to the shell's dark ventral surface. ***Xepe quih iti cöiixi iiqui izooxolcam*** 'eighth tide line', literally 'the eighth place where the sea finished', is a phrase sometimes used in casual speech to refer to the high tide line. The phrase occurs in the folkloric account of the leatherback sea turtle, in which a person standing on shore at the "eighth tide line" talks to the turtle as it comes into shore. The snail's association with the name is unclear.

Recently the shell has been strung into shell jewelry, though rarely, since it is not common. One shell in the Arizona State Museum collections is strung together with shells of *Zonaria annettae* and *Trivia solandri*. A second necklace containing one *J. pustulata* shell was bought by Richard Felger in the 1960s.[30]

Tonnidae (tun shells)

Malea ringens

◊ ***xasecö cmaam*** 'female ***xasecö***'
xepe yamaasij 'what is rolled by the sea'

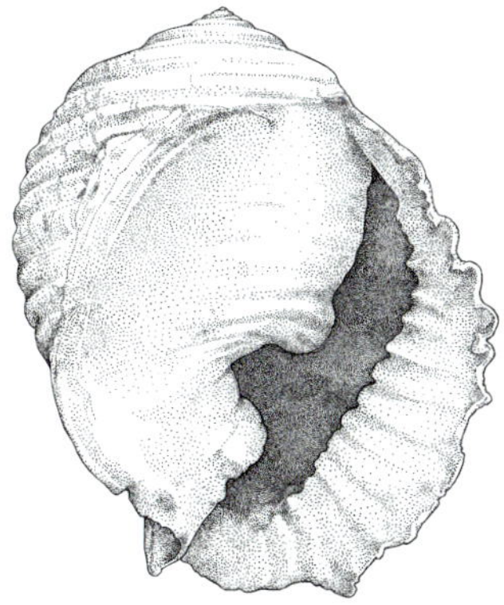

L 190 mm

The snail's large size and its shell's strongly toothed outer lip and deeply notched inner lip make it very distinctive.

The snail is uncommon; I had not seen it in the area until I was given a large shell in 2008 that had been collected from deep water. Raquel Moreno said that she had found a small one long ago and showed it to her father, who called it ***xasecö cmaam*** 'female ***xasecö***'. The name notes the shell's similarity to the ***xasecö*** (*Ficus ventricosa*), a shell with a similar round shape, but having a longer siphonal canal, considered to be a "male" characteristic.

The smooth roundness of the shell suggests an easily rolled shape, alluded to by the name ***xepe yamaasij*** 'what is rolled by the sea', a name given to other gastropods with similar inflated shells (see *Cypraecassis coarctata*, *Semicassis centiquadrata*, and *Strombus galeatus*). However, when I showed this shell to Cleotilde Morales, she exclaimed, "This is the *real* ***xepe yamaasij***!"

Cassidae (helmet shells)

Cypraecassis coarctata [= Cassis coarctata]

hant iiha quimxoj oeenec 'what the *hant iiha quimxoj* carried'
♦ ***xepe yamaasij*** 'what is rolled by the sea'

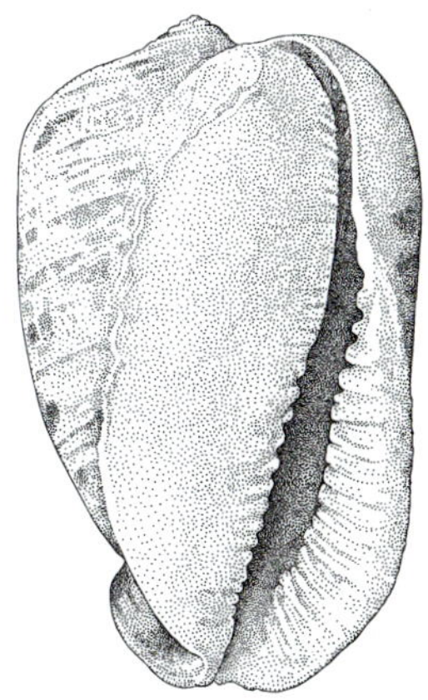

L 72 mm

This snail is somewhat uncommon and is not often found alive. Its cream-colored shell is characterized by broad lips, prominent toothed aperture, and large siphonal notch.

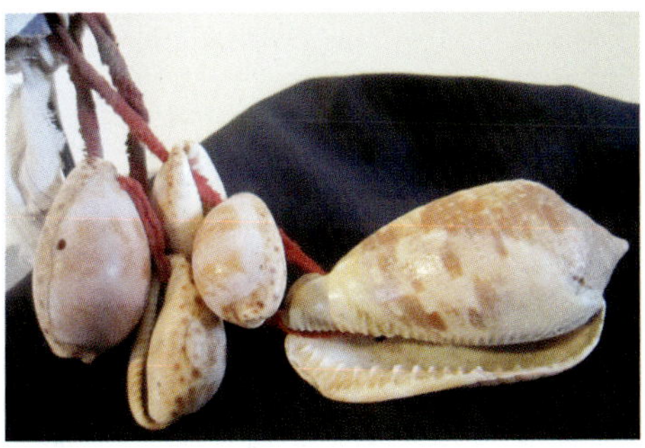

Figure 5.10. A *Cypraecassis coarctata* shell strung together with four *Zonaria annettae* shells on a cloth strip and used either as a child's rattle or amulet. Collected by William Neil Smith. Private collection.

The name *hant iiha quimxoj* ***oeenec*** 'what the *hant iiha quimxoj* carried' refers to the bearers of knowledge, those who told about the ancient things (see Mythology and Folklore). The connection to the snail is unclear.

The name ***xepe yamaasij*** 'what is rolled by the sea' is shared with other good-sized snails with shells having similar smooth inflated shapes (see *Malea ringens*, *Strombus galeatus*, and *Semicassis centiquadrata*).

I have seen one instance of the shell being used (figure 5.10).

Semicassis centiquadrata [= *Cassis centiquadrata*]

♦ ***haxölinaail ctocnij*** 'round shell'
◊ ***haxölinaail quihimca*** 'shell having ringworm [spots]'
xepe yamaasij 'what is rolled by the sea'

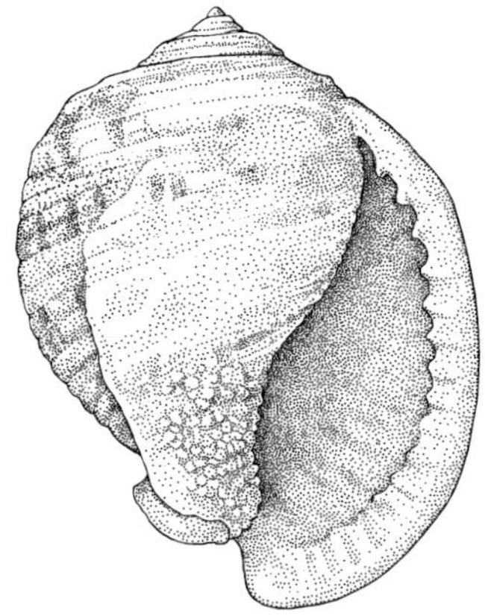

L 65 mm

This rather uncommon snail is rarely found alive, and its round, inflated shell is most often seen worn, in beach drift.

The name ***haxölinaail ctocnij*** 'round shell' reflects its shape.

The playful name ***haxölinaail quihimca*** 'shell having ringworm [spots]' reflects the shell's dappled markings. María Luisa Astorga laughed as she recalled as a child referring to the shell as *haxölinaail "cmaacoj" quihimca* '"old man" shell having ringworm [spots]' because of a man who was visiting the area. He had a somewhat spotty complexion and a drooping mouth, made all the more obvious by his frequent gum chewing. María Luisa added that the corner of his mouth reminded her of the shell's round, open siphonal canal. Such naming illustrates the way the Seris often playfully give names suggested by a person's physical characteristics or actions.

The snail shares the name ***xepe yamaasij*** 'what is rolled by the sea' with other medium-to-large snails having shells with similar smooth, inflated shapes (see *Malea ringens*, *Strombus galeatus*, and *Cypraecassis coarctata*).

Ficidae (fig snails)

Ficus ventricosa

xasecö

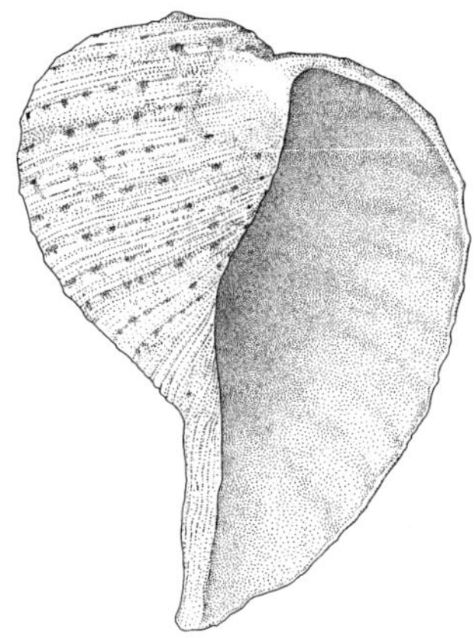

L 102 mm

The shell of this uncommon, large snail is thin, fragile, and delicately marked with brown, while its interior is a creamy pink.

Xasecö is a primary name. As the only snail of its family in the area, it is well known and easily identified by name by most people. When comparing this shell's long siphonal canal with that of *Malea ringens* (***xasecö cmaam*** 'female ***xasecö***'), the phrase ***xasecö ctam*** 'male ***xasecö***' was used to describe the *Ficus ventricosa* shell.

According to most, the snail was not eaten; in fact, it is rarely found alive.[31]

The shell was used as a dipper or cup. As described by Sara Villalobos:

> These are not eaten. They are like water dippers. The old people used them for water [from which to drink]. They also used *xapij* [reedgrass, like straws to suck up water.] The Ancestors didn't have food or drink or clothes. They used the *coton* [an early type of shirt or blouse], and they didn't have any coverings [blankets]. These shells were brought over from *Xazl Iimt* [Isla Ángel de la Guarda].

The shell was used to dip broth from the carapace of a freshly roasted sea turtle or as a funnel for filling small-mouthed clay vessels with water. The liquid was scooped into the shell and poured out via the shell's siphonal canal. In demonstrating such a use, Cleotilde Morales explained:

> It is a thing for dipping. The shell was taken care of—it is thin-edged. Also you can drink from it. It was used like a dish. [The meat] was eaten, but I haven't seen anyone eating it. The Giants ate the meat of [all] shells.

Eva López said that when young people were tattooed on their chins and on the inside of their lips, they drank water from the shell's narrow end because their newly tattooed skin was so tender. A note by Edward Moser describes such tattooing:[32]

> Tattoo the inside of the lips. Ramón López's mother and Juan Tomás had blue inside their lips from tattoo. Was a custom previously, at the age of puberty, tattooed the inside of their lips. The idea was to teach them silence and patience. Old folks would then counsel them on life, and their mouths hurt so much, they couldn't talk.

Edward H. Davis mentioned this shell by its Seri name and noted that it was used for drinking (appendix 7). Gwyneth Harrington collected the shell and recorded that it was used for drinking medicine from the shell's narrow end.[33] María Luisa Astorga confirmed this, saying that medicine (or other liquid) was given to a baby via this shell. She added that her grandmother had one of the shells in which she kept pollen for yellow face pigment, commenting that because her grandmother used the pollen, she always had nice skin.

Ranellidae (tritons)

Cymatium keenae

◊ ***hacx xah cöscaap*** 'I'll just stand over here'
◊ ***xepe yamaasij ctam*** 'male ***xepe yamaasij***'

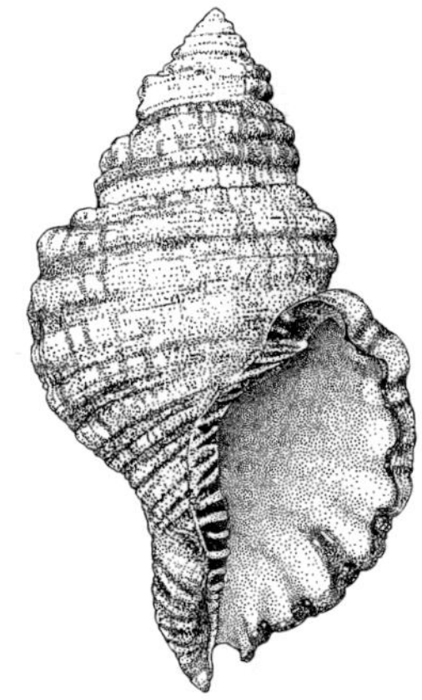

L 130 mm

Since this snail is from deep water it is seldom seen. When not beach worn, its robust shell is covered by a notable periostracum of dense bristles, called *iteems* 'its whiskers'. The shell is usually found in beach drift left by commercial shrimp dredging; the specimen illustrated was found live in the net of a fisherman, near Isla Patos.

Because it is uncommon, most people who saw the shell did not recognize it. Two different names were given to me at different times by Cleotilde Morales, both reflecting its similarity to other shells. In giving the name ***hacx xah cöscaap*** 'I'll just stand over here' she compared this shell to *Fusinus dupetitthouarsi* (having ***hacx xah cöscaap*** as one of its names), noting the high spires and long siphonal canals of both shells, characteristics she described as "male." She later provided the name ***xepe yamaasij ctam*** 'male ***xepe yamaasij***' comparing this shell's shape to the ***xepe yamaasij*** (*Strombus galeatus*), but having the more "male" characteristic of being less smooth than the latter. Both names appear to be have been somewhat improvised rather than being well-used names.

Muricidae (murex and rock snails)

Chicoreus erythrostomus [= *Hexaplex erythrostomus*]

xiica iteencoj cheel 'things with red mouths'
♦ ***xnoocat an cheel*** '***xnoocat*** with a red interior'
(var. ***xnoocat an cquihjö*** '***xnoocat*** with a red interior')

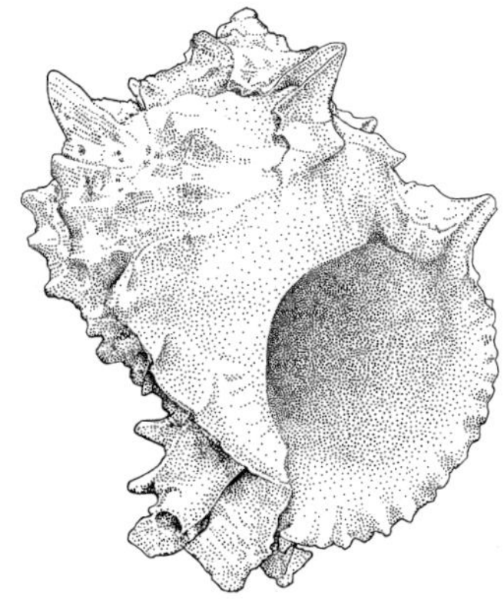

L 112 mm

The pink-mouth murex has a white shell with a polished, vivid pink interior. The live snails are collected in the very low intertidal zone.

An old name for this snail was ***xiica iteencoj cheel*** 'things with red mouths'. Today it is more commonly known as ***xnoocat an cheel*** '***xnoocat*** with a red interior' or ***xnoocat an cquihjö***, which has a variation in the color term 'red'.[34] The name ***xnoocat*** has a common variant ***noocat***.[35]

Figure 5.11. *Chicoreus erythrostomus* shells placed around a decorated tree trunk, Desemboque, 2007. Colorful strips of cloth tied to vegetation are traditional decorations.

Figure 5.12. *Chicoreus erythrostomus* opercula to be strung. Desemboque, 2007.

Figure 5.13. Victoria Astorga holding a *Chicoreus erythrostomus* shell. Punta Chueca, 2003.

The muscular foot of the snail was eaten and considered good food. Angelita Torres described the meat as tasting like ***cotopis*** (*Turbo fluctuosus*). However, one person mentioned that the snail was considered *hapáx* and something to be avoided as food by young people of both sexes.

In recent years, this snail has been gathered, boiled, and the meat removed and sold to non-Seri Mexicans, who transport it for sale elsewhere. ***Cotopis*** snails are sometimes sold like this as well.

There is no common use for the shell, but a nice specimen might be kept just because it is attractive, or the shells might be used in other ways, perhaps as a ground border (figure 5.11).

In recent years the opercula of this snail have been strung on necklaces (figure 5.12).

A childhood memory recalled by María Luisa Astorga was of her grandmother, Carmelita Encinas, singing the following song to comfort her when she was sad:

Xnoocat Cöicoos

Xe pe ca no ti ya ta gui
Xepe quih ano tíyataj
Xe pe ca no ti ya ta gui
Xepe quih ano tíyataj
Hai so ji ma cöi yo fain
Hai hisoj imac cöiyofain
Xe pe ca no ti ya ta gui
Xepe quih ano tíyataj
Hai so ji ma cöi yo fain
Hai hisoj imac cöiyofain.

The ***Xnoocat***'s Song

[My saliva] reaches the sea,
[My saliva] reaches the sea,
The wind wraps itself around me.
[My saliva] reaches the sea,
The wind wraps itself around me.

As María Luisa explained:

During a catastrophic event (*hant haaco iima*) the ***xnoocat*** was cast out of the sea and far up onto dry land. He sat there a long time; his saliva was drying up. He had a supernatural vision of the turbulent wind wrapping around and around his midsection; of the waves coming in, covering him, streaming his saliva out to sea, and carrying him out to where he longed to be. The snail sang this song.

A small desert dune camp near Desemboque carries the name *Xnoocat Coos* 'singing ***xnoocat***'. María Luisa suggested it being the place this song was sung.

Eupleura muriciformis

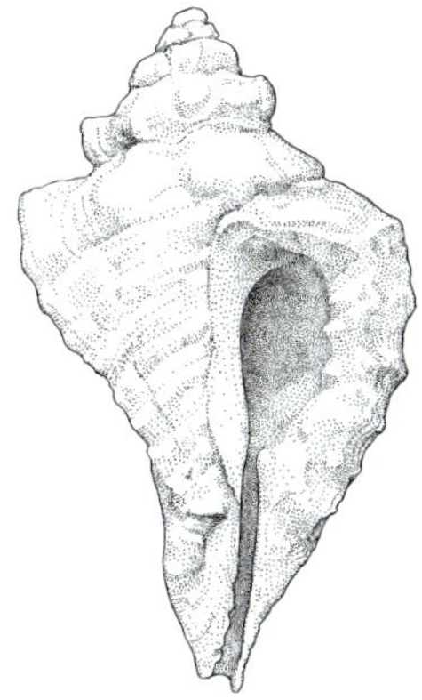

L 31 mm

hant iti yeemej coospoj 'whose slow movement marks the sand'

This small snail has an off-white shell. The snail illustrated (very worn) was found in Estero Sargento on a mud flat.

The name ***hant iti yeemej coospoj*** 'whose slow movement marks the sand' is also given to other small snails (see *Vokesimurex elenensis* and *V. tricoronis*).

Hexaplex nigritus

♦ ***xnoocat*** (var. ***noocat***)
xnoocat cmaacoj 'old man ***xnoocat***'
xnoocat coopol 'black ***xnoocat***'
xnoocat coospoj 'spotted ***xnoocat***'
xnoocat cooxp 'white ***xnoocat***'
xnocatiil '(very) large ***xnoocat***'

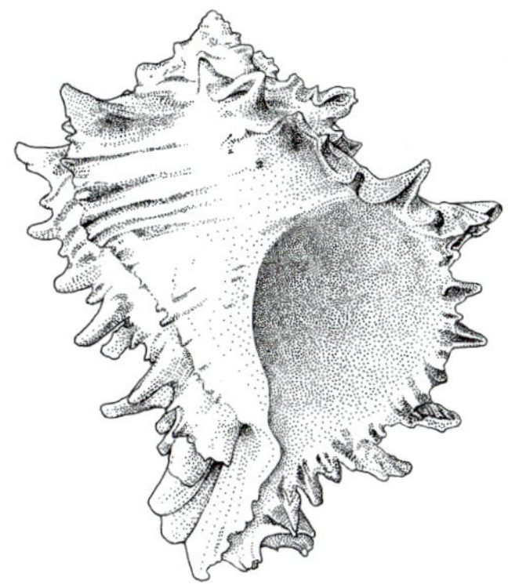

L 105 mm

The black murex is a large snail found at extreme low tide. Its robust white shell is distinctively marked with striking black designs and black-tipped spines.

The name ***xnoocat*** has a common variant ***noocat***. Other names for this shell include descriptive terms: ***xnoocat coopol*** 'black ***xnoocat***', ***xnoocat coospoj*** 'spotted ***xnoocat***', and ***xnoocat cooxp*** 'white ***xnoocat***'.

The name ***xnocatiil*** '(very) large ***xnoocat***' aptly reflects the snail's size, as it is the largest murex in the area.

Xnoocat cmaacoj 'old man ***xnoocat***' is described as having a shell that is narrower and has a longer siphonal canal.

The egg cases of the murex snail are called *xnoocat an ihiicto* '[thing] in which are ***xnoocat*** eggs'.

This snail was eaten, as especially evidenced in old camps around Estero Santa Cruz, where the broken shells are abundant, often fire-blackened (figure 5.14).[36] The snails were roasted under a brush fire, and according to one person, they were then broken open on a metate (grinding stone), using a smaller rock. María Luisa Astorga explained that if the snails were boiled, the meat popped out easily. However, if the snails were roasted, the shell was then carefully chipped with a rock (smaller than a mano, a stone used as the upper millstone for grinding foods on a metate), making a hole in its side. The meat was then removed through the hole. A shell chipped open in this way was called *hacaacp*.[37]

Figure 5.14. Fire-blackened *Hexaplex nigritus* shells in a desert camp bordering Estero Santa Cruz, 2007.

Raquel Moreno described the preparation somewhat differently. The live mollusk was placed with its aperture up and roasted under a brush fire. When the fire burned down, the juicy meat with the operculum just "popped up" and was easily removed by shaking the shell, or pried out with a sturdy twig, such as one from the creosotebush.[38] Only the muscular foot was eaten. Another woman said that sometimes the meat was thinly sliced and dipped in sea turtle oil flavored with oregano.[39]

According to Cleotilde Morales, this snail, as well as *Chicoreus erythrostomus*, was considered *hapáx* and was not eaten by young people of childbearing age.

There is no known use for the shell itself, other than a fine specimen being kept as a curiosity. Unlike those of the pink-mouth murex, the opercula of this snail are not strung on necklaces because of their larger size.

A note by Edward Moser recorded Roberto Herrera saying, "this [snail] shell feeds on jackknife clams and feeds its young—the [juvenile snail] holds on to the large [snail] shell, which gives bits to the young."

Mancinella speciosa [= *Thais speciosa*]

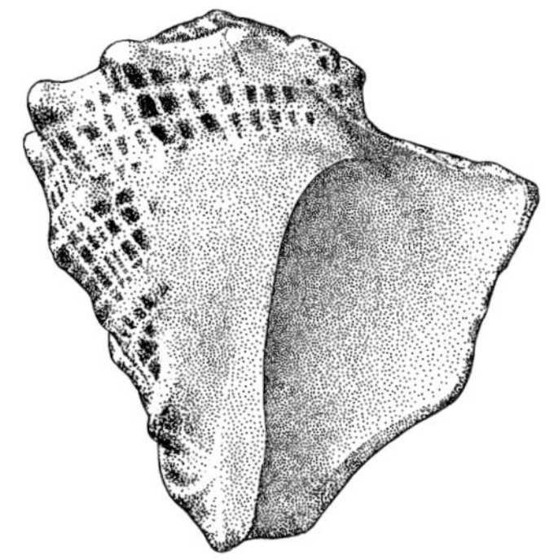

L 24 mm

This somewhat uncommon snail is found on intertidal rocks. The woman who brought the shell to me said that she had not seen one before, nor was it recognized by anyone else to whom I showed it.

Mancinella tuberculata [= *Neorapana tuberculata*]

♦ ***haxölinaail ancoj cmasl*** 'shells with yellow interiors'
honcaacoj imazaj 'the Yellow-footed Gull's clay pot'
napxeec 'the Turkey Vulture's daughter'

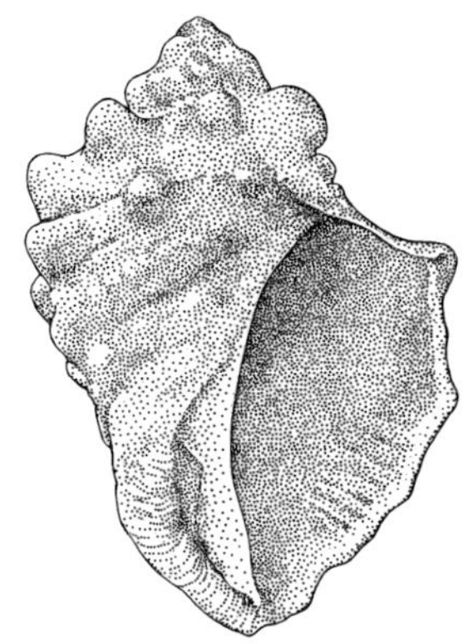

L 52 mm

Common in the rocky intertidal zone, this medium-sized snail has a shell with a distinctive, yellowish-orange interior.

The name ***haxölinaail ancoj cmasl*** 'shells with yellow interiors' is descriptive. When giving the name ***honcaacoj imazaj*** 'the Yellow-footed Gull's clay pot', Eva López related how the gull (*Larus livens*) pecks at the interior of the shell, and noted the similarity between the color of the shell's interior and the gull's beak. ***Napxeec*** 'the Turkey Vulture's daughter' is a name most often given to *Cerithium* spp. snails, although two much larger snails (*Strombus galeatus* and *Melongena patula*) are included by some Seris.

According to a few people, the snail was eaten in the past, boiled in the shell, and the meat eaten with rice and cilantro or other such condiments.

Mexacanthina lugubris angelica [= *Acanthina angelica*]

iquihimz iic cöihiipe 'ringworm medicine'
hast imocl cooxalca 'things that sit under a rock'

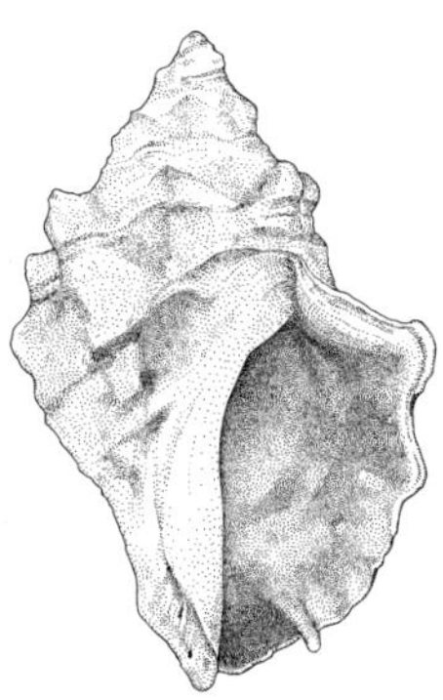

L 36 mm

This small snail is found in large numbers on, and clustered beneath, intertidal rocks. The shell is often covered by encrustations, making it rather nondescript, while its banded interior is a muted purple. The curious "tooth" on its aperture makes the shell distinctive.

The name ***iquihimz iic cöihiipe*** 'ringworm medicine' is literally 'what is good for ringworm'.[40] The snail's secretion was used to treat ringworm and other skin disorders. The raw snail was removed after breaking the shell and rubbed on the skin. Raquel Moreno recalled seeing the snails having a colored secretion.

According to Juanita Herrera the name ***hast imocl cooxalca*** 'things that sit under a rock' describes how groups of these snails congregate "like little families."

The snail was not eaten.

Pascula ferruginosa [= *Morula ferruginosa*]

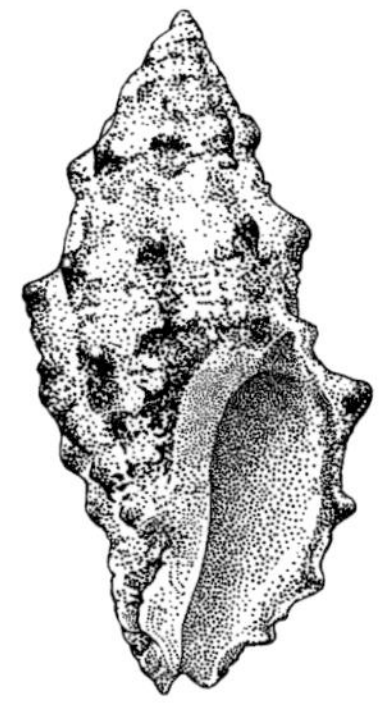

L 19 mm

This small snail is common under intertidal rocks. Its shell has a rather knobby dark sculpture.

The shell has been recently used in shell jewelry. No Seri name was provided for the snail.

Plicopurpura pansa

iquihimz iic cöihiipe 'ringworm medicine'

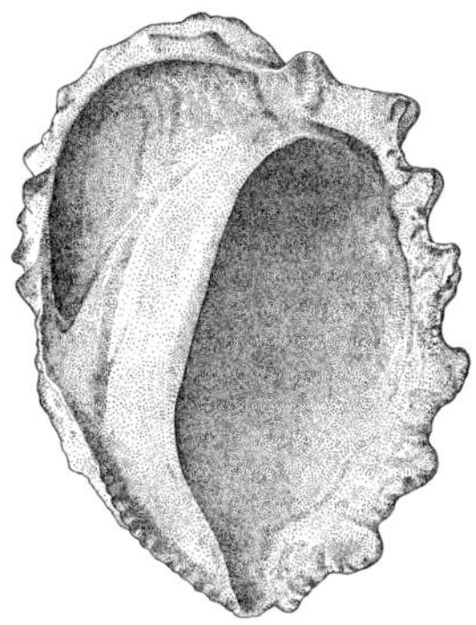

L 29 mm

Since the snail is rare in the present-day Seri territory, found at its southern limit, only a few people said that they had seen one. The snail lives on rocky shores in the higher intertidal zone where it is exposed to waves. When the live snail is handled and disturbed, its secretion leaves a vivid purple stain on one's skin—no doubt the snail's most notable characteristic.

The name ***iquihimz iic cöihiipe*** 'ringworm medicine' (literally 'what is good for ringworm'), also given to *Mexacanthina lugubris angelica*, alludes to the use of the snail's secretion in treating ringworm and other skin disorders. The meat of the live animal was described as being slippery and having a strong odor.[41]

Pteropurpura centrifuga

xepe án icös 'spines of the sea area'

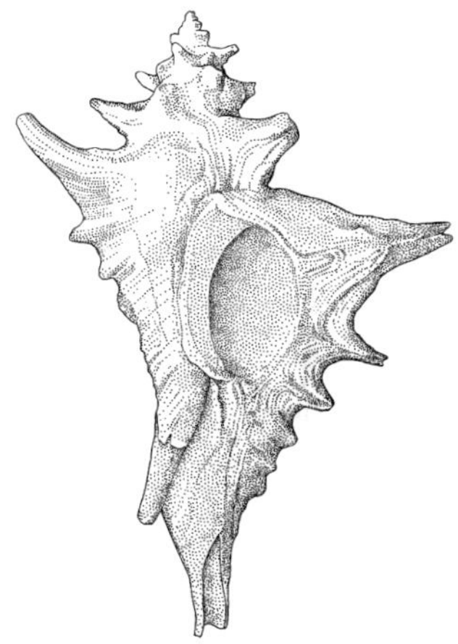

L 70 mm

This uncommon murex has a cream-colored, ornate shell. Its shell was given to me by Lorenzo Herrera, who retrieved it from a shrimp boat along with a *Vokesimurex tricoronis* shell. Lorenzo recalled his father providing the name ***xepe án icös*** 'spines of the sea area' for the *Pteropurpura centrifuga* snail. He added that his father said that the name was told to him by Porfirio Díaz, one of the last Seris to have lived on Isla San Esteban long ago.

Pteropurpura erinaceoides

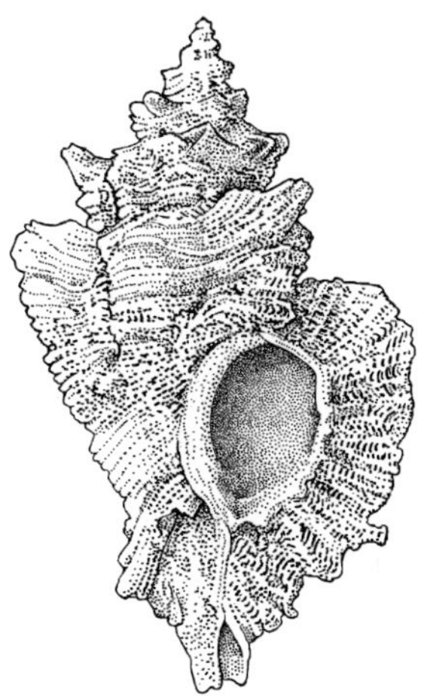

L 50 mm

This small snail has a delicate, cream-colored shell with ornate webbed sculpture. Its somewhat flat spines end in curved tips. No Seri name was provided for the snail.

Stramonita biserialis [= *Thais biserialis*]

haxölinaail ancoj cmasl 'shells with yellow interiors'

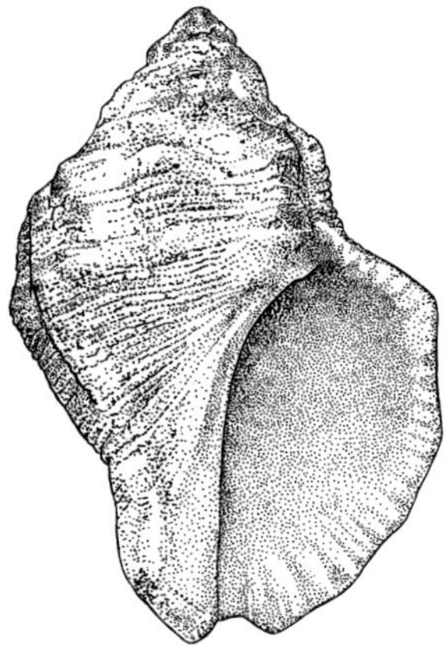

L 55 mm

The snail is common on intertidal rocks; its shell has a distinctive yellowish-orange interior. Though apparently less common than the similar *Mancinella tuberculata*, both are found on the rocky shores near Desemboque.

This snail and *M. tuberculata* are descriptively called *haxölinaail ancoj cmasl* 'shells with yellow interiors'.

This snail was eaten in the recent past and the meat sometimes sold to outsiders. The snail was boiled in the shell, and its muscle eaten with rice and cilantro or other condiments.

Vitularia salebrosa

This snail is quite uncommon. Its shell has a white interior, and its reddish-brown exterior has a somewhat knobby, webbed sculpture. The shell was found in beach drift south of Punta Chueca by Alfredo López, who said that he had not seen one before.

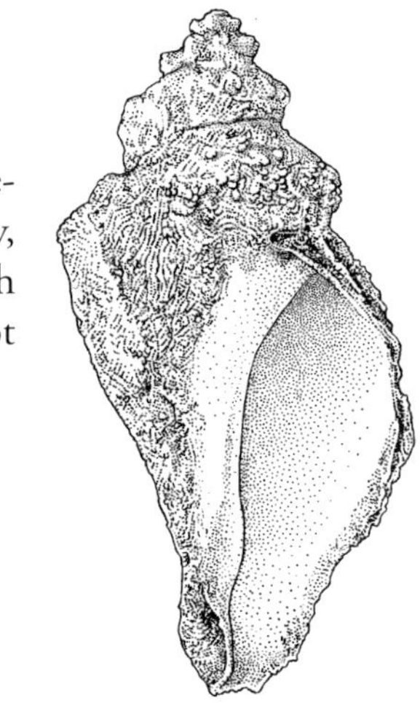

L 82 mm

Vokesimurex elenensis [= *Murex elenensis*] (a) and *Vokesimurex tricoronis* [= *Murex tricoronis*] (b)

hant iti yeemej coospoj 'whose slow movement marks the sand'
hitcz ihizcoofizlc 'the firewood my younger sister gathered'
motcz
quitxísotaj 'whose body whorls are narrow'
xahícosa
xpacaao iiha 'the mermaid's possession'

Since these small, delicate shells are most often found in beach drift with their spines broken and worn, undamaged shells are sometimes valued as keepsakes. The *Vokesimurex elenensis* shell is a white and buff color, while the white shell of the less common *V. tricoronis* has a more delicate sculpture, marked with fine reddish bands. Not often found alive, the snails were not eaten.

For such attractive little shells, it is somewhat surprising that most people could not identify them by name, and when named, there was much variability.

Hant iti yeemej coospoj 'whose slow movement marks the sand', also given to *Eupleura muriciformis*, apparently describes the shell's trail. According to Angelita Torres, the name originated with Loreta Marcos, who also named another shell (see Names and Naming, p. 25).

The name ***hitcz ihizcoofizlc*** is unclear linguistically. René Montaño attempted to clarify the meaning, but was trumped by the women in his household. Though not quite making grammatical sense, the name was very clear to them, and was explained by his mother-in-law, Victoria Astorga: "This shell was there when things were being named. It said, '*Hitcz ihizcoofizlc*' [Younger sister, let's go gather firewood]. So the Name Giver said, 'That will be your name.'"

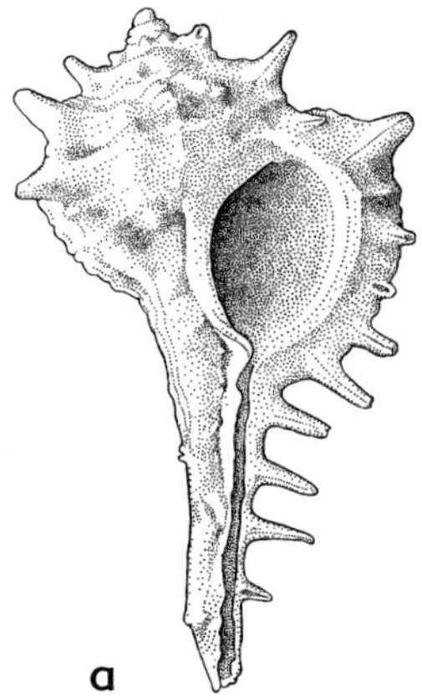

a L 56 mm

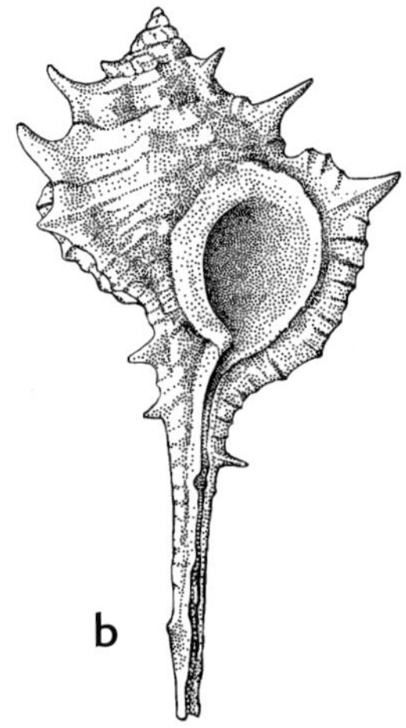

b L 50 mm

Motcz is a primary name.

Quitxísotaj 'whose body whorls (literally, buttocks) are narrow' describes the narrow body whorls and siphonal canals of the shells.

The name ***xahícosa*** is probably related to *cahícosa*, a type of mule deer found on Isla Tiburón, described as having "bumpy" antlers. Perhaps the name suggests the similarity in the shell's spines to the deer's antlers.

It is unclear how the name ***xpacaao iiha*** 'the mermaid's possession' relates to the shell.

I was given a perfect little *Vokesimurex tricoronis* shell kept in tissue paper by a friend whose husband had found it in a crab trap. It was obviously special to her.

Buccinidae (whelks)

Cantharus elegans

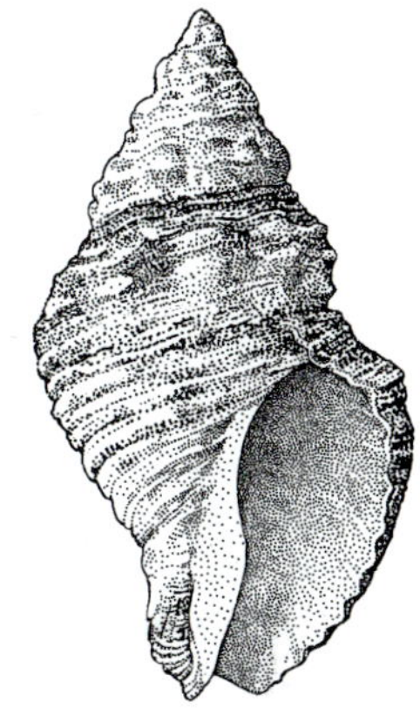

L 39 mm

The high-spired, ridged shell of this snail is marked with brown designs. The snail is common on intertidal rocks north of Bahía de Kino.

No one recognized the shell or provided a name.

Cantharus macrospira [= *Solenosteira macrospira*]

hant itcoj 'bases of the land'
haxölinaail caaoloj 'grooved shell'

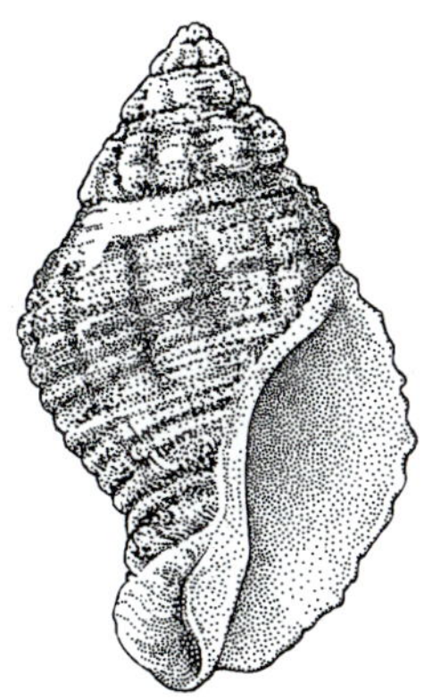

L 25 mm

The snail, its white-ridged shell usually covered with a tan periostracum, is very common on intertidal rocks in the Desemboque area. A distinctive characteristic is the way in which the male bears a mass of the snail's stalked egg capsules attached to its shell.

For such a common snail it is surprising that not many people recognized the shell or named it.[42] Only one person supplied the name ***hant itcoj*** 'bases of the land'. The connection is unclear.

The name ***haxölinaail caaoloj*** 'grooved shell' was given by Alfredo López, and reflects the grooved surface of the whorls.

Columbellidae (dove snails)

Anachis scalarina

haxölinaail coopl caacöl 'large black shells'

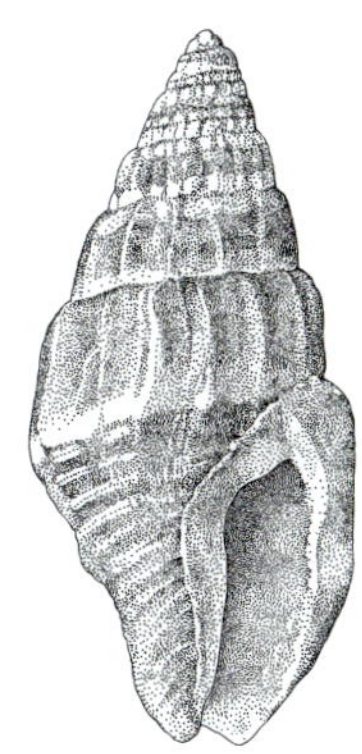

L 18 mm

This small, dark-shelled snail is found in large numbers beneath intertidal rocks.

The name ***haxölinaail coopl caacöl*** 'large black shells' is a plural form, used even when a single shell is referred to; this is true of some other shell names (see *Olivella dama*, for example). Though not black, the shell is dark and small, so the shell's design is not obvious. The inclusion of *caacöl* 'large' distinguishes the snail from the smaller snail *Mitrella ocellata*, which has the name ***haxölinaail coopl*** 'black shells'.

The shell is one of the most commonly strung in shell jewelry, often used as a filler with other more striking shells.

Columbella fuscata

hast iti caatolam 'what crawl on rocks'
iteel it hacaptxö 'what is pierced on its edge'
♦ ***mosnaact*** Common Ground-Dove

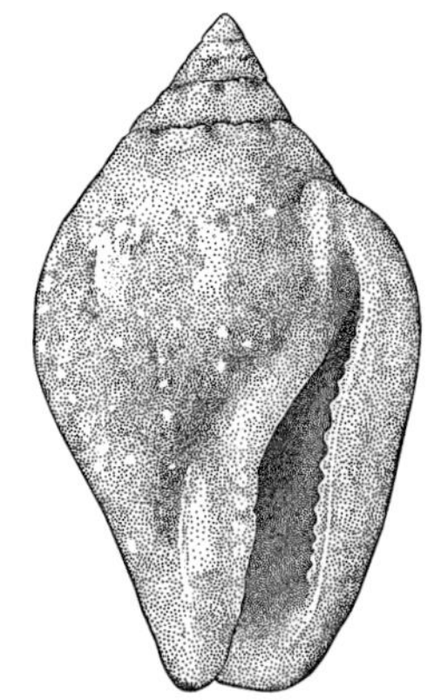

L 18 mm

This snail's delicately patterned, buff-colored shell is less vividly marked than that of the similar *Columbella strombiformis*.

The name ***hast iti caatolam*** 'what crawl on rocks' is based on the snail's movement. The name uses a plural form of the verb, even when one snail is referred to. The name ***iteel it hacaptxö*** 'what is pierced on its edge' is also given to *C. strombiformis*. These snails, unlike most others, are sometimes strung by piercing an area near their lip rather than their apex. The name ***mosnaact*** reflects the muted color and delicate markings of the Common Ground-Dove (*Columbina passerina*).

The shell, usually gathered containing a hermit crab, is common in shell jewelry.

Columbella strombiformis

hast iti caatolam 'what crawl on rocks'
iteel it hacaptxö 'what is pierced on its edge'
mosniil itaajc 'the *mosniil* sea turtle's saliva'
♦ ***mosniil oozt*** 'what the *mosniil* sea turtle tattooed'

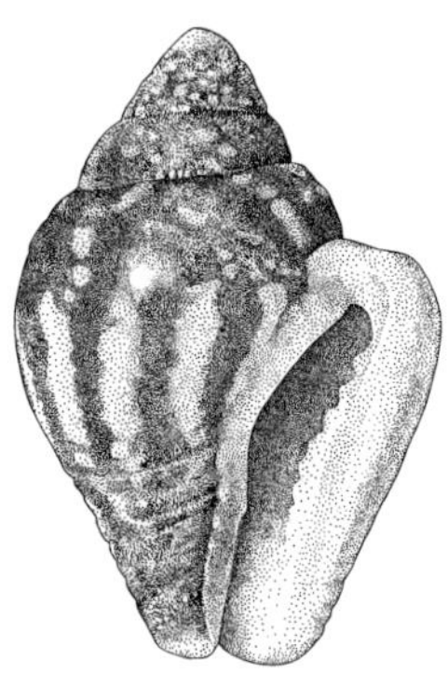

L 25 mm

The sturdy and attractive shell of this small snail is vividly marked with zigzag brown and white designs.

The name ***hast iti caatolam*** 'what crawl on rocks' is based on the snail's movement. The name uses a plural form of the verb, even when one snail is referred to. The name ***iteel it hacaptxö*** 'what is pierced on its edge' is also given to *C. fuscata*. These snail shells, unlike others, are sometimes strung by piercing an area near their lip rather than their apex. The name ***mosniil itaajc*** 'the *mosniil* sea turtle's saliva' alludes to the shell's markings. The name ***mosniil oozt*** 'what the *mosniil* sea turtle tattooed' focuses on the shell's striking zigzag designs. According to some, at the time of the flood Catastrophe this shell was a woman, and the *mosniil*, a variety of the black sea turtle, was a man. He tattooed the woman, who changed into this shell, and the man became the sea turtle.

The shell, usually gathered containing a hermit crab, is often strung in shell jewelry.

Costoanachis coronata [= *Anachis coronata*]

haxölinaail cooscl 'speckled shell'
♦ ***haxölinaail quimaaxat*** 'gray-brown shell'
◊ ***zojiime***

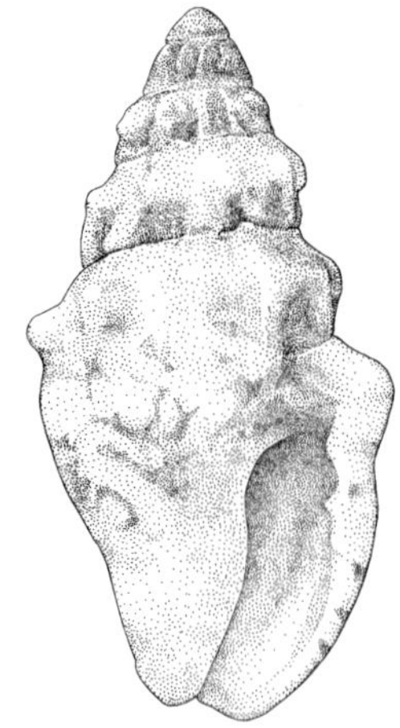

L 12 mm

This very common, small snail with a brown speckled shell is found in large numbers beneath intertidal rocks.

Both of the names ***haxölinaail quimaaxat*** 'gray-brown shell' and ***haxölinaail cooscl*** 'speckled shell' note the shell's muted color and fine markings. The latter name is also given to the less common *Echinolittorina aspera* snail.

In 2009 one family named this snail ***zojiime***, because it is abundant at *Zoj Iime*, a seaside camp north of Punta Chueca, even though the family was aware of its other, more common, names. They told me they enjoyed using their own, personally given name for the snail.

The shell is often used in shell jewelry as a filler between more distinctive shells. Sometimes the shell is whitened before stringing by toasting it in hot sand and soaking it in a commercial bleach solution.

Mitrella ocellata [= *Mitrella guttata*]

hastisol 'yellow rock'
♦ ***haxölinaail coopl*** 'black shells'

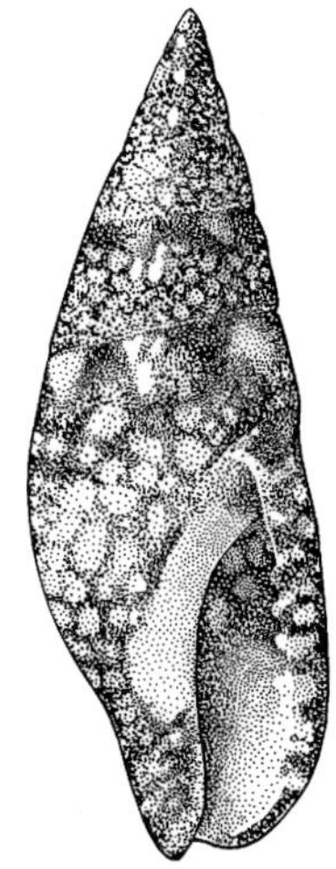

L 14 mm

This small, smooth-shelled snail is found in great numbers beneath intertidal rocks.

The name ***hastisol*** 'yellow rock' (from *hast* 'rock, mountain' and *-isol* 'yellow'), also a place name, was given by one person. The name is somewhat puzzling, since the shell is not yellow and contrasts with the shell's other name referencing black. No explanation was given.

The name ***haxölinaail coopl*** 'black shells' is shared with *Anachis scalarina* and notes the dark appearance of the small snail. The name is plural, used even when a single shell is referred to.

A shell easily gathered in quantity, it is often used as a filler in shell jewelry with other, more decorative, shells.

Parametaria dupontii

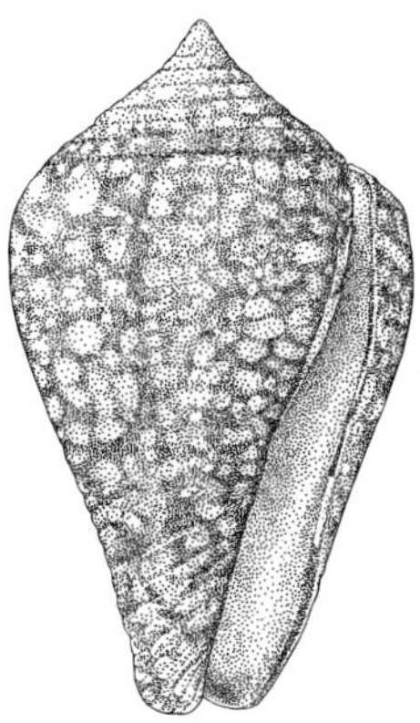

L 15 mm

The sturdy shell of this snail is spotted brown and has a distinctive spiraled and finely grooved surface. I have not seen the shell used in shell jewelry, and no Seri name was given.

Melongenidae

Melongena patula

napxeec 'the Turkey Vulture's daughter'

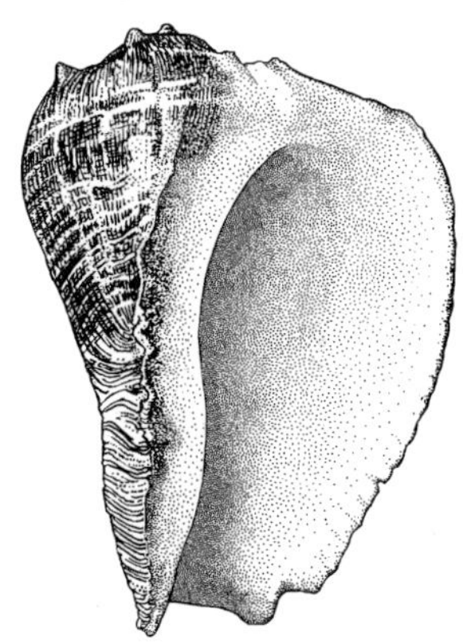

L 160 mm

The shell of this large, uncommon snail has a striking reddish-brown exterior marked with cream-colored bands.

The snail is rarely seen alive today, and most people who saw the shell did not recognize it. A few identified it as ***napxeec*** 'the Turkey Vulture's daughter', a compound of *naapxa* (*Cathartes aura*, Turkey Vulture,) and *eec* 'his daughter'. The name is most commonly given to smaller *Cerithium* spp. snails, but is also shared by the large *Strombus galeatus* and the *Mancinella tuberculata* snails. The connection to the bird is unclear.

The meat was not eaten in the recent past. However, notably around Estero Tastiota (an estero between Bahía de Kino and Guaymas) this shell, broken, occurs on middens together with other food shells.

I have seen this attractive shell kept as a curiosity displayed in several houses.

Nassariidae (nassa mud snails)

Nassarius brunneostomus [= *Nassarius moestus*]

◊ ***haxölinaail anxö icaticpan*** 'shell that is a lot of work'
♦ ***haxölinaail cnoosc*** 'rough shell'
haxölinaail quicös 'spiny shell'
xpaḻeemelc quicös 'spiny ***xpaḻeemelc***'

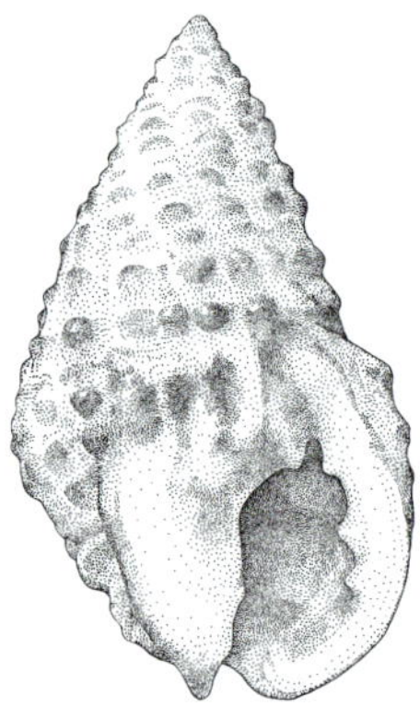

L 15 mm

This snail is commonly found on sandy substrates and on estero mud flats. Its finely knobbed shell is a light tan.

A name recently given, ***haxölinaail anxö icaticpan*** 'shell that is a lot of work', alludes to the difficulty in collecting and working with these small snails.

Haxölinaail cnoosc 'rough shell', ***haxölinaail quicös*** 'spiny shell', and ***xpaḻeemelc quicös*** 'spiny ***xpaḻeemelc***' are descriptive names. The latter names single out the snail's very sharp apex; when many small ones are strung together this is especially evident—the necklaces are rather uncomfortable to wear.

The shell is very common in necklaces, used in its natural state or dyed with commercial dyes, and often used as a filler with more unusual shells.

Nassarius tiarula

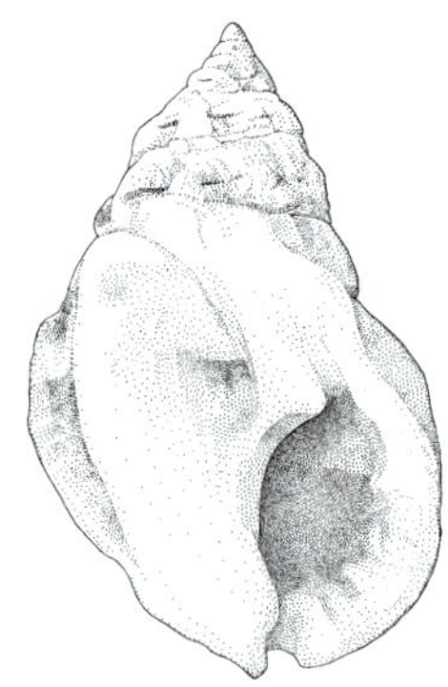

L 17 mm

haxölinaail cnoosc 'rough shell'

This small snail with its cream-colored shell occurs in great numbers on sand or estero mud flats—often massed on dead sea creatures—making it easy to collect.

The name ***haxölinaail cnoosc*** 'rough shell' describes the shell's knobby sculpture.

The shell is very common in necklaces, used in its natural state or colored with commercial dyes.

Figure 5.15. *Nassarius* spp. snails feeding on a small dead stingray, eastern shore of Isla Tiburón, 2009.

Fasciolariidae (tulip and spindle snails)

Fusinus ambustus

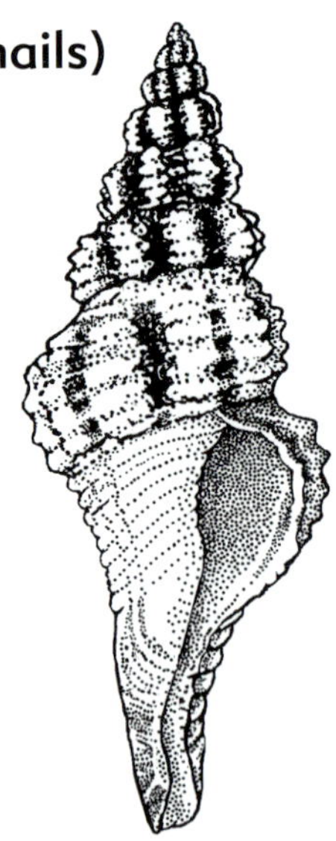

L 50 mm

The shell of this small spindle snail is distinctive because of its dark markings and small size; otherwise it is very similar to the much larger *Fusinus dupetitthouarsi*. No name was supplied for this snail.

Fusinus dupetitthouarsi

♦ ***hacx xah cöscaap*** 'I'll just stand over here'
siip isoj 'real young man'
sipisot 'narrow young man' (var. ***hasipisot***)
◊ ***xasecö ctam*** 'male ***xasecö***'
xpaḻeemelc cacösxaj 'tall ***xpaḻeemelc***'

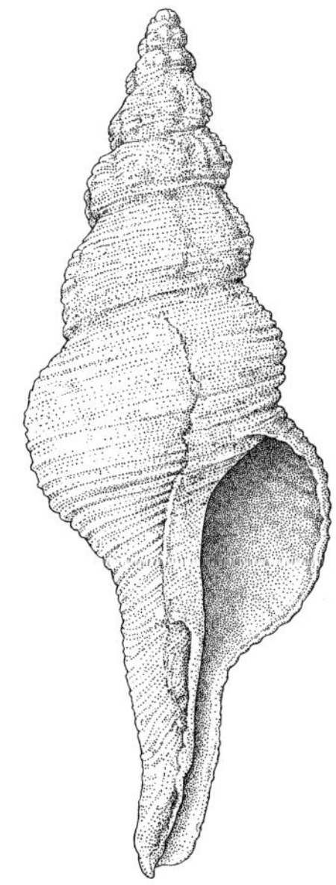
L 180 mm

The large size and long, narrow shape of this snail's white, heavily grooved shell single it out from other shells in the area. Somewhat uncommon, the empty shells are typically found in beach drift, which in recent years is often the result of dredging by commercial shrimp boats.

The name ***hacx xah cöscaap*** 'I'll just stand over here' was provided by the shell's own words, as related by Ernestina Morales:

> The Name Giver was there, and was giving different names to all of the shells, and the one called ***hacx xah cöscaap*** was there [standing outside the group because he arrived late]. He didn't know what his name would be, and he was about to go away and leave the ones that had names. He said: "*Hatee hacx xah cöscaap aha*" 'I think I'll just stand over here'. So when he said that and was leaving, the Name Giver said, "Well, you will be called ***hacx xah cöscaap***—you have that for your name." He said that to him, and then he had that name.

The shell's long narrow shape suggests the name ***sipisot***, derived from *siip* 'young man' and *-isot* 'narrow'. In a note by Edward Moser, the word for 'boy (vocative)' in Giant speech was *asipo kíisil* 'small young man'. Older people confirmed this archaic expression (now written as *hasiipo quiisil*).

The name ***siip isoj*** 'real young man' is probably a variation of the name ***sipisot***. One person identified this shell as ***xasecö ctam*** 'male ***xasecö***' noting the shell's more "male" characteristics in comparison to those of the ***xasecö*** (*Ficus ventricosa*) shell, which has the more round and squat "female" attributes. The descriptive name ***xpaḻeemelc cacösxaj*** 'tall ***xpaḻeemelc***' notes its height in addition to its "typical" ***xpaḻeemelc*** shape.

The snail was rarely found alive. Angelita Torres, who spent much of her childhood on Isla Tiburón, said that she had never seen anyone eating it, adding that she did not know of any use for the shell.

María Luisa Astorga said that this shell was sold or traded to non-Seris on nearby ranches, along with ***xtiip*** (*Laevicardium elatum*) shells. She added that she was told that her grandmother played with this shell when she was a girl. It was considered a male doll because of its shape.[43] Her grandmother called it *hapequet siip* 'young man doll'. María Luisa described another male doll made from the entoplastron of a sea turtle that was said to go to the dolls' houses at night to have sexual intercourse with female dolls. The shell ***hacx xah cöscaap*** was placed upright in the sand by the door of a doll's house to guard the other dolls from the bone doll.

Victoria Astorga explained how she had played with the shell, tucking seaweed into its aperture for hair and standing it in the sand, apex down.

Pleuroploca princeps [= *Fasciolaria princeps*]

◊ ***xasecö caacoj*** 'large ***xasecö***'

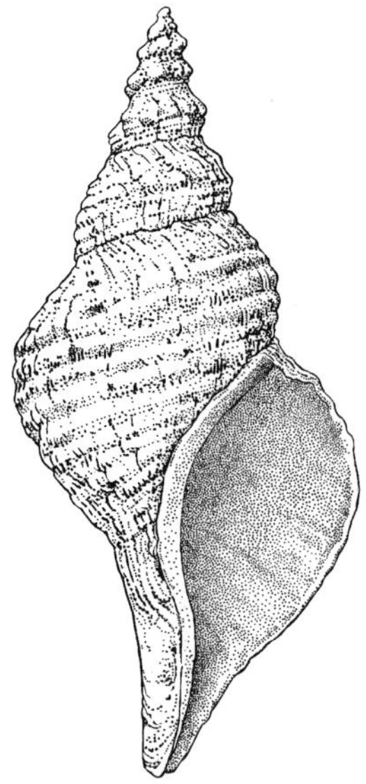

L 307 mm

The impressive size of this snail makes it noteworthy. As it is usually only known from specimens dredged by commercial shrimp boats or brought up by divers, many people I asked had not seen it before. Cleotilde Morales provided the name ***xasecö caacoj*** 'large ***xasecö*** (*Ficus ventricosa*)', which seemed more of an improvised description than a well-known name.

The shell, a light orange color under a brown periostracum, is kept only as a curiosity.

Volutidae (volutes)

Enaeta cumingii [= *Lyria cumingii*]

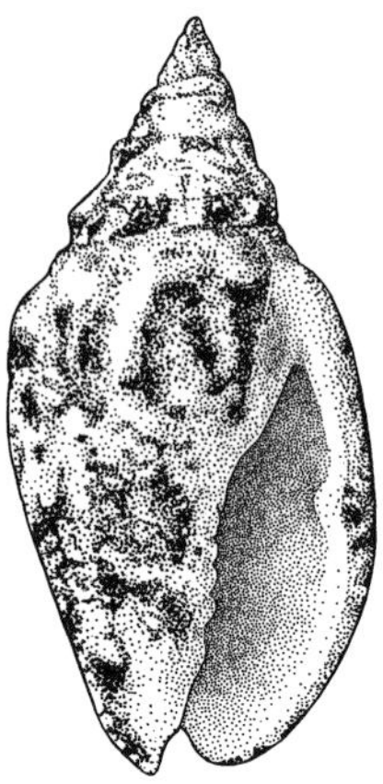

L 24 mm

The sturdy shell of this snail is cream-colored and spotted with dark brown designs.

Only recently brought to my attention, the snail has no Seri name. María Luisa Astorga brought me a few worn shells that she had gathered in beach drift at a beach north of Desemboque, saying that she had never seen them before. She chuckled and added that she had given this shell a made-up name, *haxölinaail captasa* 'unlucky shell', because she found it rather unattractive.

Harpidae (harp shells)

Morum tuberculosum

haxölinaailc imahcaptxö 'shell that is not perforated'
haxölinaailc impafc 'shell that is not pounded'
♦ ***oot yácmolca*** 'the coyote's fetish'

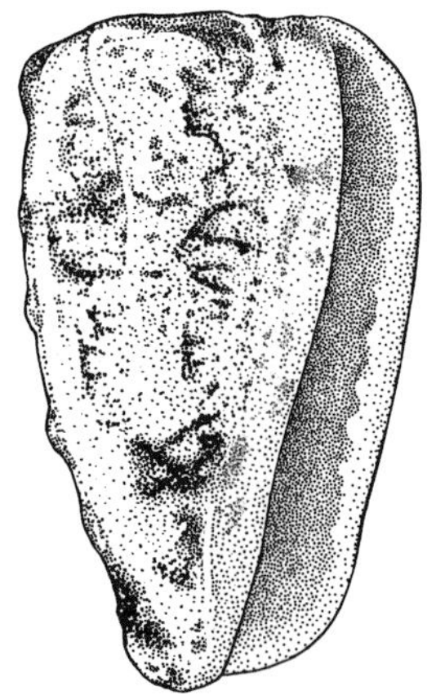

L 37 mm

The snail is somewhat uncommon in the area; its shell's distinctive flat spire makes it appear somewhat beach worn.

The names ***haxölinaailc imahcaptxö*** 'shell that is not perforated' and ***haxölinaailc impafc*** 'shell that is not pounded' were said to allude to the shell's flat spire that made it difficult to pierce for stringing. The names were given by a woman who was told by her mother that her own mother, María, had provided the names.

The name ***oot yácmolca*** 'the coyote's fetish' seems appropriate for this somewhat drab shell, as the coyote foolishly finds it attractive.[44] Cleotilde Morales suggested the shell's raised dark spots looked like the tracks left by a coyote trotting along the beach.

Usually found as empty shells in beach drift, this snail was not a food.

I have recently seen several shells hung as pendants on a shell necklace, and two men recalled using the shell, with a large bird feather inserted in its apex, as a throwing toy (see *Oliva incrassata*).

Olividae (olive snails)

Because of their attractive highly polished exteriors, the olive snails are favorites for use in shell jewelry. Additionally, most are common and easy to gather and prepare for stringing. They were not eaten.

The snails of the Olividae incorporate the term ***xpa̱leemelc*** in their names, as do those of several other families (see table 3.2, p. 24). The name is probably a compound of *xepe* 'sea' and *ipl mateemij* 'it moves slowly with its tongue'.

Agaronia testacea

iplmateemij (var. ***ipl queemij***) ‘it moves slowly with its tongue’
♦ ***mayoocö ixpal̲eemelc*** ‘the Yellow-crowned Night Heron’s ***xpal̲eemelc***’
oot ixpal̲eemelc ‘the coyote’s ***xpal̲eemelc***’

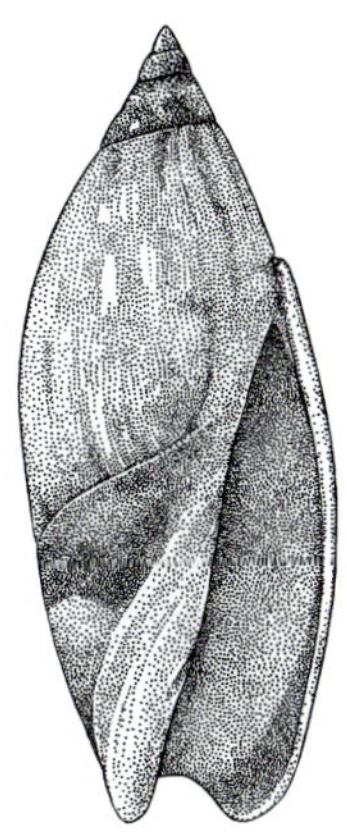

L 40 mm

The shell of this medium-sized olive snail is marked with bluish-gray and tan markings. The snail is common on sand flats.

The name ***iplmateemij*** ‘it moves slowly with its tongue’ is a contracted form of *ipl* ‘tongue’, *hamt* ‘dirt’, and *queemij* ‘to move slowly’. An old name for this snail, it aptly describes the snail’s movement as it glides with its large foot over the submerged sand. A variation of the name is ***ipl queemij*** ‘whose tongue moves slowly’. ***Xpal̲eemelc*** is possibly a contraction of ***iplmateemij*** (see Classification and Naming).

Mayoocö ixpal̲eemelc ‘the Yellow-crowned Night Heron’s (*Nyctanassa violacea*) ***xpal̲eemelc***’ is the snail’s most common name and suggests the snail’s resemblance to the bird’s plumage in markings and color (figure 5.16). When talking about the shell, Eva López described the bird fishing at night, demonstrating the sound it makes that one can hear from the beach: when it misses a crab it is trying to spear, it makes a soft, grunting sigh of disappointment; when it finally does catch a crab, it gives a satisfied, contented sigh.

Figure 5.16a–b. An *Agaronia testacea* shell and a Yellow-crowned Night Heron. Photograph courtesy of Fulvio Eccardi.

The coyote is said to own strange, old, or otherwise unusual items. The connection to this snail, implied in the name ***oot ixpal̲eemelc*** ‘coyote’s ***xpal̲eemelc***’, is unclear.

The snail was given a spontaneous personal name by Juanita Herrera: when left with one unsold necklace (tourists had bought all but the one made with this shell) she named the shell *captasa* ‘unlucky’.

The shell was used as a shell/feather toy made by inserting a large sea bird feather into its broken-off tip (see *Oliva incrassata*). José Ángel Montaño mentioned that this shell worked well because it was sleeker and more streamlined than other shells, adding that as a boy he and his playmates discovered that a tail feather of the *ziic coopol* (Double-crested Cormorant, *Phalacrocorax auritus*) was best—when used with this shell, the toy flew straight and true when thrown, shell-first, at its target. José Ángel further recalled the boys sometimes notching the feathers for decoration.

Today, small *A. testacea* shells are occasionally strung on necklaces.

Oliva incrassata

♦ ***xpal̲eemelc caacoj*** ‘large ***xpal̲eemelc***’
xpal̲eemelc comihj caacoj ‘large ***xpal̲eemelc comihj***’
xpal̲eemelc coospoj ‘spotted ***xpal̲eemelc***’
xpal̲eemelc quipcö ‘thick ***xpal̲eemelc***’

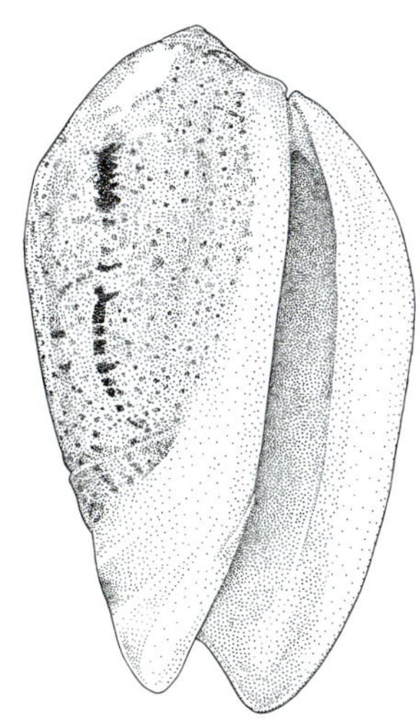

L 93 mm

The attractive, large heavy shell of this olive snail is cream-colored and spotted with deep brown. Somewhat uncommon, the snail is found on sandbars at extreme low tide.

The snail's most common name is ***xpaḻeemelc caacoj*** 'large ***xpaḻeemelc***', a name shared by *Oliva porphyria* and *Strombus galeatus*. The name ***xpaḻeemelc comihj caacoj*** 'large ***xpaḻeemelc comihj*** (*Oliva spicata*)' contrasts the shell to that smaller snail. The names ***xpaḻeemelc coospoj*** 'spotted ***xpaḻeemelc***' and ***xpaḻeemelc quipcö*** 'thick ***xpaḻeemelc***' are descriptive of the shell's markings and relative thickness.

Good-sized, sturdy, and smooth, the shell was used in a shell/feather toy (figure 5.17). The quill of a large feather from a bird such as a pelican, gull, or cormorant was inserted into a hole made in the apex of the shell. The toy was either held by the shell and thrown far, in a game to see who could retrieve it, or it was held by the feather and flipped, aimed at a target circle drawn in the sand. The toy was called *ziic ina xpaḻeemelc ano cöhazám* '***xpaḻeemelc*** into which a bird feather has been inserted'. Other snail shells of a similar size and shape were also used.

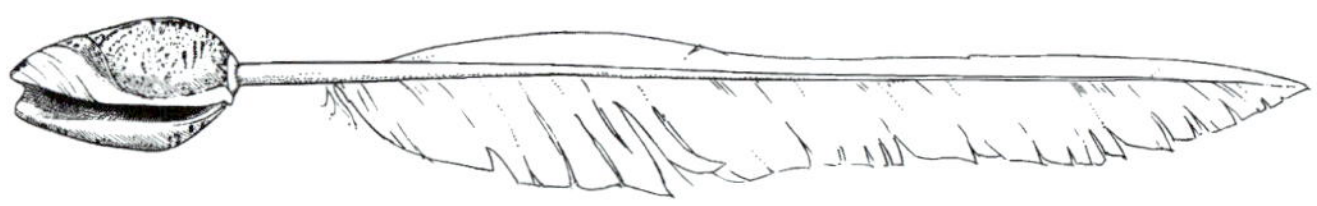

Figure 5.17. A toy made from a large bird feather inserted into a shell.

Figure 5.18. The shell of an *Oliva incrassata* snail at an inland desert camp.

Oliva porphyria

xpaḻeemelc caacoj 'large ***xpaḻeemelc***'

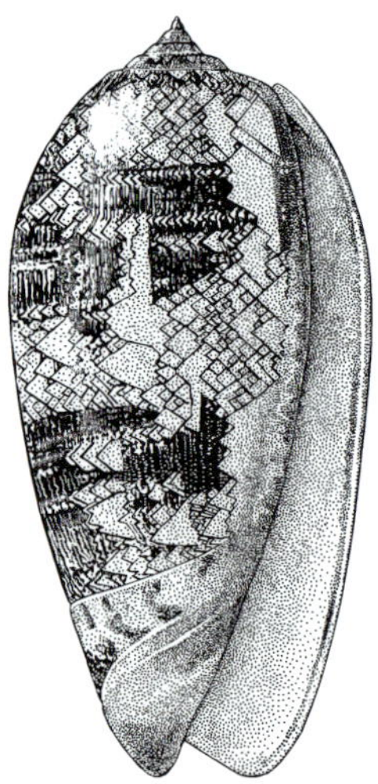

L 103 mm

The creamy white shell of this very uncommon and striking large olive snail is beautifully patterned with rust-colored designs.

The snail was brought up in a net in a sandy bay near Cerro Tepopa and was not recognized by name by anyone who saw it then. Later, the descriptive name ***xpaḻeemelc caacoj*** 'large ***xpaḻeemelc***' was associated with the shell after another was found. The name is shared by *Oliva incrassata* and *Strombus galeatus*.

Oliva spicata

haxölinaail comihj 'polished shell'
xpaḻeemelc comihj 'polished ***xpaḻeemelc***'

This common snail is often collected alive on intertidal sand flats. Its attractive, buff-colored, delicately marked and highly polished shell is a favorite for stringing in shell jewelry.

The names ***xpaḻeemelc comihj*** 'polished ***xpaḻeemelc***' and ***haxölinaail comihj*** 'polished shell' are clearly descriptive.

The shell of this snail was used as part of a shell/feather throwing toy, in the same way as other shells of similar size and shape (see *Oliva incrassata*).

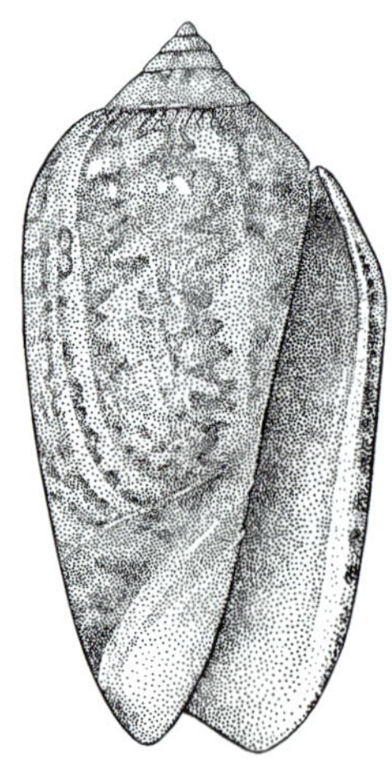

L 50 mm

Olivellidae (dwarf olive snails)

The most heavily used shells in jewelry were those of the *Olivella* snails, most often the very common *Olivella dama* and the smaller *O. steveni*. Used much less were the tiny *O. alba*. Another, unidentified, snail described by Juanita Herrera as being similar to the ***xiica cooxp*** (*Olivella dama*) but with a more elaborately marked shell, was collected in the past from an estero on Isla Tiburón. The snail's name, ***xiica cooxp coospoj*** 'spotted ***xiica cooxp***', alludes to the shell's more prominent markings.

Olivella alba

xiica cooxp heecto 'small ***xiica cooxp***'

shell L 4 mm

This tiny, white-shelled olive snail was known only from a few beaches.

In the past, the shells were strung into necklaces, though rarely, since gathering and preparing the snails was tedious. When I showed a necklace made with these shells to a group of women, one recalled by name a woman who long ago especially liked to work with such tiny shells.

Olivella dama

mas (var. ***nas***)
sayana
♦ ***xiica cooxp*** 'white things'
xiica hant cosítyalca 'things that scratch (slit) the land'

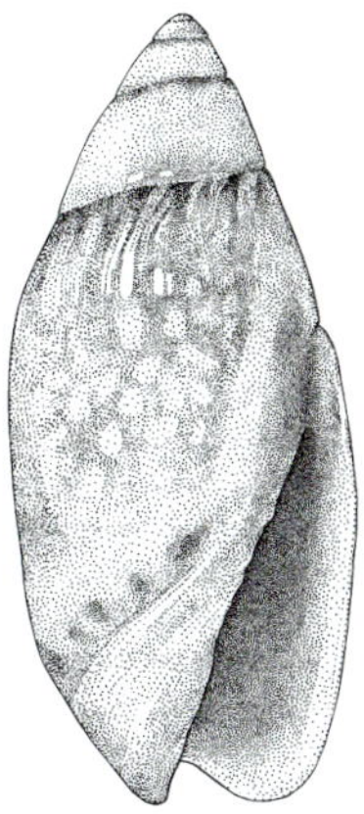

L 19 mm

Olivella dama is probably the most common olive snail in the area. Widespread and abundant on intertidal sand flats, these sturdy-shelled snails are collected in large numbers at low tide when their slit-like tracks are easily seen.

The primary name ***mas***, with a variant ***nas***, is an old name for the snail. According to Cleotilde Morales and others, when a Seri man named *Mas* died and his name was no longer spoken, the descriptive name ***xiica cooxp*** 'white things' was then given to the snail.[45] An anecdote perhaps referring to the same individual describes a man named *Mas Quiho* 'who finds/sees ***mas***' because of his penchant for gathering the snail—a very unusual thing for a man to do (see Gathering Methods, p. 52).

Almost forgotten today is the shell's very old name, ***sayana***.[46]

Xiica cooxp 'white things' is the snail's most common name. The plural name is used even when referring to a single specimen (see Names and Naming, p. 25). An older name, ***xiica hant cosítyalca*** 'things that scratch (slit) the land', aptly describes the snails' tracks in the sand (figure 5.19).[47]

The snail was not eaten; as one person commented, eating its meat caused blindness. However, others remarked that eating it was said to help with weight loss.

Figure 5.19. 'Things that scratch the land': *Olivella dama* snails leave their tracks at low tide. Desemboque, 2013.

Figure 5.20. Bleaching *Olivella dama* shells in hot sand. Desemboque, 2003.

From early collections and photographs it is clear that *Olivella dama* shells were some of the most commonly used for personal adornment in the past; the Arizona State Museum collections show the shells often included in necklaces and other shell objects. Victoria Astorga recalled that her grandmother, Angelita, told her that long ago the people sold necklaces made with the shell in the Sonoran state capital of Hermosillo.[48]

Both natural and bleached shells are strung. When used in their natural state (although they are boiled in order to remove the animal), the shells are referred to as ***xiica cooxp*** *quis* 'raw ***xiica cooxp***'. Shells are bleached by toasting in hot sand (figure 5.20) and are then referred to as ***xiica cooxp*** *cmam* 'cooked ***xiica cooxp***'. After the shells are toasted, sometimes they are whitened further in a commercial bleach solution. The animal of either the bleached or the unbleached shell is then extracted with a needle. The shell's apex is removed by tapping it with a rock or filing it with a metal file or sand paper; the apices of tiny snails are snipped off with nail clippers.

Olivella steveni

mas
♦ ***xiica cooxp*** 'white things'
xiica coxpísotoj (var. ***xiica cooxp coosotoj***) 'narrow ***xiica cooxp***'
xiica hant cosítyalca 'things that scratch (slit) the land'

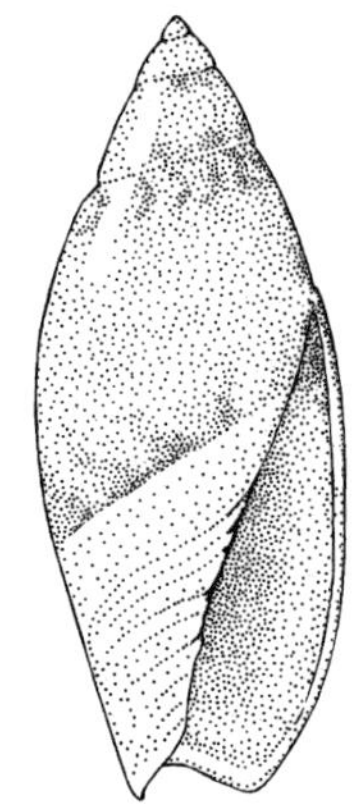

L 10 mm

The shell of this small, delicate olive snail has fine light-brown markings.

The snail shares the names ***mas***, ***xiica cooxp*** and ***xiica hant cosítyalca*** with the similar *Olivella dama*. However, the name ***xiica coxpísotoj*** 'narrow ***xiica cooxp***' distinguishes *O. steveni* from the former, which is larger. A variation of the name, holding the same meaning, is ***xiica cooxp coosotoj***.

This shell is prepared and strung on necklaces in the same way as *O. dama*, but the finished necklaces are much finer because of the shell's smaller size.

Mitridae (miter shells)

Mitra belcheri

◊ ***haxölinaail cpotyoj caacoj*** 'large ***haxölinaail cpotyoj***'

L 125 mm

The high-spired shell of this distinctive, large snail is covered with a thin dark periostracum and has a cream-colored aperture. The snail is from deeper water and is seldom collected.

The name ***haxölinaail cpotyoj caacoj*** 'large ***haxölinaail cpotyoj***' was given to this snail because of its resemblance to the ***haxölinaail cpotyoj***, a smaller and more common (unidentified) intertidal snail. The term *cpotyoj* describes the ridged sculpture on the shell's whorls.

The shell was given to me by Francisco Morales in 2008 after his son brought up the snail in a net offshore from Isla Tiburón. I have since seen another shell displayed outside of a Seri house, kept as a curiosity.

Mitra tristis

The shell of this snail is a dark chocolate brown, banded with white. It is not common and is usually found worn, in beach drift.

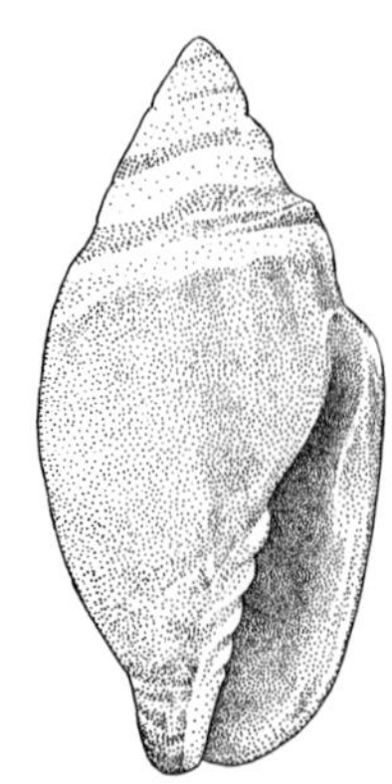

L 24 mm

Conidae (cone snails)

Any cone snail is most often referred to as ***xpa<u>l</u>eemelc***. Other snails that include the term in some of their names are *Strombus galeatus*, some of the olive snails, and a few other gastropods (see table 3.2, p. 24). The snails were not eaten.

A few people playfully call the cone snails *naapxa iif* 'the Turkey Vulture's beak', humorously noting the similarity in shape. This is not a commonly used name.

It is interesting to note that no one I asked was aware of the toxin carried by the cones, nor of the radular dart that delivers it.[49] One might assume that no Seri in recent memory has been harmed by the snail, even though live cones are collected for stringing.

Conus princeps

hant iiha cooml 'what investigate the land'
iicj 'fine sand'
mosniil yaaspoj 'the *mosniil* sea turtle's design'
♦ ***xpa<u>l</u>eemelc***

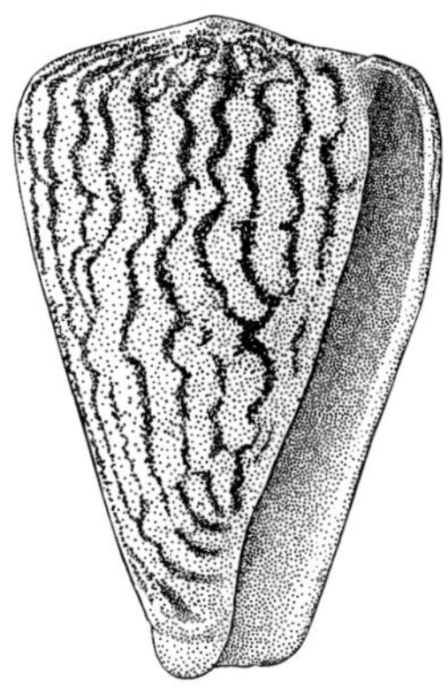

L 42 mm

The attractive shell of this good-sized cone is orangish-pink, marked with brown wavy designs. The shells are usually found only in beach drift; larger less-worn shells are sometimes washed onshore as part of the bycatch from commercial shrimp boats.

Although sharing the name ***xpa<u>l</u>eemelc*** with other cone snails, this less common cone has the distinction of having three names of its own, perhaps because it stands out by its markings and color.

The shell acquired its name ***hant iiha cooml*** 'what investigate the land' from the time when living things were being named. After the land was formed, it was still soft. The *Hant Iiha Quimx* sent out certain creatures to test the land to see if it had hardened. The first was the daddy longlegs, since it was so light on its feet, and the second was the prince cone, because it glided along slowly and carefully on the land's surface.

The name ***iicj*** is also the word for fine sand. The connection is unclear.

Mosniil yaaspoj 'the *mosniil* sea turtle's design' alludes to the patterns on the shell. *Mosniil* is a variety of black sea turtle that is no longer seen.

Like other less common shells, the shell is occasionally hung as a pendant on necklaces. It was also recalled as being used in a shell/feather throwing toy (see *Oliva incrassata*).

Conus regularis

xpa<u>l</u>eemelc

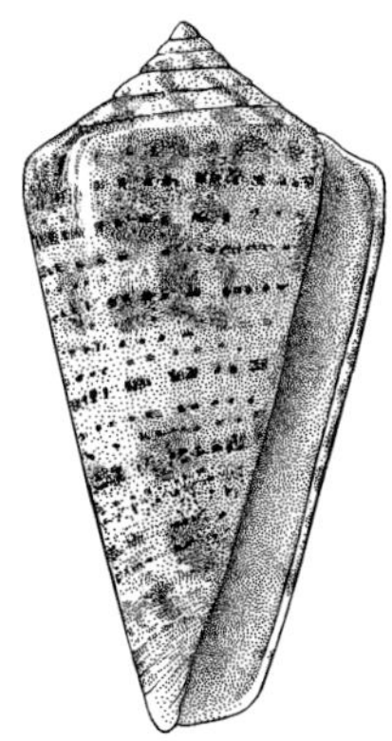

L 45 mm

The cream-colored shell of this medium-sized cone is marked with buff and brown spotted designs. It is common in beach drift.

The snail is just referred to by the generic name ***xpa<u>l</u>eemelc***.

Shells of medium-sized cones such as this one are sometimes hung as pendants on necklaces.

Conus ximenes

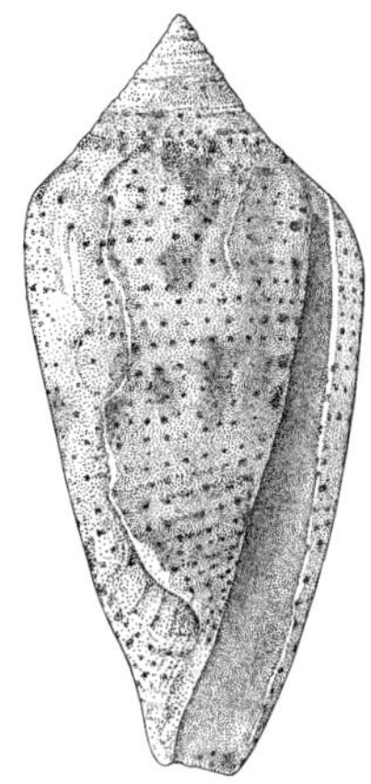

L 43 mm

mosnipol oopl 'what the leatherback sea turtle licks'
♦ ***xpal̲eemelc***

This small cone's shell is white with delicate purple markings and a light purple aperture. Commonly collected both in beach drift and live on sand shores at low tide, in recent years the shell has become popular for stringing in shell jewelry.

Besides sharing the generic name ***xpal̲eemelc*** with other cone snails, the snail was given the name ***mosnipol oopl*** 'what the leatherback sea turtle licks' by Cleotilde Morales, who said that her aunt and mother told her the name. Pointing out the shell's spots, Cleotilde explained, "The *mosnipol* goes down onto the ocean floor and licks this snail—maybe it's because of its markings!"

Terebridae (auger shells)

Terebra ornata

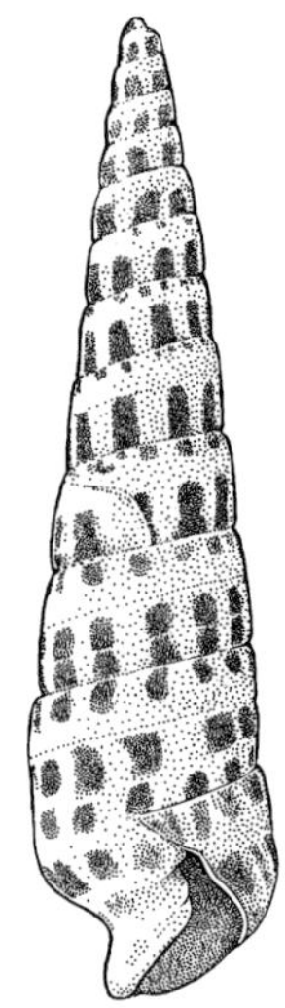

L 83 mm

◊ ***xtapacaj coospoj*** 'spotted ***xtapacaj***'

This snail with its distinctively spotted white shell has been recently rediscovered on shores bordering the Seri region to the north.

The name ***xtapacaj coospoj*** 'spotted ***xtapacaj***' notes the auger shell's similarity in shape to the ***xtapacaj*** (turret snail), but also reflects its more ornate shell.

Because of its large size and being found in fewer numbers, the shell is not often used for stringing on necklaces, although larger and distinctive shells such as this are sometimes hung as pendants.

Terebra variegata

◊ ***xtapacaj heeque*** 'small ***xtapacaj***'

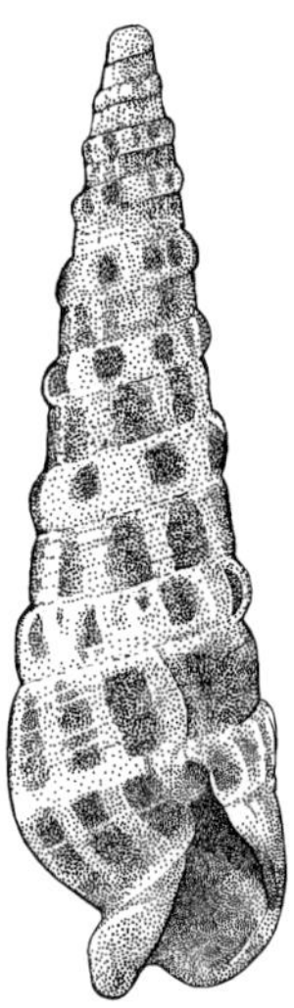

L 33 mm

This small snail with its attractively marked shell is found on shores bordering the Seri region to the north. The shells are strung as pendants on necklaces.

The name ***xtapacaj heeque*** 'small ***xtapacaj***' is shared by other snails of similar size and shape (see Cerithiidae, Potamididae, and Turritellidae), and is a recently given name.

Turridae (turrids)

Crassispira incrassata

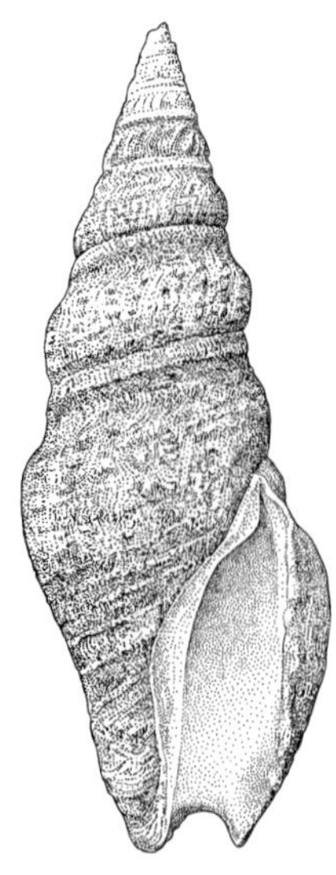

L 52 mm

oot ixpal̲eemelc 'the coyote's ***xpal̲eemelc***'
ziic coopol ixpal̲eemelc 'the Double-crested Cormorant's ***xpal̲eemelc***'

This snail has a dark, grooved shell with a white interior.

Not a common snail, it was given the name ***oot ixpal̲eemelc*** 'the coyote's ***xpal̲eemelc***' by one person, a name shared by *Agaronia testacea*. Most people, however, did not recognize the snail by name.

When Eva López supplied the name ***ziic coopol ixpaḻeemelc*** 'the Double-crested Cormorant's ***xpaḻeemelc***', she explained, "The *ziic coopol* (*Phalacrocorax auritus*) eats this snail; he gets heavy, and dives down under the water and eats fish!"

Knefastia dalli

hant yapol ipapl 'necklace of the dark place'

L 55 mm

The shell of this snail is tan with brown markings; the aperture is stained with a light orange-brown. The spiral ribs are somewhat knobby.

The shell was found in beach drift on the eastern shore of Isla Tiburón, and Cleotilde Morales and Alfredo López provided the name ***hant yapol ipapl*** 'necklace of the dark place' at that time.[50] As they explained, *hant yapol* 'dark place' refers to a dark area in the sea, as a rocky sea floor would appear from the surface. The connection to the shell's name is unclear.

Shells of this snail and other similar shells are sometimes hung as pendants on necklaces with other more common, smaller shells.

Architectonicidae (sundials)

Architectonica nobilis

itx imonaaaij 'whose body whorl doesn't return'

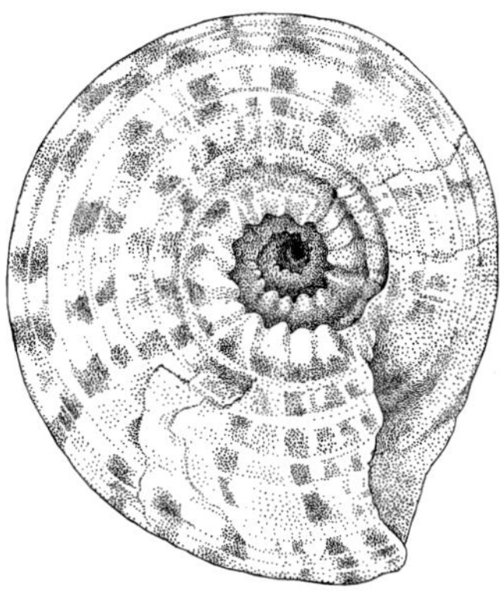

L 20 mm

The low-spired shell of this very uncommon snail has a conspicuous, deep, and beaded umbilicus. The shell is white, marked with light brown designs.

The shell's distinctive umbilicus suggests the snail's name, ***itx imonaaaij*** 'whose body whorl (literally, buttock) doesn't return'.

This was a very special shell. When I showed one to an older woman, she exclaimed over it, not only for being uncommon, but because of what it meant to her as a mother. The shell was strung through its center and hung on the neck of the *caal*, the youngest child of a pregnant mother. The amulet was said to protect the *caal* from *Coen*, a malevolent spirit who especially targeted such a child (see Beliefs and Practices, p. 41).

In describing types of amulets, María Luisa Astorga recalled being told that her grandmother had strung ***itx imonaaaij*** shells on the necks of her children.

Heliacus bicanaliculatus

This small uncommon snail has an attractive shell beaded with dark and light brown.

Found on a remote, little-disturbed beach, the snail's shell is a recent component in necklaces, though rarely. When strung, it resembles tiny Seri olla-shaped baskets. I had not seen the shell used until it was given to me on a necklace made by María Luisa Astorga in 2008. María Luisa had no name for it, but likened it to an ***haapx*** (*Tegula rugosa*) snail.

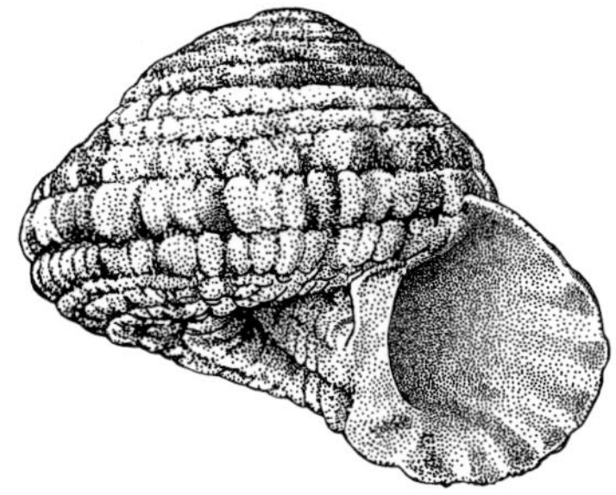

L 12 mm

Bullidae (bubble snails)

Bulla gouldiana

cacaapxom 'what fattens [something]'

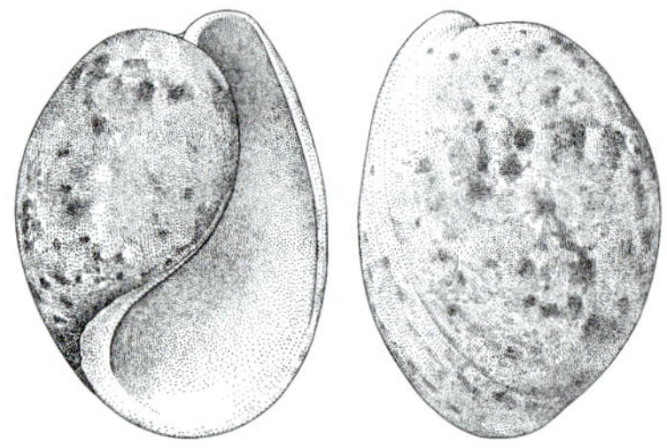

L 32 mm

The shell of the bubble snail is thin, delicate, and smoothly globular. Its interior is white and its exterior a light brownish-violet, splotched with darker patches.

According to Eva López, this snail received its name because of what it said:

> The Name Giver was there. Lots of people were there—shells, animals, large land mammals, they were all there. This thing named itself. It said, "*He hyacaapxom iha,*" 'I fattened them [referring to the others present]'. That is what it said. So [the Name Giver] said, "*Me cacaapxom inspáh aha.*" 'You will be called ***cacaapxom***'.

Although most people said that the meat of this snail was not eaten, one person said that it was considered to be good food by the Ancestors, but was not eaten now: "Those who ate it grew fat." This is perhaps connoted by the snail's name.

According to Cleotilde Morales, the bubble snail was "just played with." Children would flick the live snails at each other with thumb and finger, "like shooting marbles," and because the animal itself was sticky, the shells would adhere to their target.

The shells are occasionally strung in necklaces. Eva López described a past use:

> The Ancestors wore these as earrings. They would gather lots of these and make ear ornaments. They are light. They would pierce them and make ear ornaments. When they were made into ear ornaments it seems that they were really pretty. They were strung on threads; the wind would blow them and they would really make a sound.

When talking about interesting things he has seen in the sea, Efraín Estrella mentioned that he had more than once seen a *copsiij cooil* 'blue jellyfish' (*Physalia* sp.) with its tentacles in contact with a live *Bulla* snail.

Aglajidae (aglajas)

Navanax inermis

hant iti queemij 'what moves slowly on the land'

This odd little creature was found in the shallow waters of Estero Santa Cruz. In explaining the name ***hant iti queemij*** 'what moves slowly on the land', Cleotilde Morales described how the creature glides across the submerged sand, its wing-like appendages waving slowly as it moves.

approx. L 120 mm

Aplysiidae (sea hares)

Aplysia californica

hatx cöcazoj

approx. L 300 mm

The sea hare is an internally shelled gastropod (or "naked snail") with flaccid grayish-green flesh that, when disturbed, releases a violet ink. The creature is a graceful thing to watch as it moves slowly across sandy, seaweed-covered areas in shallow water.

The etymology of the name ***hatx cöcazoj*** is not clear, and no one I asked could provide an explanation for the name. The word *hatx* is clearly the unpossessed form of 'rump' or 'buttocks'. The second part is an inflected form of the word *cazoj*, meaning 'one' or 'alone'. The name ***hatx cöcazoj*** was also given to a small sea slug, *Oxynoe panamensis*.[51]

Probably because the sea hare is generally left alone, few people I questioned were aware of its small internal shell. Those who did know of it gave it the name *iixquij* 'its cartilage', a term also used for the internal pen of the squid.

The people did not eat the sea hare; as someone explained, it was not considered to be made of flesh, because it was just full of the reddish *ix* 'its liquid'. Seeing them eaten was remarkable enough that one consultant recalled hearing that long ago a Yaqui man the Seris called *Cmaacoj Cmaaitj* 'Soft Old Man' lived with the people and enjoyed eating the animal. It is unclear if there is a connection between the physical characteristics of the creature and the man's name.

The sea hare was used as bait when no other was available, although it is not preferred (see Bait). The animal is eaten by sea turtles; in the 1970s, one woman, when butchering a sea turtle, found seventeen sea hares in its gut. She opened one of them, saw that it had eaten eelgrass (*Zostera marina*) and quipped, "Oh look, they eat eelgrass too!" (as did the Seris, who used the seeds).[52]

The color violet or purple is *hatx cöcazoj ix cmis* 'like the liquid of the ***hatx cöcazoj***'. Because the color of its ink is similar to blood, a girl or woman of childbearing age was cautioned against handling or stepping on this animal, as it would cause her to have abnormal menstrual flow or to hemorrhage during childbirth.[53]

In two instances of which I am aware, the ink was used to dye limberbush fibers used for making baskets. One woman was known to have once dyed some of the fibers by boiling them in water colored with the ink, and incorporated them in a basket design. Such use was noteworthy, since other dyes used were plant-derived. More recently, Raquel Moreno told me that she had once experimented with the ink and had made a small basket with the dyed fibers.

Angelita Torres sang about the *honcaacoj* (Yellow-footed Gull, *Larus livens*) and the ***hatx cöcazoj***:

Honcaacoj yas

Hatx cöa zó ma pa txí
Hatx cöcazoj immaptx
Hatx cöa zó ma pa txí
Hatx cöcazoj immaptx
I qui ya ma pa txí
Iiqui ihyamaptx
I qui ya ma pa txí
Iiqui ihyamaptx.

Song of the *honcaacoj*

The sea hare bursts,
The sea hare bursts,
I burst it!
I burst it!

As Angelita explained, the gull is pecking at the sea hare, whose violet ink stains the gull's feet. Another gull comes and wants to take the sea hare away (to eat), and the first gull sings this song.

Pleurobranchidae

Berthellina ilisima

xepenozaah 'sun in the sea'

approx. L 25 mm

This striking orange slug is found on the underside of rocks in the low intertidal zone.

The name ***xepenozaah*** 'sun in the sea' alludes to the slug's glossy, vivid orange color.[54] The name is shared with *Limaria pacifica*, a swimming clam that has bright orange tentacles.

This uncommon slug was not eaten. When I showed María Luisa Astorga a photograph of one, she laughed and said that it reminded her of canned apricots and made her hungry.

According to several women the slug was used medicinally.[55] The slug was dried and later boiled and the liquid drunk to stop hemorrhaging.

In an anecdote from Seri oral history, long ago some of the people traded for food from a vessel passing through the Gulf—food that the people had never seen. There were sacks of white things, which they referred to as *potaat cmis* '[things] like maggots' (most likely rice) and other objects they called *xepenozaah cmis* '[things] like a ***xepenozaah***'. Among several nonnative fruits that have been suggested (including the persimmon and tomato), the most likely candidate for being carried on such a voyage is oranges. As a matter of fact, on a map of Isla Tiburón an intriguing handwritten note made by Edward Moser near the island's southwest shore reads "rock underwater, 'oranges' wreck." The Seri word for the orange is *sahmees*, a term not found in the vocabulary of any other group in the area. It is possible that through time, a shortened version of ***xepenozaah cmis*** might have been *zaah cmis* 'like the sun', from which it is no great leap to arrive at the word *sahmees*. In fact, one Seri family still pronounces the orange's name as *zahmees*.

Plakobranchidae

Usually seen only at extreme low tide, these slugs are not recognized by name today. When I showed a photograph of the *Elysia diomedea*, a slug I found in Desemboque, to Angelita Torres, she recalled seeing such creatures on the shores of Isla Tiburón during her childhood, but was not aware of a name for them. María Luisa Astorga had seen them, but also said that she could not recall a name. She added that some people might say that such things were from somewhere else, like the deeper sea, or might be grouped with other small creatures under the name *xepe án ihicoomz* 'pinworms of the sea area' or *xepe án imiipla* 'the badness (worthless things) of the sea area' (things that are not used nor eaten—the "junk or leftovers").

Raquel Moreno said that her father, José Juan Moreno, had once told her of a creature he called *xpanaams ano coom* 'what lies in the seaweed', and she suggested that perhaps he had referred to the slug.

Elysia diomedea

approx. L 25 mm

This is the most commonly found slug in the area. The beautiful creature's colors and delicate frilled mantle give it a rather flamboyant appearance, evoking the source of its common English name, Mexican dancer.

Nudibranchs

The nudibranchs are not named by the Seris, but are grouped with other such small creatures as the "leftovers."

When talking about odd creatures, María Luisa recalled being at a beach north of Desemboque at low tide, many years ago. She found a light-colored creature about three inches in length that she had never seen before. She described it as having wavy filaments on its back and a firm texture that reminded her of a cantaloupe. She said she was curious enough to dissect the creature and added that she never saw another. Since her description was clear, I showed her photographs of different sea slugs and nudibranchs. When she saw the nudibranch *Cadlina luarna*, without hesitating she identified it as the one she had found.

Onchidiidae

Onchidella binneyi

tamax

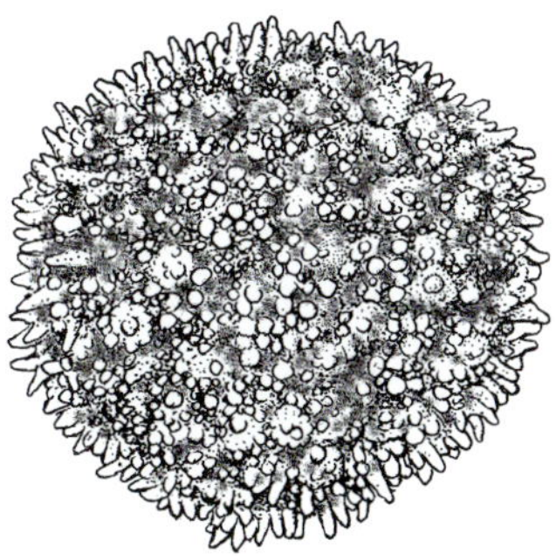

approx. L 20 mm

This small, somewhat warty-looking slug is found beneath rocks in the low intertidal area, often together with sea cucumbers (*Holothuria* sp.) and chitons. When disturbed, the papillae on its perimeter exude a bluish-white secretion.

Tamax is a primary name. As expected, since the creature does not have a shell, it is not considered as *haxölinaail* 'shell'.

According to the oldest people today, these were collected at low tide, boiled, and eaten. While recalling her aunt cooking them, Victoria Astorga described how a milky fluid dripped from the creature when squeezed. In a handwritten note Mary B. Moser recorded being told by an older woman, Chona Burgos, that the slugs were common on the shores of Isla Tiburón, and were eaten raw by the Ancestors "when there was nothing else to eat."[56]

Figure 5.21. *Onchidella binneyi* slugs on an intertidal rock. Desemboque, January 2007.

The name ***tamax*** occurs in the name for the Bufflehead Duck (*Bucephala albeola*): ***tamax*** *capóc* 'what pries off ***tamax***'. A note by Edward Moser records that the people say that the duck eats this slug.[57]

Hydrobiidae

xojmás

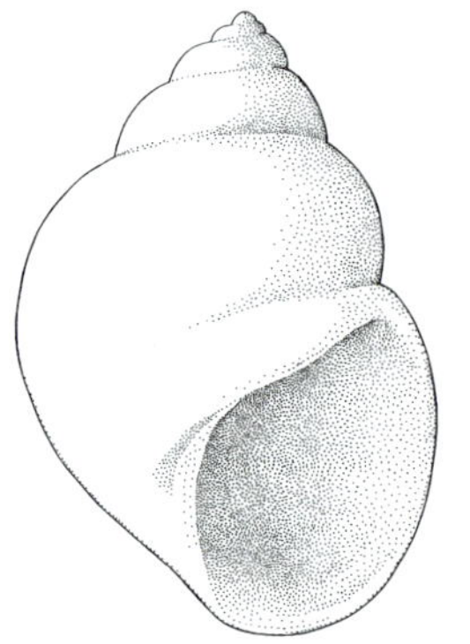

L 5 mm

This tiny snail has a fragile, translucent shell with inflated whorls. The snail inhabits fresh or brackish water environments, uncommon in the Seri area.

An unusual necklace made with the shells is in the Seri collection at the Arizona State Museum (figure 5.22). I have not seen the snails elsewhere. It is intriguing to wonder about their origin and the fact that someone took the time and care to string these tiny shells, although I have been told that certain women favored working with shells of this size, such as those of *Olivella alba* (see Olivellidae).

Figure 5.22. Detail of a necklace made with the tiny snails of the Hydrobiidae family. Collected by Wilma Kaemlein in 1969. Arizona State Museum, catalog no. E-8294.

Although the snail is not recognized today, the name ***xojmás*** was given by those who saw its photograph, as they commented that it did not look like a marine snail.[58]

Ellobiidae

Melampus mousleyi

◊ ***haxölinaail comihj*** 'smooth shell'
◊ ***xepe yamaasij heeque*** 'small ***xepe yamaasij***'
◊ ***xnaa caaa ano haxölinaail*** 'shell in the *Salicornia*'

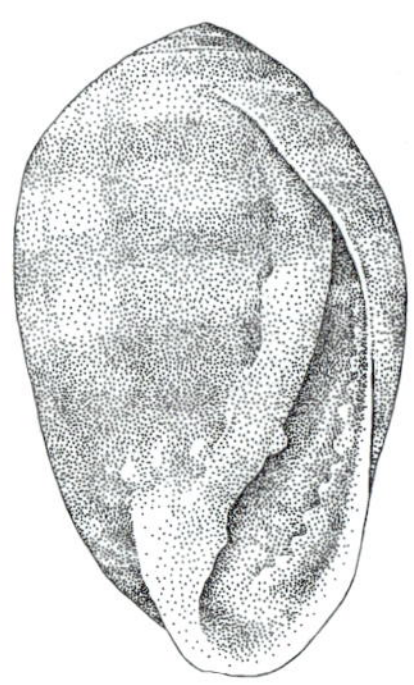

L 15 mm

These mud, or mangrove, snails are abundant on sandy mud flats bordering estero channels, where their empty shells are found heaped together with shells of *Cerithidea mazatlanica* among salt-tolerant plants.

This shell is not readily recognized, and the names given are not well used. The name ***haxölinaail comihj*** was given by one person, although that name is more commonly used for *Oliva spicata*. The term *comihj* often implies 'polished', which describes well the *O. spicata* shell, but it can also mean 'smooth', a more appropriate translation in this case, as the *Melampus mousleyi* shell can hardly be described as polished.

One woman recalled seeing heaps of these shells near Estero Sargento and referring to them as ***xnaa caaa ano haxölinaail*** 'shell in the *Salicornia*'. *Xnaa caaa* (pickleweed, *Salicornia bigelovii*) is one of the common salt-tolerant plants in the area.

Cleotilde Morales provided the name ***xepe yamaasij heeque*** 'small ***xepe yamaasij***', noting its similarity to the much larger shells of that name, notably *Strombus galeatus*.

Figure 5.23. A hatband that includes *Melampus* sp. shells. Collected by James Manson in the mid-twentieth century. Arizona State Museum, catalog no. E-8506.

Today the shell is only occasionally used, in shell jewelry. One woman was remembered as having worn a necklace made of the shells; apparently it was unusual enough to stand out. A hatband in the Arizona State Museum collections includes this shell strung with *Liocerithium judithae*, *Trivia solandri*, and *Zonaria annettae* shells (figure 5.23).

Succineidae

an unidentified terrestrial pulmonate snail, possibly *Succinea* sp.
xojmás

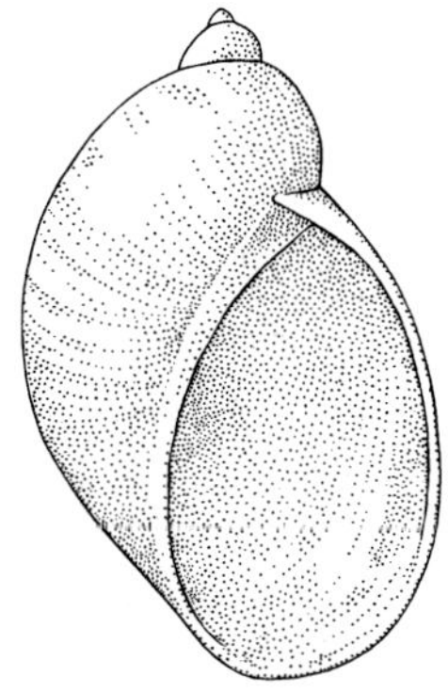

L 19 mm

The terrestrial pulmonate snails are associated with some moisture. Since a stable supply of fresh water is scarce in the area, the snails are sometimes found where water has pooled for a time after rain, such as dry lake beds. The shells are fragile and easily crushed. Since such snails are not seen as often as marine snails, they are sometimes kept as curiosities.

The name ***xojmás*** is probably a compound of *hax ano* ***mas*** '***mas*** (*Olivella dama*) in fresh water'. The name relates the snail to the marine snail *Olivella dama*, which has ***mas*** as one of its names.

Bulimulidae

an unidentified terrestrial pulmonate snail, possibly *Rabdotus* sp.
xojmás

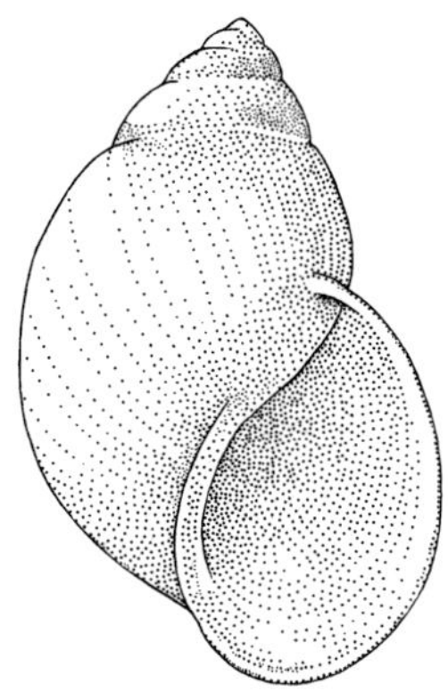

approx. L 30 mm

The snail's shell was found on Isla Tiburón and carefully kept by a Seri woman.[59] The name ***xojmás*** reflects its similarity to *Olivella dama* (see Succineidae).

Helminthoglyptidae

Sonorella sp.

a talus snail
xojmás

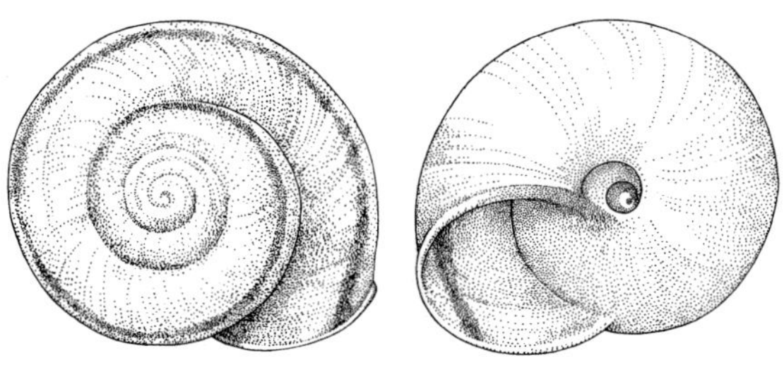

L 9 mm

This is a terrestrial pulmonate snail associated with rocky terrain. The shell of the live snail is described as tan or gray, with white markings; the shell illustrated is marked with a reddish-brown band.

The name ***xojmás*** is evidently a compound formed from the name *hax ano* ***mas*** '***mas*** in fresh water' (see Succineidae).

The snail, considered unusual since it was not found near the sea, was said to occur in moist places at higher elevations. A high mountain on Isla Tiburón is named *Xojmás Queaaf* 'waist cord of the ***xojmás***' because the live snail was found in areas encircling the mountain, evoking a waist cord made of shells. Raquel Moreno remembered her grandmother once collecting the shells from a mountain north of Desemboque and stringing them to sell to tourists: "When they weren't sold, people would wear them to a fiesta [as if they were special]." Raquel explained that when the snails were found alive, they were boiled to remove the animal. The shells were pierced and strung with *Olivella dama* shells and stem sections of reedstem milkweed.

Victoria Astorga recalled walking part of the way up Cerro Tepopa long ago with her young daughter, Ramona, and finding a deep narrow canyon in which a huge mesquite tree was growing and from which a great number of doves flew. There they found a moist area with a lot of green moss or algae and many ***xojmás*** snails. The shiny trails of the snails covered the rocks, and Victoria and Ramona were amazed to see snails that were not in the ocean and that squirted liquid that was not salty. Victoria gathered some live snails that she described as having white and light violet shells, and put them in her waist-bag to take back to camp to show her husband. When she arrived, the snails' shells were all crushed, they were so fragile. She and her daughter were afraid of that strange place because Ramona suggested that a large snake lived there, and they never went back.

POLYPLACOPHORA

Chitonidae (chitons)

Chitons are distinguished by their eight shell plates and leathery girdle. In the intertidal zone they are often associated with sea cucumbers and the pulmonate slug *Onchidella binneyi*, attached to the underside of intertidal rocks.

Although the chitons are very different in form, they are included with other shelled mollusks as *haxölinaailc* (see Classification and Naming).

Figure 6.1. Chitons on a *Pinna rugosa* shell, Isla Tiburón, 2009.

Chiton virgulatus

♦ ***capoclim***
capoclimiil 'blue/green ***capoclim***'
tamax capoclim

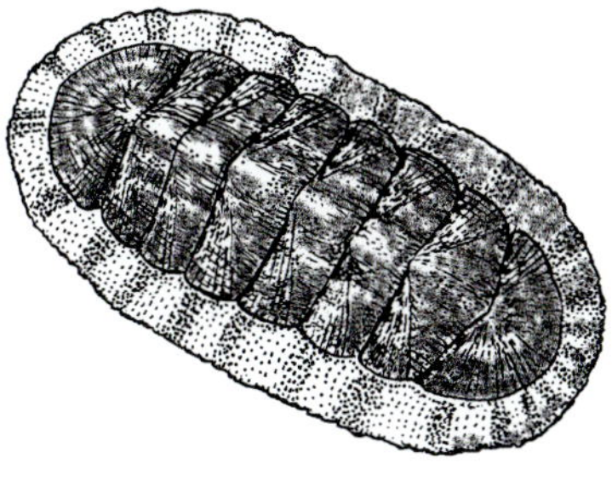

L 53 mm

This greenish chiton is perhaps the most common in the area.

All species of the family are most commonly known by the name ***capoclim***. The name is related to the word *capocj*, which means to pry off something, such as a cactus fruit or the shell of a roasted sea turtle. However, the actual form of the name is anomalous.

The name ***capoclimiil*** 'blue/green ***capoclim***', given to *C. virgulatus*, describes the chiton's greenish color.[1]

An older name for any chiton, rarely used today, is ***tamax capoclim***. Other creatures that include *tamax* in their names are the clam *Codakia distenguenda*, the cup-and-saucer snails *Crucibulum* spp., the pulmonate slug *Onchidella binneyi,* and the glass-hair chiton *Acanthochitona exquisita*.[2] It is interesting to note that with the exception of the clam, all occur in the same habitat and are somewhat similar in form.

According to Sara Villalobos, chitons were eaten in the past, "when there was no food." More recently a few other older people recalled having eaten them. Victoria Astorga described their taste as similar to ***satoj*** (mussels, *Modiolus capax*). María Luisa Astorga remembered seeing her grandmother boiling and eating chitons at a

Figure 6.2. A chiton shell plate with other food shells in a wind-exposed hearth on a seaside dune. Desemboque, 2007.

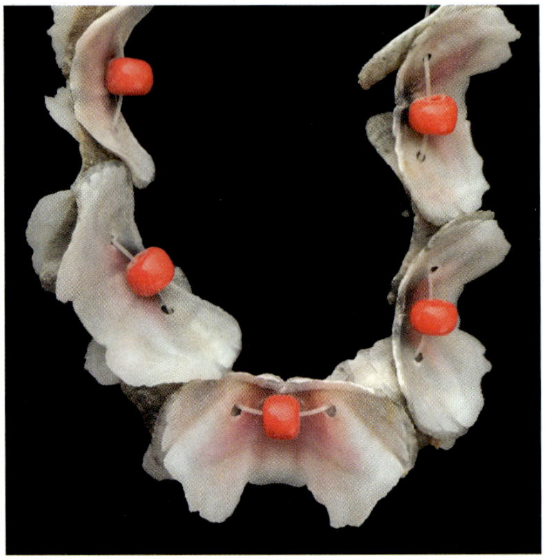

Figure 6.3. Detail of a necklace made with chiton shell plates.

camp at *Hast Itliitxo* (Las Cuevitas, between Desemboque and Puerto Libertad). Chiton plates are common in shell middens in the area (figure 6.2).

The chitons' plates are strung in shell jewelry. After the animal is removed by boiling, the separated plates, said to resemble butterflies, are pierced and strung. Such use is relatively new, perhaps begun around 1998. According to Raquel Moreno and others, the shells of the common *Chiton virgulatus* are not used for stringing, as they are harder and more narrow, making them more difficult to pierce. Instead, those of a pink-shelled chiton are used.

Acanthochitonidae (glass-hair chitons)

Acanthochitona exquisita

haxz heeque 'puppy'
♦ ***tamax cosi*** 'spiny ***tamax***'

approx. L 35 mm

The distinctive tufts of golden spines are what make this glass-hair chiton notable. It is found adhering to the underside of rocks in the low intertidal zone.

The name ***haxz heeque*** 'puppy' was supplied by Juanita Herrera, and is not a well-used name. She recalled as a child showing the chiton to her father, Antonio Herrera, who told her this name.[3]

The descriptive name ***tamax cosi*** 'spiny ***tamax***' references the chiton's spines. The name is also given by some to the spiny cup-and-saucer snail *Crucibulum spinosum*, which has a vaguely similar spiny surface.

The glass-hair chitons, as other chitons, were eaten in the past. They were cooked by boiling, which apparently caused their spines to become erect. This was one of the intriguing characteristics several people used to describe this creature to me before I had seen one.

I was first told about this elusive creature in 2005 by Alfredo López, who described it as being similar to ***tamax*** (*Onchidella binneyi*), but having spines. After a long search for *Onchidella binneyi* and a positive identification of ***tamax***, I set out looking for its counterpart, the ***tamax cosi***, assuming it to be another slug. When peering under stones at low tide for one last look on the final day of our stay in Desemboque in the fall of 2007, I chanced on an odd-looking grayish-green chiton attached near a larger *Chiton virgulatus*. I pried it off its rock for a closer look and was entranced when some formidable tufts of golden spines suddenly stood up around its edges as it curled up in typical chiton fashion. I found another, stowed them in a jar of seawater, and my husband and I drove the long bumpy coast road to Punta Chueca. When we arrived we found Alfredo López and his wife at home with a group of women standing around after selling necklaces to some just-departed tourists. I promptly placed one of my two spiny creatures in Alfredo's hand for identification. He just as promptly plopped it into his mouth and appeared to chew with great relish! While I was of course horrified that I would so quickly lose it, everyone else greatly enjoyed the humor. After removing the chiton and showing it to me whole, Alfredo identified it—a ***tamax cosi***, pure and simple. After one more foray into Alfredo's mouth and more laughter all around, the animal was put back into seawater, and I left very pleased with a final answer to my puzzle.

Other people to whom I showed the chiton did not recognize it by name. In fact, many said that they had never even noticed the creature—not surprising perhaps, as its inconspicuous shell plates are not used for stringing on necklaces.

Figure 6.4. Two *Acanthochitona exquisita* chitons (left) with a *Chiton virgulatus*. Desemboque, 2007.

SCAPHOPODA

Little recognized and not used, the tusk shells are known by one Seri name.

Dentaliidae

Dentalium sp.

cozaaij 'what makes a circular cut'

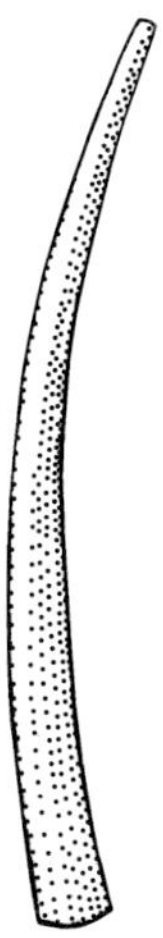

L 34 mm

Various species of scaphopods occur in the area. The tusk-shaped shells are white, with some having a finely ribbed sculpture. The mollusk lives buried vertically in subtidal sediments.

The people say that when this shell is stepped on and breaks off, it makes a circular and very painful wound in the sole of one's foot. The name ***cozaaij*** 'what makes a circular cut' is shared with small Vermetidae (worm snails), as their sturdy shells are said to leave a similar circular wound when stepped on.

Distinctive and uncommon in much of the area, the tusk shells were not recognized by many of the people to whom I showed them, and the few who did recognize the shells were unaware of any past use for them.[1] Angelita Torres described the mollusk as occurring together with pen shells and mussels on sand beaches at *Ziipxöl Iifa* (Las Víboras) and *Sacpatix* (a camp on the eastern shore of Isla Tiburón). The shells are most obvious in beach drift, though being very light and easily rolled away by the wind, they are not found in quantity.

Although tusk shells seem to be relatively unrecognized today, I know of one instance in which they were cached: around fifty of the shells had been saved in a woman's personal storage box collected by William Neil Smith in the mid-twentieth century (see Preliminaries).

CEPHALOPODA

Neither squids nor octopuses are considered as *haxölinaailc* 'shells', as they are not externally shelled. The squids were rarely eaten and are recognized by only one name. On the other hand, octopuses were a common food; four species have Seri names, and while others are described, the actual species referred to have not been identified.

Teuthida

Squid

pajaas

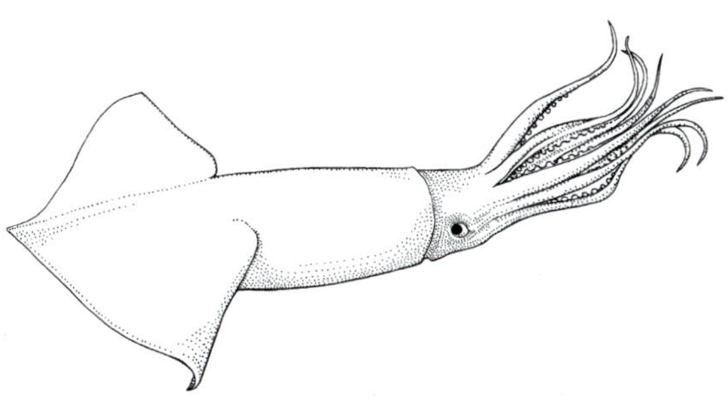

Any squid is called ***pajaas***; common squid species in the Gulf are *Lolliguncula diomedeae* and the large *Dosidicus gigas*.

The terms for the squid anatomy are for the most part those of the octopus (see Anatomy, p. 32) with the exception of the squid's fins, *íseloj* 'its pectoral fins'; the mantle, *ilít* 'its head'; the internal pen, *iixquij* 'its cartilage' or *itac* 'its bone'; and the raspers on the sucker rings, *inooslca* 'its claws'. When talking about the raspers and the beak (*itaast* 'its teeth') Efraín Estrella matter-of-factly informed me that, "When a boat sinks and [its occupants] aren't found, squids have eaten them." Unlike the suction cups of the octopus, the suckers of the squid are not strung for necklaces.

Live squids are not a common sight. Raquel Moreno laughed as she recounted a time as an adult when she saw a very strange creature swim by as she stood on the shore; she ran away screaming—it was a large squid, the first she ever saw.

Until commercial fishing in the area began in the early twentieth century, the Seris did not deep sea fish and did not fish for squid.[1] Other meat, such as venison and that of sea turtle, was abundant and favored; however, squids were occasionally eaten as well.

Figure 8.1. Squid raspers on a beached squid. Desemboque, 2010.

At certain times of the year, squids swimming close to shore were easily caught by hand. As Lorenzo Herrera explained:

> The *Comcaac* used to eat squids, but now lots of people don't know about that. They weren't hunted. On the west side of Isla Tiburón at a camp called *Heeme*, sometimes groups of large squids would swim to shore during the months of March and April. They were easily caught—some were too big to be carried by one man. The fins were eaten, but not the body or arms. The meat was [filleted horizontally] into thin slices. Sea salt could be collected on large rocky areas at that camp. The meat was salted and dried in the sun on racks [*hapatxam*] for four or five days. It could be stored and later cooked by boiling. After being cooked, the meat was squeezed to remove the liquid because it had a strange flavor. The meat was then eaten with sea turtle oil and flavored with salt and oregano.

The strange flavor or sensation referred to is also associated with the meat of ***cahiixona*** (*Crucibulum* spp.), ***oot isatoj*** (*Modiolus tumbezensis*),[2] and ***capoclim*** (chitons). The comment is made that "*Hateen hai cöimahoiitot*," 'It knocks the air out of the mouth' (see Food).[3]

Many people today do not eat squid, referencing its odd taste and odor. Others do enjoy it; one family was singled out as reportedly liking to eat the meat sliced thinly, dipped in flour, and fried, "like potato chips."

Octopodidae (octopuses)

Figure 8.2. An octopus sculpted in rock by José Ramón Torres, 2009.

Octopus spp.

hapaj
ziix cotopis 'thing that attaches by suction'

The generic name ***hapaj*** is given to any octopus. An archaic descriptive name is ***ziix cotopis*** 'thing that attaches by suction'.

Octopus meat was considered a good food. The octopuses were easily found during the cooler months in the rocky low intertidal area. When talking about octopus eggs, Lorenzo Herrera described the spent female guarding her eggs as being "languid, lightweight, and lacking muscle tone" and noted that she would not try to escape if disturbed. He added that such an octopus was not eaten, as its meat was rather "mushy."

The people say that octopuses eat small fishes such as *canaaajöa* (Gobiidae), clams, and small crabs. Several added that during times of high winds and heavy seas when sand fills the octopus lair and there is nothing for it to eat, the octopus eats its own arms (see *Octopus alecto*). One consultant noted that he has seen a large *Octopus bimaculatus* eating a smaller octopus of the same species.

The lairs were easily identified by the surrounding shell rubble (figure 8.3). When the octopus was not visible in its cave, the regular upward pulsing of the water's surface identified the occupant as an octopus rather than something to avoid, such as *Eurythoe complanata*, a stinging fire worm.

One's arm was inserted into the cave as far as possible, touching the octopus. With the other hand, the surrounding rocks were quickly removed until the animal could be grasped, with care taken to avoid its beak. When the octopus was secured, it was pulled out and killed by repeatedly throwing it onto the rocks. The mantle containing the visceral mass was then turned inside out and the internal organs (including the ink sac, which was not used) removed and discarded.[4] If many octopuses were carried together, they were strung through the muscle area at the base of their arms onto a loop of wire, or merely hung by their hooded mantles on one's fingers.

Octopuses were occasionally hunted for sale, and the people would sometimes travel by boat to distant beaches where larger octopuses were found, and in greater numbers. On February 15, 1950, such a trip ended in tragedy. Two boats were returning to Desemboque from a beach to the north, heavy with people and containers of octopuses. One boat returned safely, but wind and high waves of the incoming tide capsized the other, and nine of the ten people on board lost their lives.

Lorenzo Herrera described fishing by boat at night, when the shallow sea floor would be illuminated by a lantern hung from a boat's prow, and large fish were visible and easily speared. He recalled spearing good-sized octopuses moving along the sea floor, specifically mentioning *Octopus bimaculatus*, and noted that this was the only octopus he ever saw during such excursions.

When the meat was to be promptly consumed rather than sold, the octopuses were either boiled or, in the past, baked (*hapát*) under coals.[5] When baked, the meat could then be washed off (with either fresh water or in

Figure 8.3. An octopus in its cave (eye, center). Desemboque, 2013.

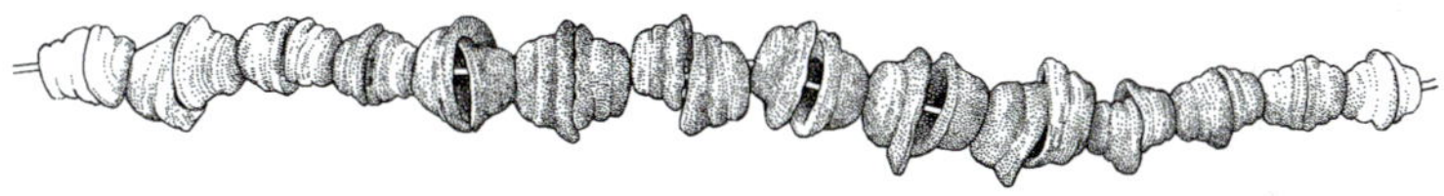

Figure 8.4. Dried strung octopus suction cups.

Figure 8.5. Detail of a photograph showing Angelita Díaz, wife of Manuel Encinas (right), with a young woman, who is wearing a necklace of octopus suction cups. The boy (left) is identified as Roberto Herrera, Angelita's grandson. Bahía de Kino, 1924. Photograph by Edward H. Davis, courtesy of the National Museum of the American Indian, Smithsonian Institution (N23800).

the ocean) and the dark outer skin removed. If not eaten immediately, the meat could be sliced and dried; salt was not used for preservation. It could then be stored and eaten like jerky or boiled again—when rehydrated by boiling, it was said to be just like the fresh meat.

The suction cups were removed after cooking and could be strung into necklaces,[6] often as an amulet for a child. Sometimes the child would chew and eat the dried suction cups directly from the necklace, or the suction cups could be rehydrated like the body meat. As one person quipped, "Dried octopus was the Seri *Maruchan* [a dehydrated noodle soup]!"

Octopuses were usually not consumed raw. However, someone remarked that it was known that "very long ago when people didn't have a fire, they ate octopus raw," perhaps referring to the "peculiar" Seris who lived on Isla San Esteban, who were said to have eaten much of their meat raw.[7]

Victoria Astorga talked about enjoying octopus meat and described how it was cooked. When discussing octopus and squid, she said that she had never seen a large squid, but recounted a time when something very large and dark swam by the shore, and her husband Miguel Barnett said it was the *itni* (the hump-shaped mantle containing the visceral mass) of an enormous octopus. Victoria laughed as she recalled that it scared them so much, they all ran away from the shore. In another instance relating to such a creature, it was rumored in the past that an enormous octopus lurked near *Hast Xnoois*, a small island rock (Morrito de Turner) near Isla Turners (see Mythology and Folklore). However, no one I spoke to recently said that they had ever seen such an octopus.

The raw meat of octopuses was cut up and used to bait hooks. According to Lorenzo Herrera, when fishing with a large hook in deep water, the complete animal was used, but with the *itni* removed, as the fishes would not attack the bait if it were whole. The fishes caught with octopus bait were *caanj* (*Mycteroperca jordani*), *zixcám coospoj* (*Epinephelus analogus*), and *zixcám caacoj* (*Stereolepis gigas*).

Borys Malkin describes the *Octopus bimaculatus* used as bait:

> The octopus, the informants said, is consumed by the baya, pinta, sardinera and pescara [*sic*], but not by the lisa, culon or birds. The reply suggests clear distinction between predatory and non-predatory fishes. How the Seris have arrived at this idea about the octopus is not clear, but they do use these animals to bait the above-mentioned predatory fishes.[8]

According to older people, the term *xpanaamalc* is 'octopus' in "coyote talk."[9] Victoria Astorga described the coyote trotting along the beach looking for food, especially octopus, and singing a song:

Oot quih Xpanaamalc Quícotim

Ha té o to, ha té o to, té sa cáo ti
Hatée oot o, hatée oot o, tiix sahcaao
Ha té o to, ha té o to, té sa cáo ti
Hatée oot o, hatée oot o, tiix sahcaao
Ha té xpa ná mal quí co tim
Hatée xpanaamalc quícotim
Ha té xpa ná mal quí co tim
Hatée xpanaamalc quícotim
Ho ho hó
Ho ho ho
Cmí que sma hái co tim
Cmiique smaháaicotim.

Coyote Hunts Octopuses

As for me, I'm a coyote; as for me, I'm a coyote. Go eat that one!
As for me, I'm a coyote; as for me, I'm a coyote. Go eat that one!
As for me, I'm hunting octopuses.
As for me, I'm hunting octopuses.
Oo Oo Oo
Nobody's going to have a share.

As Victoria explained, the coyote gleefully sings to himself, "I'm going to get an octopus, I'm going to get an octopus! And I'm not going to share it!"

In 2008 Angelita Torres recounted a time when she was young:

> I have seen a coyote hunting octopuses. When I was still a young girl, we were at *Xpanoháx* [Puerto Libertad]. When the tide was really low, a coyote was hunting octopuses. It dug into an octopus lair, put its face close and stuck its nose into the lair, and the octopus attached itself [by suction] to the coyote's nose. [The coyote pulled it out and] used its paws to scrape it off, and he ate it.

A translation of a text written by Xavier Moreno in 2010 relates a personal account that took place when he was a young man:

> One afternoon I said, "Tomorrow I am going hunting." So that same afternoon I got ready. The next morning before the sun came up, I left for the desert. I got far away, came to the mountain called *Hoinajeecoj* [a large hill north of Desemboque], went around it and kept walking.
>
> I wasn't alone; I had a companion with me. We went to lots of places, and every once in a while my companion would say, because he was tired, "Let's rest, then." We were returning to our camp, but it was still far away. We kept walking and came to the mountain called *Hast Yaxaajoj*, where we rested. We took advantage of some rocks there, and sat down on them to rest.
>
> I then saw that under the rock where my companion was resting, there was something that looked like a snake coming out from beneath the rock. I said to my companion, "Don't move! There's a snake coming out from beneath the rock you are sitting on. Don't move or it will bite you!" My companion just laughed and said, "You're just kidding, there's no snake!"
>
> I saw another thing come out that looked like a snake, and out spurted a bit of water from under the rock. Then I knew what it was.
>
> The animal was something from the ocean! It was an octopus. We got it and took it back to our home. But no one believed us. They didn't believe that we could have found an octopus in the desert and brought it home.[10]
>
> I started to think about it—how could an octopus have come to be in the desert, and I could have brought it to my house?!
>
> I asked my grandfather, *Cmiique* Roberto [Roberto Herrera] how it was that this could have happened. He said to me: "I too, have seen this. When I was with the people in the past, living on Isla Tiburón, [one time] I was hunting deer in the desert. I shot a deer and said, 'I hope it's fat!' I cut open the deer [along its breast bone] and saw that its fat was so thick and white that it resembled a mass of maggots. I cut its meat off and hung it on my *peen* [carrying yoke] to carry it.
>
> "I went around a mountain. In the distance I saw something running so fast I couldn't see its shape. I watched and waited to see what it was. When it got close I could see that it was a coyote. There was an octopus attached to its head/face. The coyote couldn't see where it was going—it was shaking its head back and forth, like it didn't know what to do. The coyote was so far into the desert that the octopus would then have dropped off in the desert, and would be found there. It's because of the coyote that one can find an octopus in the desert."
>
> This is what my grandfather explained to me. Now I understood how I could find the octopus in the desert.

In an instance related by Cleotilde Morales, her father Jesús Morales was far into the desert, near a mountain called *Hast Heen*, when he watched a coyote running crazily through the desert, an octopus attached to its head. The coyote couldn't see where it was going and kept running into cactus. Jesús made up a song,[11] sung by Cleotilde:

Há xo ja nó xo cá ma jöí
Haxoj án ano caamjö iha
Há xo ja nó xo cá ma jöí
Haxoj án ano caamjö iha
Há ti tá xo có co zi ya
Haa ntita x, coocöz iha
He ya he mo ca ta nó ta ma xo có co zi
Hehe án com ano ntaaca x, xocoocöz
He ya he mo ca ta nó ta ma xo có co zi
Hehe án com ano ntaaca x, xocoocöz.

On the [rocky] shore it is hunting.
On the [rocky] shore it is hunting.
There it goes trotting.
In the desert I am going, I am trotting!
In the desert I am going, I am trotting!

Since octopuses are rather odd creatures, it is a bit surprising that references to them are rare in Seri folklore today. There is one instance of a place name incorporating the word for octopus: *Hapaj Itni* 'octopus umbilical hernia' is a mountain near Desemboque having that shape.

According to oral accounts related by Lorenzo Herrera, long ago a group of Seris lived on Baja California and were called the *Hant Ihiini Comcaac* 'Baja California Seris' (see *Euvola vogdesi*). The people navigated the sea in large balsas, each carrying many men, all paddling. The people on shore had a saying when the balsas were approaching, "*Ox hapaj cola ccótonim immís!*" 'Look how they are just like octopuses jetting [through the sea]!'

María Luisa Astorga had a bundle of *najcaazjc* (*Asclepias subulata*) twigs wrapped with cloth strips, tucked into the wall of her ramada. The bundle, called *siipoj imal* 'what doesn't accompany the Osprey' (meaning unclear), was often placed in houses for good luck. María Luisa commented that when such a bundle was poked into the cave of a large octopus, it would cause the octopus to emerge quickly. She said that perhaps it was because of the supernatural power of the bundle, but then added that "maybe it was because it just disturbed the octopus!"

Of the different octopuses described by the Seris, only four have been linked to recognized species.

Octopus alecto

hapaj inloj caact 'octopus that cuts off its arms'
hapaj inloj cacöla 'long-armed octopus'
hapaj inloj cozla 'octopus with arms that easily detach'

The names of this uncommon red-colored octopus are descriptive and reference the creature's most notable characteristic of having long, thin arms that "just drop off" (as a defense mechanism, some octopuses can autotomise [detach] their limbs when under attack). Said to occur around the Tepopa peninsula, the octopus was not eaten; in fact, any woman I asked strongly remarked that she did not like it and was afraid of it because it was so strange. According to one person who said he had witnessed it, and confirmed by others, the octopus eats its own arms when no other food is available.

Octopus bimaculatus

hapaj áa 'true octopus'
hapaj isoj 'real octopus'
hapaj quimaaxat 'gray-brown octopus'

Easily recognized by its two ocelli, or false eyespots, this is the most frequently encountered octopus in the area, something reflected by the names ***hapaj áa*** 'true octopus' and ***hapaj isoj*** 'real octopus'. The octopuses' dens are easily found in the rocky low intertidal zone.

The rather nondescript color of this octopus gives it the name ***hapaj quimaaxat*** 'gray-brown octopus'.

Octopus fitchi

hapaj cosni

This tiny red octopus, found beneath rocks in the low intertidal zone, is rather feisty and has a toxic bite. People have observed that instead of ink, the octopus squirts seawater, something that was apparent when I caught and photographed one.

The meaning of the name ***hapaj cosni*** is not clear. The word *cosni* only occurs in the name for this octopus and the name of a mountain, *Hast Cosni.*

María Luisa Astorga informed me of a common belief: when a person was bitten by this octopus, the bite would swell and be painful at high tide, and when the tide was low, the swelling and pain would go away. This would continue for several days.[12] Such ebb and flow is also said to be true of the pain caused by the stinging bristles of the fire worm, *Eurythoe complanata.*

This octopus was known for its medicinal properties. According to Felger and M. Moser:

> [It] was dried, crushed and cooked with hierba mansa (*Anemopsis californica*) and the resulting tea was given to a person or horse to make him run fast. Sara Villalobos said that this tea was taken by someone fleeing from his enemies. Either the dried octopus or the prepared, dried mixture was kept until needed.[13]

One of several archaic practices used to stop a strong wind that had been blowing for a long time involved placing a live sea creature in a fire. An early note by Edward Moser describes the practice:

> To stop the wind: wave sea life in air—cooked sea life. Put a live sea animal—shellfish, fish, *pulpo* [octopus]—crabs, etc.—in fire. When it is dying—(before it is dead)—take it from fire and wave it in the wind and say: ʔái cʔinséemeetinxo [early transcription—CMM] (repeat) "stop the wind!" (talking to the animal) Then put it back in fire. Do it especially during long windstorm. Old folks still do it.

In confirming this, Juanita Herrera recounted a childhood memory, when she and her family were living near Estero Santa Rosa in an ocotillo-frame structure. It was very cold; a strong wind had blown for many days and it looked like it was going to rain. Her father made a fire and put a live ***hapax cosni*** into the fire.[14] He removed the octopus when it was dead and placed it by the fire. As Juanita recalled, the wind stopped and the sky cleared.

Angelita Torres also confirmed the custom, saying that her uncle, Antonio Morales, had practiced it. Efraín Estrella added that a ***satoj*** (*Modiolus capax*) could be used as well.

Octopus hubbsorum

hapaj cheel 'red octopus'

This octopus is said to be large; according to one person it could weigh two kilos. It is caught in the same areas as *Octopus bimaculatus*.

The name ***hapaj cheel*** 'red octopus' describes the octopus's color.

Octopus sp.

hapaj caacoj 'large octopus'

No one I asked had any information about such an octopus, which exists only as a brief mention accompanying the name of a small island, *Hast Xnoois* (see Mythology and Folklore).

Octopus sp.

hapaj cmaaitj 'tender [meat of] octopus'

This good-sized unidentified octopus is described as having tender meat.

Octopus sp.

hapaj cooil 'blue/green octopus'

Manuel Monroy described this octopus as greenish and about one foot in total length, with short, thick arms. Lorenzo Herrera reported that the octopus has been taken in traps from an area off the coast of Cerro Tepopa.

Paroctopus digueti [= *Octopus digueti*]

This small octopus is found inhabiting empty sea shells in the low intertidal area. The shells named are those of *Laevicardium elatum*, *Dosinia ponderosa*, *Codakia distinguenda*, and the larger murexes. The Seris I asked consider this octopus to be an immature *Octopus bimaculatus*, and do not give it a separate name.

Figure 8.6. Manuel Monroy with an octopus he has just extracted from its cave, 2008.

Figure 8.7. María Luisa Molina holding freshly killed octopuses, 2008.

An octopus hunt

During the cooler months of the year octopus hunting is at its best. María Luisa Astorga observed that February is especially good, before the months turn warm, but added that, "there are always octopuses found around Desemboque!"

On February 3, 2008, my husband Steve and I accompanied Manuel Monroy and his wife María Luisa, a childhood friend, on an octopus hunt at a beach area north of Desemboque. I noticed they were both wearing tennis shoes, while I for some reason still wore sandals (a step down from my confident barefoot childhood days, I later found out on some slippery seaweed-covered rocks). In about an hour and a half Manuel and María Luisa had taken six good-sized octopuses; the first was *Octopus bimaculatus* and the next *O. hubbsorum*—all taken that afternoon were of these two species.

Manuel and his wife each carried a metal rod with one end hooked and barbed, a small bottle of commercial bleach, and a net bag. They hunted separately in the partially submerged low intertidal zone, finding the occasional boulder standing out in contrast to the more common, smaller, beach rocks—the boulders harbored larger octopuses. A space beneath each boulder was probed with the rod; the rod meeting with something soft was an indication that an octopus was present. A few drops of the bleach (a pine-scented cleaner is sometimes used) were sprinkled in the water by the rock, and the octopus emerged very quickly, at which time it was hooked and secured. The octopus was then thrown against rocks to kill it; in one case María Luisa killed one by pounding its visceral area with a rock. The octopuses were then put into the net bag and taken back to Desemboque where a local Seri man was paying twenty pesos a kilo (at the current exchange rate, about two dollars) for the octopus meat; he then trucked it on ice for sale outside of the area.

Argonautidae (paper nautilus)

Although scientifically documented as occurring in the Gulf, no member of this family is recognized by name today, and no one to whom I showed a photograph of the creature had seen one in its natural habitat. In 2005, Cleotilde Morales was given a brood case of a Pacific argonaut in Baja California, which she carefully kept as a curiosity. Alfredo López, her husband, provided a name with which he had recently dubbed the creature: *haxöl hapaj* 'clam octopus'.

Figure 8.8. Cleotilde Morales with the brood case of an *Argonauta argo.* Punta Chueca, 2005.

Figure 9.1. A man spearing a lobster near the southern end of Isla Tiburón, 1939. Photograph by Edward H. Davis, courtesy of Robert Fuller Davis.

OTHER MARINE INVERTEBRATES

Although mollusks were the most heavily used marine invertebrates in the Seri culture, they were by no means the only ones noticed and named. Many of the commonly encountered creatures living in the intertidal zone or shallow water have Seri names; some were important foods, such as the good-sized swimming crab *Callinectes bellicosus*.

The invertebrates are not named as a group, although the term *imítajoj* 'things without bones' might be the best way to characterize them. However, the term is more descriptive and is not commonly used.

A broad term for marine life, *xepe án icám* 'life of the sea area', apparently refers to creatures actually inhabiting the sea. According to Efraín Estrella, included in that group are such things as whales, sharks, and fishes, but not sea lions or shore crabs.

Small creatures found in the intertidal zone are sometimes collectively referred to as *canoomalc*.[1] More commonly used is *xepe án imiipla* 'the badness (worthless things) of the sea area', a phrase that describes sea life lacking apparent nutritional or cultural value—the "inconsequential"—such as worms, sea cucumbers, small crabs, and, suggested by one person, the organisms causing red tide. Small mollusks are included by some as well. Another name, *xepe án ihicoomz* 'pinworms of the sea area', singles out small worm-like creatures.

Although the focus of this work is on mollusks, I include in table 9.1 other marine invertebrates having Seri names, as well as a few details about them.[2]

Table 9.1. Other marine invertebrates

Sponges		
moosni ocáxalam	'what the sea turtle (*Chelonia mydas*) nibbles'	an unidentified sponge or substance eaten by the black sea turtle
tacj queemtajc	'dolphin that cuts meat to pieces'	an unidentified sponge (the name is also given to the killer whale *Orcinus orca*)
xepenozatx	'glochids in the sea'	sponge (a name shared by the fire worm *Eurythoe complanata*)
Sea anemones, corals, jellies		
hast iti coocp	'what emerges on the rock'	calcified seaweed; sea fan
najoo ixpanaams	'seaweed of the spotted sand bass (*Paralabrax maculatofasciatus*)'	calcified seaweed; sea fan
xepe án impós caacöl	'large *xepe án impós*'	black coral
xepe ooil 		sea fan
comcaii him oocatx 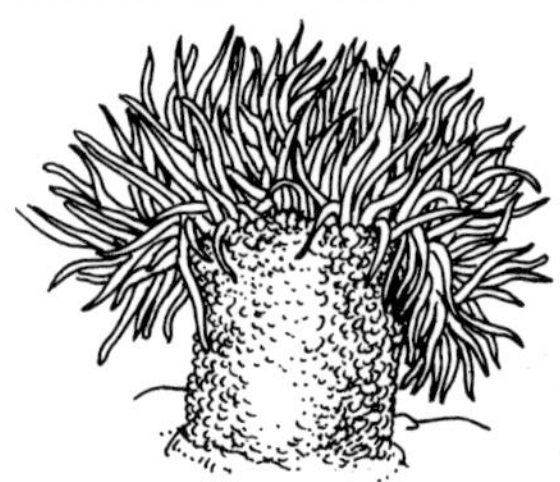	'what the old woman discarded'	sea anemone
copsiij 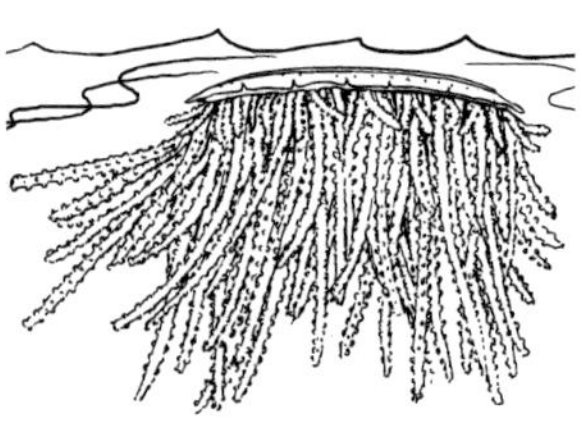		*Porpita pacifica*, jellyfish
copsiij cooil	'blue jellyfish'	*Physalia* sp.
copsiij cooxp	'white jellyfish'	*Porpita* sp.
copsiij hapám	'jellyfish that is swallowed'	*Pleurobrachia* spp., sea gooseberry (a comb jelly)

ipxom haaco iimj quih oiitoj	'what the *ipxom haaco iimj* (large sea turtles) eat'	*Porpita pacifica*
xpanaams cocoopxa	'stinging seaweed'	*Aglaophenia* sp., a colonial hydroid

Worms

hast iixz 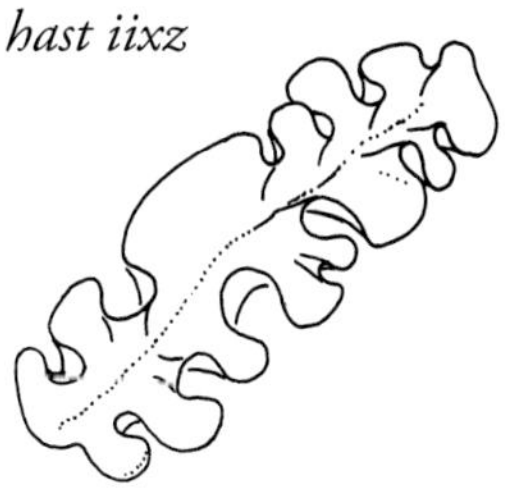	'the rock's pet' (reflecting the manner in which the flatworm hugs rocks or other surfaces)	a flatworm
hant itoozj	'intestines of the land'	a worm
xepenopotaa	'maggot in the sea'	a small reddish intertidal worm
xat		possibly *Phascolosoma* sp., a peanut worm
hant yapxöt	'flower of the land'	possibly *Themiste* sp., a peanut worm
xepenozatx 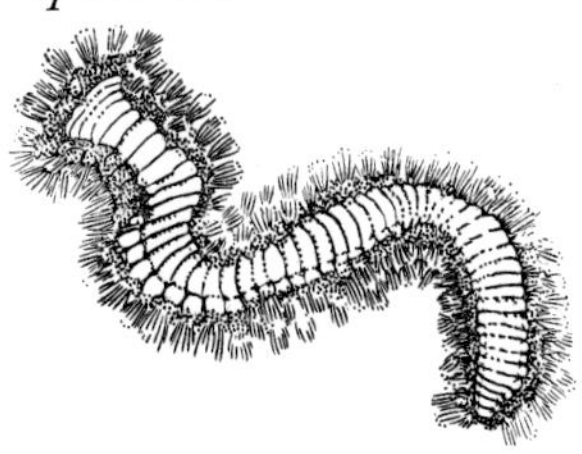	'glochids in the sea'	*Eurythoe complanata*, a fire worm
hant ipos 	'throat of the land'	Onuphidae, *Diopatra* sp., a tube worm

Arthropods

cmootc (var.: *mootc, moocj, mooc, cmooct, smoocp*) 		*Tetraclita* spp.; *Megabalanus* spp., thatched or acorn barnacles

moosni ihaapxaj 	'sea turtle's bead'	*Chelonibia testudinaria*, sea turtle barnacle
ziix haquim ihaapxaj 	'whale's bead'	*Cryptolepas rhachianecti*, whale barnacle
haxoj cahiift cöcapaseeeja	'*haxoj cahiift* with a rudder'	an amphipod
haxoj cahiift (var.: *haxoj cahiifp*, *xojcahiift*)		*Tylos* sp., a beach isopod
cmaam iixz	'the woman's pet'	*Ligia* sp., a beach isopod
zixcám iixz	'the fish's pet'	Cymothoidae, a parasitic isopod
oot icamn	'the coyote's *camn*'	*Squilla* sp., mantis shrimp
seeten iixz	"the *seeten*'s pet'	*Pontonia* sp., commensal shrimp
haxoj án ihiifz	'the shallow sea area's snapping'	Alpheidae, pistol shrimp, snapping shrimp
montaj 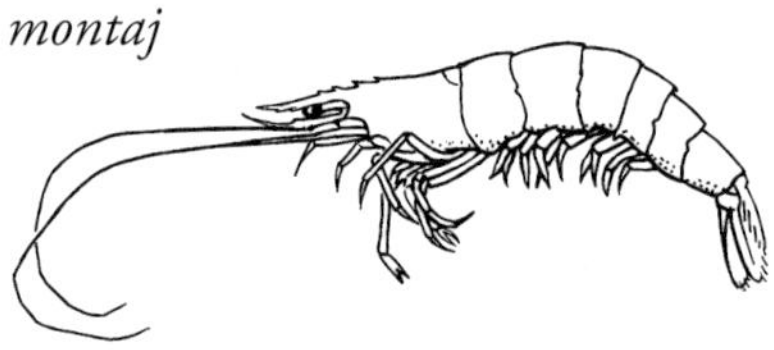		Penaeidae, shrimp
montaj caacoj	'large *montaj*'	Penaeidae
montaj cooil	'blue/green *montaj*'	Penaeidae
montaj quimaaxat	'gray-brown *montaj*'	Penaeidae
hant quixaa 	'what has roots in the earth'	ghost or mud shrimp
ptcamn 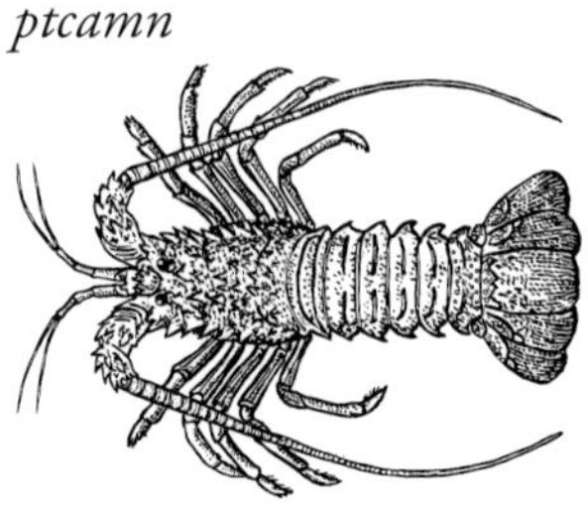		*Panulirus inflatus*, Cortez spiny lobster

ziix hast imocl quiij	'thing that sits under a rock' (obsolete name)	*Panulirus inflatus*, Cortez spiny lobster
tacamn (var. *xptacamn*)		slipper lobster
tacamn cmaam	'female *tacamn*'	*Evibacus princeps*, Panamic slipper lobster
tacamn ctam	'male *tacamn*'	*Scyllarides astori*, rock slipper lobster
hast ctopl 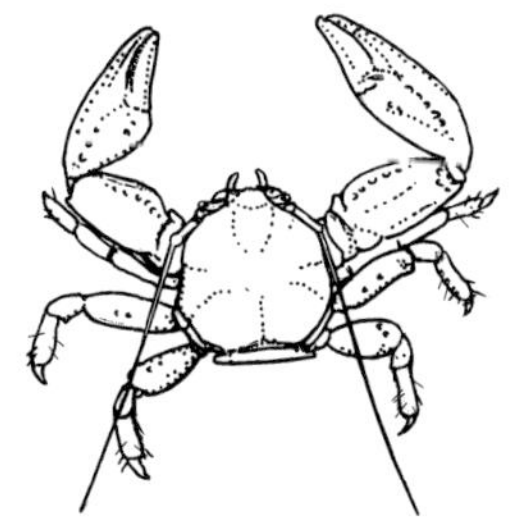	'what clings to a rock'	*Petrolisthes armatus*, a rock slider
hast iiqui cxatlc	'what is thin against a rock'	*Petrolisthes armatus*
oot iqueepl	'the coyote's *queepl*'	*Petrolisthes armatus*
hant quixoaa 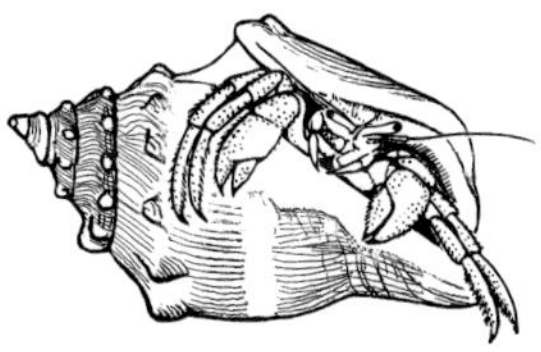	'what plans to fight'	Paguroidea, hermit crab
iicj ano moosni	'black sea turtle (*Chelonia mydas*) in fine sand'	*Hippa pacifica*, mole crab
queepl 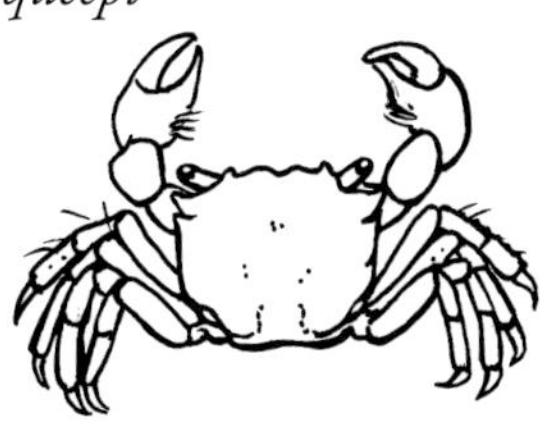		*Tetragrapsus jouyi* and others, grapsoid crabs
inl quiixaz	'whose arms (claws) clack'	Grapsidae, various shore crabs
pnaacoj an inl quiixaz	'*inl quiixaz* in the mangroves'	*Goniopsis* sp., a mangrove crab
queepl inol cahomtxö	'*queepl* that extends its arm'	*Uca* spp., fiddler crab (synonym: *itoj caacöl* 'whose eyes are large')
queepl inol quitox	'arm-eyed *queepl*'	*Uca princeps*
xepe taacoj iixi	'where the high tide finished'	*Uca* spp., fiddler crab
queepl insiijim 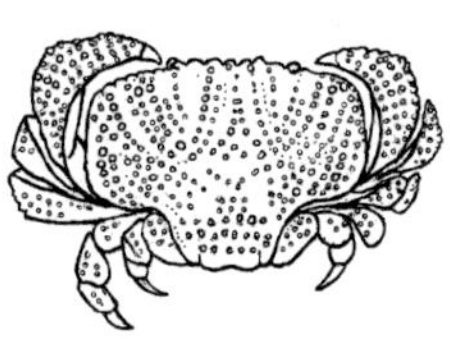	'*queepl* that doesn't move'	Xanthidae, pebble crab

queepl itoj cacöla 	'*queepl* with long eyes'	ghost crabs and fiddler crabs
ziix hanteeno quiij	'thing that sits in the earth'	an unidentified crab
hast cahooxp	'what whitens the rock'	a small white-mottled crab
zamt, zamt áa 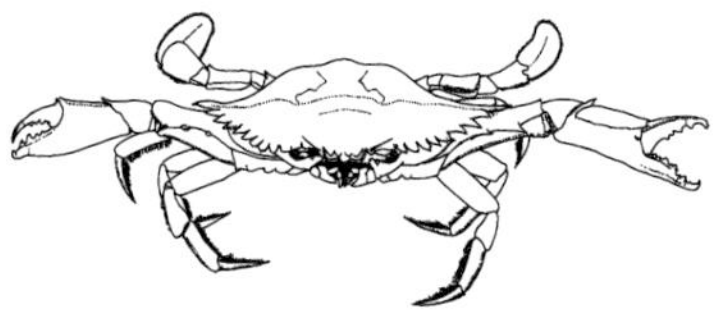	*zamt áa* 'true *zamt*'	*Callinectes bellicosus*, Cortez swimming crab
zamtooil	from *zamt* + *-ooil* 'blue'	*Callinectes arcuatus*, a swimming crab
oot izamt 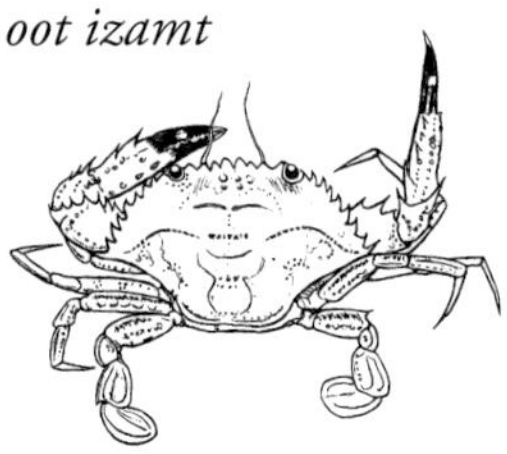	'the coyote's *zamt*'	*Cronius ruber* and other small, feisty swimming crabs
zamt hacaatol	'dangerous (in a metaphysical way) *zamt*'*	*Euphylax dovii*, a portunid crab
haxoj án iicöla	'the shore area's length'	*Stenorhynchus debilis* and others, arrow crabs
xepe án iixaz	'the sea area's tinkling'	an unidentified crab
hant ihamjoj	'land' + (unknown)	*Herbstia camptacantha* and others, spider crabs
xpacaafp 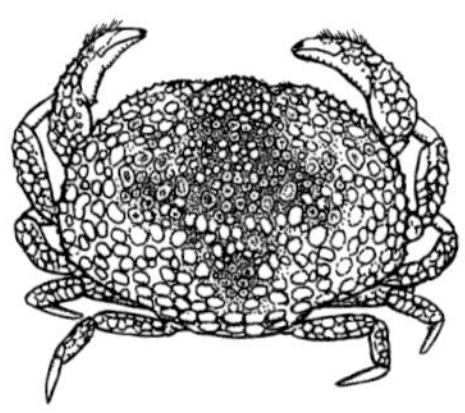		*Hepatus lineatus*, Panamic calico crab
xepe ano coiiz	'spider in the sea'	a sea spider
xepe quilp	'seawater squirter'	*Pleuroncodes planipes*

Sea stars, sea urchins, and others

azoj canoj cheel	'red *azoj canoj*'	*Pentaceraster cumingi*, Panamic cushion star

* When this seldom-seen crab was found in a trap in the Infiernillo Channel, it caused quite a stir of interest. After it was photographed, it was released because it was considered dangerous. René Montaño wrote an essay in Seri about the crab (Montaño Herrera 2012b).

azoj canoj 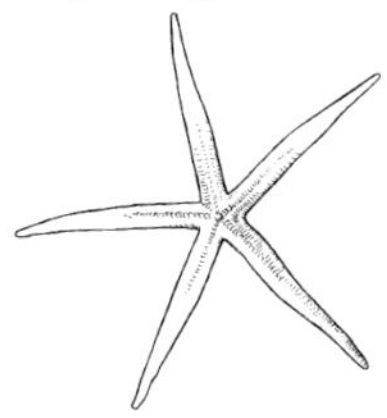	'star that shines (star)'	sea star
pyooque 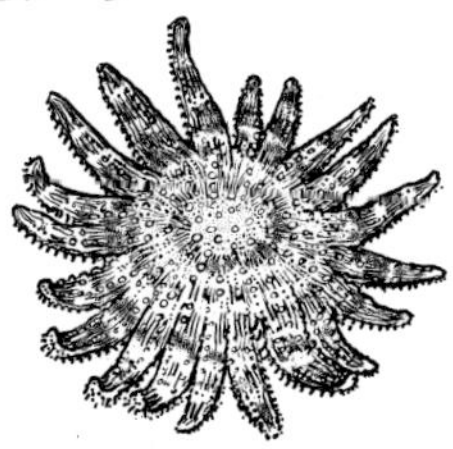		*Heliaster kubiniji*, a sun star
oot ipyooque	'coyote's *pyooque*'	an unidentified sea star
hanol cahít 	'what breaks off its arms'	Ophiuroidea, brittle star
xepenosiml quictamo (synonym: *xepenosiml caacöl* 'large *xepenosiml*')	'fierce *xepenosiml*'	*Eucidaris thouarsii*, slate pencil urchin
xepenosiml 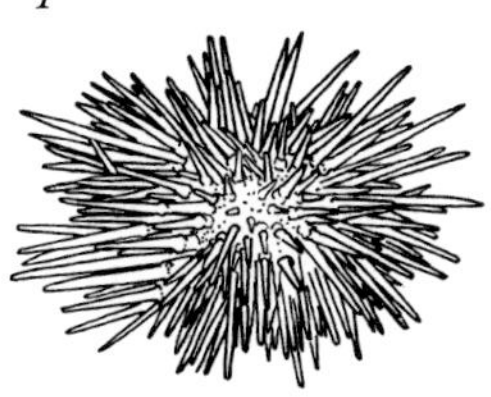	'barrel cactus in the sea'	*Echinometra vanbrunti*, purple urchin
xepeíisj 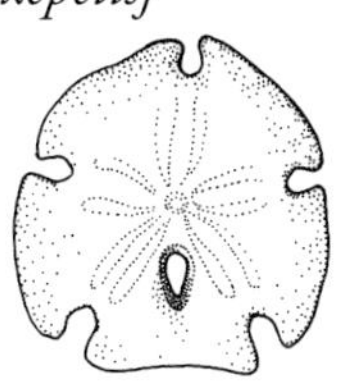	'sea's basket'	Scutellidae, sand dollar
cataamax hapx ooxquim caacoj	'large *cataamax hapx ooxquim*'	*Isostichopus fuscus*
cataamax hapx ooxquim 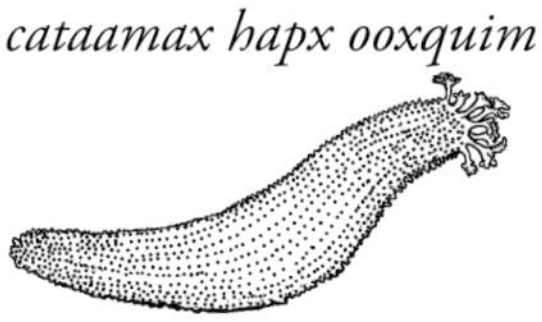	'what the person harvesting eelgrass pulls up'	*Holothuria lubrica* and other sea cucumbers
xepeiyas	'sea's liver'	Ascidiacea, tunicate

Table 9.2. Some marine invertebrate anatomical terms

Part of the anatomy	Seri name	Literal translation
antennae (arthropod)	*itaamalca*	'its horns'
antennules (lobster)	*itxaloms*	(also 'his/her loose strands of hair')
arthropod cheliped (pincer, claw)	*yacazni*	'with which it bites'
bristles (fire worm) or spicules (sponge)	*zatx*	(also 'glochids')
eyestalk, eye	*itoj*	'its eye'
legs (arthropod)	*inloj*	'its hands/arms'
lobster tail	*ptcamn yaap*	'lobster's neck'
molted exoskeleton	*ano yapox*	'with which it sheds'
mouthparts	*itaast*	'its teeth'
spines (sea urchin)	*icös*	'its spine(s)'
swimming leg (swimming crab)	*yeesc*	(also 'its rattle' of rattlesnake; 'his/her little toe')
tail fan (shrimp or lobster)	*hapaspoj yaa*	'its paper (hand of cards)'

APPENDIX 1
Seri names

Names with phonological variants are marked with an asterisk. The inclusion of the family name implies that the members of that family can be referred to by that name; the individual species that follow are those illustrated in the species accounts.

caaitmoj	*Pinctada mazatlanica*
cacaapxom	Bullidae; *Bulla gouldiana*
cahiixona	Calyptraeidae; *Crucibulum scutellatum, C. spinosum*
camazjij	*Megapitaria squalida*
camazjij caacoj	*Megapitaria aurantiaca*
camonlc	*Trivia solandri*
camós caacoj	*Macrocypraea cervinetta*
capoclim	Chitonidae; *Chiton virgulatus*
capoclimiil	*Chiton virgulatus*
cascanoohcö	*Euvola vogdesi* (convex, inflated valve)
cascanoohcö ooix	*Euvola vogdesi* (concave, almost flat valve)
cascapeetij	*Euvola vogdesi* (concave, almost flat valve)
casooca	*Crucibulum scutellatum*
casopaj	*Pinctada mazatlanica*
casquim quih iti ihiij	*Crepidula onyx, C. striolata.*
catamax	*Crucibulum scutellatum, C. spinosum*
cateesic	an unidentified oyster
cmaam	'woman', an unidentified shelled mollusk
cmacjij*	*Lottia dalliana*
cmatjij oot quih yaa	*Hipponix antiquatus*
comcaii inoosj	*Nerita scabricosta*
comcaii inoosj heeque	*Nerita funiculata*
copas	*Pinctada mazatlanica*
copas hapapl	*Isognomon janus, I. recognitus*

cotitla oot quih yaa	*Strombus granulatus*
cotopis	*Turbo fluctuosus*
cotopis quislitx	*Tegula rubroflammulata*
cozaaij	*Dentalium* spp. and small worm snails
ctam yazexe	*Tagelus peruvianus*
haan	*Chionista fluctifraga*
haapx	*Tegula rugosa*
hacx xah cöscaap	*Cymatium keenae, Fusinus dupetitthouarsi*
haitepl	an archaic name for *Hast Quipac* (Campo Dólar); an unidentified shelled mollusk
halít cahooxp	*Dosinia ponderosa*
hanl	*Nodipecten subnodosus*
hant haaco iima ano haxölinaail	*Janthina prolongata*
Hant Ihiini stacj	*Placunanomia cumingii*
hant iiha cooml	*Conus princeps*
Hant Iiha Quimx itaajc ootp	*Theodoxus luteofasciatus*
hant iiha quimxoj oeenec	*Cypraecassis coarctata*
hant itcoj	*Cantharus macrospira*
hant iti queemij	*Navanax inermis*
hant iti yeemej coospoj	*Eupleura muriciformis, Vokesimurex elenensis, V. tricoronis*
hant yapol ipapl	*Knefastia dalli*
hapaj	Octopodidae (octopus)
hapaj áa	*Octopus bimaculatus*
hapaj caacoj	*Octopus* sp.
hapaj cheel	*Octopus hubbsorum*
hapaj cmaaitj	*Octopus* sp.
hapaj cooil	*Octopus* sp.
hapaj cosni	*Octopus fitchi*
hapaj inloj caact	*Octopus alecto*
hapaj inloj cacöla	*Octopus alecto*
hapaj inloj cozla	*Octopus alecto*
hapaj isoj	*Octopus bimaculatus*
hapaj quimaaxat	*Octopus bimaculatus*
hapos cahaaxat	*Chama buddiana*
hapxaacoj	an unidentified gastropod, similar to *Tegula rugosa*, but much larger
hapz	an unidentified shelled mollusk
hasajitij	*Crucibulum scutellatum*
hasit camós	*Trivia solandri*
hasla	*Argopecten ventricosus*
Hast Án ano haapx	*Astraea unguis*
hast imocl cooxalca	*Mexacanthina lugubris angelica*
hastisol	*Mitrella ocellata*

hast iti caatolam	*Columbella fuscata, C. strombiformis*
hast iti cöcooyaj	Cerithiidae, Columbellidae, Olivellidae
hast iti cöcooyam	*Lottia dalliana, L. stanfordiana*
hast iti cöquiipax	Cerithiidae, Columbellidae, Olivellidae
hast iti cpaainalca	Cerithiidae, Columbellidae, Olivellidae
hast iti queemij	*Diodora inaequalis*
hatajeen camós	*Zonaria annettae*
hatajeen camós cahtxima	*Jenneria pustulata*
hatajeen camós coopol	*Jenneria pustulata*
hatj ihaxöl	*Sanguinolaria tellinoides, Tellina regia*
hatj ihaxöl cooxp	*Psammotreta pura*
hatx cöcazoj	*Aplysia californica*
hax cöoaal	*Hipponix antiquatus*
haxöl	*Leukoma grata*
haxöl áa	*Leukoma grata*
haxöl aapa	*Chione tumens*
haxöl icaai	*Sanguinolaria tellinoides, Simomactra dolabriformis*
haxöl icaai ctam	*Pholas chiloensis, Tagelus peruvianus*
haxöl icaai sleecoj iti yazám	*Psammotreta cognata, Tagelus peruvianus*
haxölinaail an icoopxoj	*Strombus galeatus*
haxölinaail ancoj cmasl	*Mancinella tuberculata, Stramonita biserialis*
haxölinaail anxö icaticpan	*Nassarius brunneostomus*
haxölinaail caaoloj	*Cantharus macrospira*
haxölinaail cheel*	*Trivia solandri*
haxölinaail cheel caacoj	*Macrocypraea cervinetta*
haxölinaail cmasl	*Anomia peruviana*
haxölinaail cnoosc	*Nassarius brunneostomus, N. tiarula*
haxölinaail comihj	*Melampus mousleyi, Oliva spicata*
haxölinaail cooil	*Lottia dalliana, L. stanfordiana*
haxölinaail cooil caacoj	*Lottia dalliana*
haxölinaail cooil ctopis	*Lottia dalliana, L. stanfordiana*
haxölinaail cooil cooxi	*Lottia strigatella*
haxölinaail coopl	*Anachis scalarina, Mitrella ocellata*
haxölinaail cooscl	*Costoanachis coronata, Echinolittorina aspera*
haxölinaail cotázita	*Strombus gracilior*
haxölinaail cotitla cmaam	*Strombus gracilior*
haxölinaail cotitla ctam	*Strombus granulatus*
haxölinaail cotítzilca	*Strombus gracilior, S. granulatus*
haxölinaail cpotyoj	'shell with small ridges'; an unidentified gastropod with a ridged shell
haxölinaail cpotyoj caacoj	*Mitra belcheri*
haxölinaail ctocnij	*Semicassis centiquadrata*
haxölinaail cxatlc	*Anomia peruviana*

haxölinaail hamisj istj cmis	*Anomia peruviana*
haxölinaail imiteeno	'shell without a mouth'; an unidentified gastropod
haxölinaail isoj	'real shell'; an unidentified gastropod found on rocky shores of Isla Tiburón
haxölinaail quicös	*Nassarius brunneostomus*
haxölinaail quihimca	*Semicassis centiquadrata*
haxölinaail quihócaloj	*Placunanomia cumingii*
haxölinaail quimaaxat	*Costoanachis coronata*
haxölinaail quítajij	*Cerithidea mazatlanica, Liocerithium judithae*
haxölinaailc imahcaptxö	*Morum tuberculosum*
haxölinaailc impafc	*Morum tuberculosum*
haxölísotoj	*Cardites laticostatus*
haxt	*Crassostrea columbiensis, C. corteziensis*
haxz heeque	*Acanthochitona exquisita*
heexoj	*Pholas chiloensis*
hehe hasoaaj	'cross'; an unidentified bivalve described as having red cross-shaped viscera
hexot	an unidentified shelled mollusk
hitcz ihizcoofizlc	*Vokesimurex elenensis, V. tricoronis*
honcaacoj imaazax	*Mancinella tuberculata*
honquipla	'immature gull'; an unidentified shelled mollusk
iicj	*Conus princeps*
iimox*	*Spondylus limbatus*
ipac casa ihaxöl	*Tagelus peruvianus*
iplmateemij*	*Agaronia testacea*
iquihimz iic cöihiipe	*Mexacanthina lugubris angelica, Plicopurpura pansa*
isliicot itaail	*Arca pacifica*
iteel it hacaptxö	*Columbella fuscata, C. strombiformis*
itx imonaaaij	*Architectonica nobilis*
Juan Tomás ihaxöl	*Raeta undulata*
lexoj	an unidentified gastropod
mas*	*Olivella dama, O. steveni*
matcz ital coofizlc	*Crepidula onyx*
mayoocö ixpa̲leemelc	*Agaronia testacea*
miizj hant quixquim quih iti ihiij	*Crepidula onyx, C. striolata*
mooxon iixz	*Cardites laticostatus*
mosnaact	*Columbella fuscata*
mosniil inoosj	*Diodora inaequalis*
mosniil itaajc	*Columbella strombiformis*
mosniil oozt	*Columbella strombiformis*
mosniil yaaspoj	*Conus princeps*
mosniiquet	'the sea turtle's child'; an unidentified shelled mollusk

mosnipol oopl	*Conus ximenes*
mosniscl inoosj	*Diodora inaequalis*
motcz	*Vokesimurex elenensis, V. tricoronis*
mox	*Hipponix antiquatus*
naapxa yaat	*Glycymeris gigantea*
naj	*Lottia stanfordiana*
najmís	possibly from *naj* and *cmis* 'like a *naj*'; an unidentified shelled mollusk
napxeec	*Cerithium maculosum, C. stercusmuscarum, Mancinella tuberculata, Melongena patula, Strombus galeatus*
oot ano yapxoj	'what the coyote blows into'; an unidentified gastropod
oot icamazjij	*Tivela byronensis*
oot icopas	*Isognomon janus, I. recognitus*
oot ihaxöl	*Chione tumens*
oot iquiit	*Carditamera affinis* (an odd growth form)
oot isatoj	*Entodesma pictum, Modiolus tumbezensis, Mytilus californianus*
oot iscám	*Crepidula onyx, C. striolata*
oot iseeten	*Atrina maura, A. oldroydii*
oot iteexoj	*Spondylus leucacanthus*
oot ixpa̲leemelc	*Agaronia testacea, Crassispira incrassata*
oot izacz	*Trachycardium consors*
oot yácmolca	*Morum tuberculosum*
pahotmazexe*	*Anomia peruviana*
pajaas	Teuthida (squid)
pohaapx camoz	'what thinks it's an ***haapx***'; an unidentified gastropod, similar to *Tegula rugosa*
queeex	*Chionopsis gnidia*
quiimosim xepenoyaafc	*Anadara multicostata, Cardites laticostatus*
quiit	*Carditamera affinis*
quitxísotaj	*Vokesimurex elenensis, V. tricoronis*
saacj	*Modiolus capax*
satim	*Glycymeris gigantea*
satoj	Mytilidae; *Modiolus capax*
satoj aapa	possibly a large *Modiolus capax, Mytella charruana, Mytilus californianus*
saxáp	*Glycymeris gigantea*
sayana	*Olivella dama*
seeex	*Glycymeris gigantea*
seeten	Pinnidae
seeten cmaam	*Atrina tuberculosa*
seeten cmaam ctam	*Atrina maura*
seeten comihj	*Atrina oldroydii*
seeten cooil	*Atrina oldroydii*

seeten ctam	*Pinna rugosa*
seeten iizax coocp	*Atrina maura*
seeten inaail cquihjö	*Atrina maura*
seeten itaqueecöj caacoj	'**seeten** with a large adductor muscle'; an unidentified penshell
siml ihataxoj	*Trachycardium consors, T. procerum*
siip isoj	*Fusinus dupetitthouarsi*
sipisot*	*Fusinus dupetitthouarsi*
sleecoj cooxp ihaxöl	*Psammotreta cognata*
slenapzj ihaxöl	*Tellina regia*
snoomtxö	*Papyridea aspersa*
soocajam	*Pteria sterna*
spiitquim*	*Chione californiensis*
spiitquim coospoj	*Chione undatella*
spiitquim cpaxz	*Chionopsis gnidia*
spiitquim ctamoo	*Chionopsis gnidia, Periglypta multicostata*
spiitquim quictamo	*Chionopsis gnidia*
stac ano coom	*Leiosolenus spatiosus*
stacj	*Saccostrea palmula*
stacj caacöl	*Crassostrea columbiensis, Myrakeena angelica*
stacj coiicöt	*Saccostrea palmula* (an odd growth form)
stacj hapahit	*Saccostrea palmula*
stacj quítajij	*Saccostrea palmula* (an odd growth form)
taijitiquiixaz	*Heterodonax pacificus*
tamax	*Onchidella binneyi*
tamax capoclim	Chitonidae; *Chiton virgulatus*
tamax cosi	*Acanthochitona exquisita, Crucibulum spinosum*
tásona	*Crucibulum scutellatum*
teexoj	*Spondylus limbatus*
teexoj ctam	*Spondylus leucacanthus*
toerc iixz	'the Willet's pet'; an unidentified gastropod, possibly a limpet
xahícosa	*Vokesimurex elenensis, V. tricoronis*
xasaacj	an unidentified bivalve described as "red, pretty, found on west side of Isla Tiburón"
xasecö	*Ficus ventricosa*
xasecö caacoj	*Pleuroploca princeps*
xasecö cmaam	*Malea ringens*
xasecö ctam	*Fusinus dupetitthouarsi*
xat	*Crassostrea corteziensis*
xatós	an unidentified gastropod, possibly a moon snail
xatoscö ihaxöl	*Barbatia reeveana*
xepe an hapafc	*Chione undatella*
xepe án icös	*Pteropurpura centrifuga*

xepe ano xojmás	*Janthina prolongata*
xepe quih iti cöiixi iiqui izooxolcam	*Jenneria pustulata*
xepe quilp	*Theodoxus luteofasciatus*
xepe quilp cooxp	*Polinices uber*
xepe quis	*Theodoxus luteofasciatus*
xepe taacoj iixi	'where the high tide finished'; an unidentified mollusk
xepe yamaasij	*Cypraecassis coarctata, Malea ringens, Semicassis centiquadrata, Strombus galeatus*
xepe yamaasij ctam	*Cymatium keenae*
xepe yamaasij heeque	*Melampus mousleyi*
xepeiinoj	*Neverita reclusiana*
xepeiinoj cooxp	*Polinices uber*
xepeiinoj heecto	*Polinices bifasciatus*
xepeiinoj mayoocö quih yaa	*Natica chemnitzii*
xepenozaah	*Berthellina ilisima, Limaria pacifica*
xepenyaaan	*Astraea unguis*
xiica cooxp	*Olivella dama, O. steveni*
xiica cooxp heecto	*Olivella alba*
xiica cooxpísotoj	*Olivella steveni*
xiica hant cosítyalca	*Olivella dama, O. steveni*
xiica iteencoj cheel	*Chicoreus erythrostomus*
xiica quiictoj	*Pinctada mazatlanica, Pteria sterna*
xlehaxöl	*Tagelus peruvianus*
xnaa caaa ano haxölinaail	*Melampus mousleyi*
xnahaail	*Anadara multicostata, A. tuberculosa*
xnahooil	*Arca pacifica*
xnoocat*	*Hexaplex nigritus*
xnoocat an cheel	*Chicoreus erythrostomus*
xnoocat cmaacoj	*Hexaplex nigritus*
xnoocat coopol	*Hexaplex nigritus*
xnoocat coospoj	*Hexaplex nigritus*
xnoocat cooxp	*Hexaplex nigritus*
xnocatiil	*Hexaplex nigritus*
xojmás	terrestrial pulmonate and freshwater snails
xomcahaail	*Arca pacifica, Barbatia reeveana*
xpacaao iiha	*Vokesimurex elenensis, V. tricoronis*
xpajísotoj*	*Pholas chiloensis, Tagelus peruvianus*
xpaleemelc	Conidae
xpaleemelc caacoj	*Oliva incrassata, O. porphyria, Strombus galeatus*
xpaleemelc cacösxaj	*Fusinus dupetitthouarsi*
xpaleemelc comihj	*Oliva spicata*
xpaleemelc comihj caacoj	*Oliva incrassata*

xpa̲leemelc coospoj	*Oliva incrassata*
xpa̲leemelc quicös	*Nassarius brunneostomus*
xpa̲leemelc quipcö	*Oliva incrassata*
xpanooil*	*Glycymeris gigantea*
xpeecol*	an area in Estero Santa Cruz; an unidentified mollusk
xtamaxiictoj*	*Codakia distinguenda*
xtapacaj	Turritellidae; *Turritella banksi*, *T. gonostoma*, *T. leucostoma*
xtapacaj coopol	*Cerithidea mazatlanica*
xtapacaj coospoj	*Terebra ornata*
xtapacaj heeque	*Cerithidea mazatlanica*, *Liocerithium judithae*, *Terebra variegata*, *Turritella banksi*
xtiip	*Laevicardium elatum*
xtiip heeque	*Laevicardium substriatum*
xtipisot	*Tagelus peruvianus*
xtoozajö	Vermetidae; *Tripsycha tripsycha*
zaah	*Megapitaria aurantiaca*
zacz*	*Trachycardium consors*, *T. procerum*
zacz ctam	*Trachycardium consors*
zcaczj quictamo	*Trachycardium consors*
zeel	*Chama buddiana*
ziic ccap cooxp ihaxöl	*Pholas chiloensis*
ziic coopol ixpa̲leemelc	*Crassispira incrassata*
ziix cotopis	Octopodidae (octopus, old name)
ziix iiquet inaail	*Pinctada mazatlanica*, *Pteria sterna*
zojiime	*Costoanachis coronata*

APPENDIX 2
Seri names grouped by linguistic and various other characteristics

Primary names

casooca	*Crucibulum scutellatum*
casopaj	*Pinctada mazatlanica*
catamax	*Crucibulum scutellatum, C. spinosum*
cateesic	an unidentified oyster
cmacjij	*Lottia dalliana*
copas	*Pinctada mazatlanica*
haan	*Chionista fluctifraga*
haapx	*Tegula rugosa*
haitepl	an unidentified shelled mollusk
hapaj	Octopodidae (octopus)
hapz	an unidentified shelled mollusk
haxöl	*Leukoma grata*
haxt	*Crassostrea columbiensis, C. corteziensis*
heexoj	*Pholas chiloensis*
hexot	an unidentified shelled mollusk
iimox	*Spondylus limbatus*
lexoj	an unidentified gastropod
mas	*Olivella dama, O. steveni*
motcz	*Vokesimurex elenensis, V. tricoronis*
mox	*Hipponix antiquatus*
naj	*Lottia stanfordiana*
pajaas	Teuthida (squid)
quiit	*Carditamera affinis*
saacj	*Modiolus capax*
satim	*Glycymeris gigantea*
satoj	Mytilidae; *Modiolus capax*

saxáp	*Glycymeris gigantea*
sayana	*Olivella dama*
seeten	Pinnidae
soocajam	*Pteria sterna*
spiitquim	*Chione californiensis*
stacj	*Saccostrea palmula*
tamax	*Onchidella binneyi*
tásona	*Crucibulum scutellatum*
teexoj	*Spondylus limbatus*
xasaacj	an unidentified bivalve
xasecö	*Ficus ventricosa*
xat	*Crassostrea corteziensis*
xatós	an unidentified shelled mollusk
xnoocat	*Hexaplex nigritus*
xpa̱leemelc	Conidae
xpeecol	an unidentified shelled mollusk
xtapacaj	Turritellidae; *Turritella banksi, T. gonostoma, T. leucostoma*
xtiip	*Laevicardium elatum*
xtoozajö	Vermetidae; *Tripsycha tripsycha*
zacz	*Trachycardium consors, T. procerum*
zeel	*Chama buddiana*

Modified primary names

copas hapapl	*Isognomon janus, I. recognitus*	'strung ***copas***'
hapaj áa	*Octopus bimaculatus*	'true ***hapaj***'
hapaj caacoj	*Octopus* sp.	'large ***hapaj***'
hapaj cheel	*Octopus hubbsorum*	'red ***hapaj***'
hapaj cmaaitj	*Octopus* sp.	'tender (meat of) ***hapaj***'
hapaj cooil	*Octopus* sp.	'blue/green ***hapaj***'
hapaj cosni	*Octopus fitchi*	'(unclear) ***hapaj***'
hapaj inloj caact	*Octopus alecto*	'***hapaj*** that cuts off its arms'
hapaj inloj cacöla	*Octopus alecto*	'***hapaj*** with long arms'
hapaj inloj cozla	*Octopus alecto*	'***hapaj*** with arms that easily detach'
hapaj isoj	*Octopus bimaculatus*	'real ***hapaj***'
hapaj quimaaxat	*Octopus bimaculatus*	'gray-brown ***hapaj***'
hapxaacoj	an unidentified gastropod	'large ***haapx***'
haxöl áa	*Leukoma grata*	'true ***haxöl***'
haxöl aapa	*Chione tumens*	'huge or true ***haxöl***'

satoj aapa	possibly a large *Modiolus capax*, *Mytella charruana*, *Mytilus californianus*	'huge or true ***satoj***'
seeten cmaam	*Atrina tuberculosa*	'female ***seeten***'
seeten cmaam ctam	*Atrina maura*	'female-male ***seeten***'
seeten comihj	*Atrina oldroydii*	'smooth ***seeten***'
seeten cooil	*Atrina oldroydii*	'blue ***seeten***'
seeten ctam	*Pinna rugosa*	'male ***seeten***'
seeten iizax coocp	*Atrina maura*	'quarter-moon ***seeten***'
seeten inaail cquihjö	*Atrina maura*	'red-shelled ***seeten***'
seeten itaqueecöj caacoj	an unidentified pen shell	'***seeten*** with a large muscle'
spiitquim coospoj	*Chione undatella*	'spotted ***spiitquim***'
spiitquim cpaxz	*Chionopsis gnidia*	'rough ***spiitquim***'
spiitquim ctamoo	*Chionopsis gnidia*, *Periglypta multicostata*	'***spiitquim***-shaman power'
spiitquim quictamo	*Chionopsis gnidia*	'fierce ***spiitquim***'
stacj caacöl	*Crassostrea columbiensis*, *Myrakeena angelica*	'large ***stacj***'
stacj coiicöt	*Saccostrea palmula*	'***stacj*** that kill'
stacj hapahit	*Saccostrea palmula*	'edible ***stacj***'
stacj quítajij	*Saccostrea palmula*	'sharp ***stacj***'
tamax capoclim	Chitonidae; *Chiton virgulatus*	from *capocj* 'to pry off'
tamax cosi	*Acanthochitona exquisita*, *Crucibulum spinosum*	'spiny ***tamax***'
teexoj ctam	*Spondylus leucacanthus*	'male ***teexoj***'
xasecö caacoj	*Pleuroploca princeps*	'large ***xasecö***'
xasecö cmaam	*Malea ringens*	'female ***xasecö***'
xasecö ctam	*Fusinus dupetitthouarsi*	'male ***xasecö***'
xnoocat an cheel	*Chicoreus erythrostomus*	'***xnoocat*** with red interior'
xnoocat cmaacoj	*Hexaplex nigritus*	'old man ***xnoocat***'
xnoocat coopol	*Hexaplex nigritus*	'black ***xnoocat***'
xnoocat coospoj	*Hexaplex nigritus*	'spotted ***xnoocat***'
xnoocat cooxp	*Hexaplex nigritus*	'white ***xnoocat***'
xnocatiil	*Hexaplex nigritus*	'(very) large ***xnoocat***'
xpal̲eemelc caacoj	*Oliva incrassata*, *O. porphyria*, *Strombus galeatus*	'large ***xpal̲eemelc***'
xpal̲eemelc cacösxaj	*Fusinus dupetitthouarsi*	'tall ***xpal̲eemelc***'
xpal̲eemelc comihj	*Oliva spicata*	'polished ***xpal̲eemelc***'
xpal̲eemelc comihj caacoj	*Oliva incrassata*	'large ***xpal̲eemelc comihj***'
xpal̲eemelc coospoj	*Oliva incrassata*	'spotted ***xpal̲eemelc***'
xpal̲eemelc quicös	*Nassarius brunneostomus*	'spiny ***xpal̲eemelc***'

xpaleemelc quipcö	*Oliva incrassata*	'thick ***xpaleemelc***'
xtamaxiictoj	*Codakia distinguenda*	'red ***xtamax***'
xtapacaj coopol	*Cerithidea mazatlanica*	'black ***xtapacaj***'
xtapacaj coospoj	*Terebra ornata*	'spotted ***xtapacaj***'
xtapacaj heeque	*Cerithidea mazatlanica, Liocerithium judithae, Terebra ornata, Turritella banksi*	'small ***xtapacaj***'
xtiip heeque	*Laevicardium substriatum*	'small ***xtiip***'
xtipisot	Solecurtidae; *Tagelus peruvianus*	'narrow ***xtiip***'
zacz ctam	*Trachycardium consors*	'male ***zacz***'
zcaczj quictamo	*Trachycardium consors*	'fierce ***zcaczj***'

Descriptive names based on shape, color, size, texture, similarity to something (gross morphology), or location found

camazjij caacoj	*Megapitaria aurantiaca*	'large ***camazjij***'
camonlc	*Trivia solandri*	'puckered [shells]'
camós caacoj	*Macrocypraea cervinetta*	'large puckered [shell]'
capoclim	Chitonidae; *Chiton virgulatus*	
capoclimiil	*Chiton virgulatus*	'blue/green chiton'
cascanoohcö	*Euvola vogdesi* (convex, inflated valve)	from 'deeply rounded'
cascanoohcö ooix	*Euvola vogdesi* (concave, almost flat valve)	'what the ***cascanoohcö*** left behind'
cascapeetij	*Euvola vogdesi* (concave, almost flat valve)	from 'circular'
casquim quih iti ihiij	*Crepidula onyx, C. striolata*	'where the paddler sits'
comcaii inoosj	*Nerita scabricosta*	'old woman's [toe]nail'
comcaii inoosj heeque	*Nerita funiculata*	'small ***comcaii inoosj***'
cotopis quislitx	*Tegula rubroflammulata*	'***cotopis*** with an ear canal'
hanl	*Nodipecten subnodosus*	'fingers'
Hant Ihiini stacj	*Placunanomia cumingii*	'***stacj*** from *Hant Ihiin* (Baja California)'
hasit camós	*Trivia solandri*	'puckered vulva'
hasla	*Argopecten ventricosus*	'ear'
Hast Án ano haapx	*Astraea unguis*	'***haapx*** from *Hast Án*'
hatajeen camós	*Zonaria annettae*	'puckered genitalia'
hatajeen camós cahtxima	*Jenneria pustulata*	'rich ***hatajeen camós***'
hatajeen camós coopol	*Jenneria pustulata*	'black ***hatajeen camós***'
haxöl icaai ctam	*Pholas chiloensis, Tagelus peruvianus*	'male ***haxöl icaai***'

haxölinaail ancoj cmasl	*Mancinella tuberculata, Stramonita biserialis*	'shells with yellow interiors'
haxölinaail caaoloj	*Cantharus macrospira*	'grooved shell'
haxölinaail cheel	*Trivia solandri*	'red shell'
haxölinaail cheel caacoj	*Macrocypraea cervinetta*	'large ***haxölinaail cheel***'
haxölinaail cmasl	*Anomia peruviana*	'yellow shells'
haxölinaail cnoosc	*Nassarius brunneostomus, N. tiarula*	'rough shell'
haxölinaail comihj	*Melampus mousleyi, Oliva spicata*	'smooth or polished shell'
haxölinaail cooil	*Lottia dalliana, L. stanfordiana*	'blue/green shell'
haxölinaail cooil caacoj	*Lottia dalliana*	'large ***haxölinaail cooil***'
haxölinaail cooil cooxi	*Lottia strigatella*	'faded ***haxölinaail cooil***'
haxölinaail cooil ctopis	*Lottia dalliana, L. stanfordiana*	'***haxölinaail cooil*** that attaches by suction'
haxölinaail coopl	*Anachis scalarina, Mitrella ocellata*	'black shells'
haxölinaail cooscl	*Costoanachis coronata, Echinolittorina aspera*	'speckled shell'
haxölinaail cotitla cmaam	*Strombus gracilior*	'female ***haxölinaail cotitla***'
haxölinaail cotitla ctam	*Strombus granulatus*	'male ***haxölinaail cotitla***'
haxölinaail cpotyoj caacoj	*Mitra belcheri*	'large ***haxölinaail cpotyoj***'
haxölinaail ctocnij	*Semicassis centiquadrata*	'round shell'
haxölinaail cxatlc	*Anomia peruviana*	'thin shells'
haxölinaail hamisj istj cmis	*Anomia peruviana*	'shell like an *hamisj* (*Jatropha cinerea*) leaf'
haxölinaail imiteeno	an unidentified gastropod	'shell without a mouth'
haxölinaail quicös	*Nassarius brunneostomus*	'spiny shell'
haxölinaail quihimca	*Semicassis centiquadrata*	'shell having ringworm [spots]'
haxölinaail quihócaloj	*Placunanomia cumingii*	'shell with ruffles'
haxölinaail quimaaxat	*Costoanachis coronata*	'gray-brown shell'
haxölinaail quítajij	*Cerithidea mazatlanica, Liocerithium judithae*	'sharp shell'
haxölísotoj	*Cardites laticostatus*	'narrow ***haxöl***'
haxz heeque	*Acanthochitona exquisita*	'puppy'
hehe hasoaaj	an unidentified bivalve	'cross'
isliicot itaail	*Arca pacifica*	'wide-based upper back'
iteel it hacaptxö	*Columbella fuscata, C. strombiformis*	'what is pierced on its edge'
itx imonaaaij	*Architectonica nobilis*	'whose body whorl doesn't return'
miizj hant quixquim quih iti ihiij	*Crepidula onyx, C. striolata*	'where the navigator sits'
quitxísotaj	*Vokesimurex elenensis, V. tricoronis*	'whose body whorls are narrow'
siip isoj	*Fusinus dupetitthouarsi*	'real young man'

sipisot	*Fusinus dupetitthouarsi*	'narrow young man'
snoomtxö	*Papyridea aspersa*	'one-pelican-skin covering'
stac ano coom	*Leiosolenus spatiosus*	'what lies in coral'
xepe án icös	*Pteropurpura centrifuga*	'spines of the sea area'
xepe ano xojmás	*Janthina prolongata*	'***xojmás*** in the sea'
xepeiinoj cooxp	*Polinices uber*	'white ***xepeiinoj***'
xepeiinoj heecto	*Polinices bifasciatus*	'small ***xepeiinoj***'
xepenozaah	*Berthellina ilisima, Limaria pacifica*	'sun in the sea'
xepe quilp cooxp	*Polinices uber*	'white ***xepe quilp***'
xepe yamaasij ctam	*Cymatium keenae*	'male ***xepe yamaasij***'
xepe yamaasij heeque	*Melampus mousleyi*	'small ***xepe yamaasij***'
xiica cooxp	*Olivella dama, O. steveni*	'white things'
xiica cooxp heecto	*Olivella alba*	'small ***xiica cooxp***'
xiica cooxp isotoj	*Olivella steveni*	'narrow ***xiica cooxp***'
xiica iteencoj cheel	*Chicoreus erythrostomus*	'things with red mouths'
xiica quiictoj	*Pinctada mazatlanica, Pteria sterna*	'pregnant things'
xnaa caaa ano haxölinaail	*Melampus mousleyi*	'shell in the *Salicornia*'
zaah	*Megapitaria aurantiaca*	'sun'
ziix iiquet inaail	*Pinctada mazatlanica, Pteria sterna*	'the pearl's shell'
zojiime	*Costoanachis coronata*	found at *Zoj Iime*

Names reflecting physiological characteristics or actions of the mollusk

camazjij	*Megapitaria squalida*	from *cmazjij* 'to crack open'
cotopis	*Turbo fluctuosus*	'what attaches by suction'
hant iti queemij	*Navanax inermis*	'what moves slowly on the land'
hant iti yeemej coospoj	*Eupleura muriciformis, Vokesimurex elenensis, V. tricoronis*	'whose slow movement marks the land'
hast imocl cooxalca	*Mexacanthina lugubris angelica*	'things that sit under a rock'
hast iti caatolam	*Columbella fuscata, C. strombiformis*	'what crawl on rocks'
hast iti cöcooyaj	Cerithiidae and others	'what travel on rocks'
hast iti cöcooyam	*Lottia dalliana, L. stanfordiana*	'what passes over [high places of] rocks'
hast iti cöquiipax	Cerithiidae and others	'what climbs up rocks'
hast iti cpaainalca	Cerithiidae and others	'what tumble off rocks'
hast iti queemij	*Diodora inaequalis*	'what moves slowly on rocks'
haxölinaail cotázita	*Strombus gracilior*	'shell that pinches (small pinches)'
haxölinaail cotítzilca	*Strombus gracilior, S. granulatus*	'shells that pinch'
iplmateemij	*Agaronia testacea*	'it moves slowly with its tongue'

taijitiquiixaz	*Heterodonax pacificus*	'what tinkle in the surf (swash zone)'
xepeiinoj	*Neverita reclusiana*	'the sea's roar'
xepenyaan	*Astraea unguis*	from *xepe* and 'stirred up'
xepe quilp	*Theodoxus luteofasciatus*	'seawater squirter'
xepe yamaasij	*Cypraecassis coarctata, Malea ringens, Semicassis centiquadrata, Strombus galeatus*	'what is rolled by the sea'
xiica hant cosítyalca	*Olivella dama, O. steveni*	'things that scratch (slit) the land'
ziix cotopis	Octopodidae (octopus, old name)	'thing that attaches by suction'

Names reflecting cultural function or personal use

ctam yazexe	Solecurtidae; *Tagelus peruvianus*	'with what the man incises'
haxöl icaai	*Sanguinolaria tellinoides, Simomactra dolabriformis*	'clamshell for making pottery'
haxölinaail an icoopxoj	*Strombus galeatus*	'shell that is blown into'
haxölinaail anxö icaticpan	*Nassarius brunneostomus*	'shell that is a lot of work (to string)'
haxölinaailc imahcaptxö	*Morum tuberculosum*	'shell that is not perforated'
haxölinaailc impafc	*Morum tuberculosum*	'shell that is not pounded'
ipac casa ihaxöl	Solecurtidae; *Tagelus peruvianus*	'the outcast's ***haxöl***'
iquihimz iic cöihiipe	*Mexacanthina lugubris angelica, Plicopurpura pansa*	'ringworm medicine'
Juan Tomás ihaxöl	*Raeta undulata*	'Juan Tomás's ***haxöl***'
quiimosim xepenoyaafc	*Anadara multicostata, Cardites laticostatus*	'what the beggar pounds in the sea'
siml ihataxoj	*Trachycardium consors, T. procerum*	'barrel cactus scraper'

Names reflecting physiological effects on humans

cahiixona	*Crucibulum scutellatum, C. spinosum*	(unknown etymology but connotes throat irritation)
cozaaij	*Dentalium* spp., small vermetids	'what makes a circular cut'
halít cahooxp	*Dosinia ponderosa*	'what causes white hair'
hapos cahaaxat	*Chama buddiana*	'what irritates the throat'

Animal-related names

caaitmoj	*Pinctada mazatlanica*	'five point deer'
cmatjij oot quih yaa	*Hipponix antiquatus*	'***cmatjij*** belonging to the coyote'
cotitla oot quih yaa	*Strombus granulatus*	'***cotitla*** belonging to the coyote'

hatj ihaxöl	*Sanguinolaria tellinoides, Tellina regia*	'the Tern's ***haxöl***'
hatj ihaxöl cooxp	*Psammotreta pura*	'white ***hatj ihaxöl***'
haxöl icaai sleecoj iti yazám	*Psammotreta cognata, Tagelus peruvianus*	'***haxöl icaai*** stepped on by a heron'
haxz heeque	*Acanthochitona exquisita*	'puppy'
honcaacoj imazax	*Mancinella tuberculata*	'the Yellow-footed Gull 's clay pot'
honquipla	an unidentified shelled mollusk	'immature gull '
mayoocö ixpa̱leemelc	*Agaronia testacea*	'the Yellow-crowned Night Heron's ***xpa̱leemelc***'
mooxon iixz	*Cardites laticostatus*	'the scorpionfish's pet'
mosnaact	*Columbella fuscata*	Common Ground-Dove
mosniil inoosj	*Diodora inaequalis*	'the *mosniil* sea turtle's claw'
mosniil itaajc	*Columbella strombiformis*	'the *mosniil* sea turtle's saliva'
mosniil oozt	*Columbella strombiformis*	'what the *mosniil* sea turtle tattooed'
mosniil yaaspoj	*Conus princeps*	'the *mosniil* sea turtle's design'
mosniiquet	an unidentified shelled mollusk	'the (female) sea turtle's child'
mosnipol oopl	*Conus ximenes*	'what the leatherback sea turtle licks'
mosniscl inoosj	*Diodora inaequalis*	'the *mosniscl* sea turtle's claw'
naapxa yaat	*Glycymeris gigantea*	'what the Turkey Vulture baked in the coals'
napxeec	*Cerithium maculosum, C. stercusmuscarum, Mancinella tuberculata, Melongena patula, Strombus galeatus*	'the Turkey Vulture's daughter'
oot ano yapxoj	an unidentified gastropod	'what the coyote blows into'
oot icamazjij	*Tivela byronensis*	'the coyote's ***camazjij***'
oot icopas	*Isognomon janus, I. recognitus*	'the coyote's ***copas***'
oot ihaxöl	*Chione tumens*	'the coyote's ***haxöl***'
oot iquiit	*Carditamera affinis* (an odd growth form)	'the coyote's ***quiit***'
oot isatoj	*Entodesma pictum, Modiolus tumbezensis, Mytilus californianus*	'the coyote's ***satoj***'
oot iscám	*Crepidula onyx, C. striolata*	'the coyote's balsa'
oot iseeten	*Atrina maura, A. oldroydii*	'the coyote's ***seeten***'
oot iteexoj	*Spondylus leucacanthus*	'the coyote's ***teexoj***'
oot ixpa̱leemelc	*Agaronia testacea, Crassispira incrassata*	'the coyote's ***xpa̱leemelc***'
oot izacz	*Trachycardium consors*	'the coyote's ***zacz***'
oot yácmolca	*Morum tuberculosum*	'the coyote's fetish'

sleecoj cooxp ihaxöl	*Psammotreta cognata*	'the Snowy Egret's ***haxöl***'
slenapzj ihaxöl	*Tellina regia*	'the Little Blue Heron's ***haxöl***'
toerc iixz	an unidentified gastropod	'the Willet's pet'
xatoscö ihaxöl	*Barbatia reeveana*	'the Eared Grebe's ***haxöl***'
xepeiinoj mayoocö quih yaa	*Natica chemnitzii*	'***xepeiinoj*** belonging to the Yellow-crowned Night Heron'
ziic ccap cooxp ihaxöl	*Pholas chiloensis*	'the Tern's ***haxöl***'
ziic coopol ixpa̲leemelc	*Crassispira incrassata*	'the Double-crested Cormorant's ***xpa̲leemelc***'

Folklore-related names and those with obscure etymologies

cacaapxom	*Bulla gouldiana*	'what fattens [something]'
cascanoohcö	*Euvola vogdesi* (convex, inflated valve)	'deeply rounded'
cascanoohcö ooix	*Euvola vogdesi* (concave, nearly flat valve)	'what the ***cascanoohcö*** left behind'
cascapeetij	*Euvola vogdesi* (concave, nearly flat valve)	'circular'
cmaam	an unidentified shelled mollusk	'woman'
comcaii inoosj	*Nerita scabricosta*	'old woman's [toe]nail'
hacx xah cöscaap	*Cymatium keenae, Fusinus dupetitthouarsi*	'I'll just stand over here'
hant haaco iima ano haxölinaail	*Janthina prolongata*	'shell of the Catastrophe'
hant iiha cooml	*Conus princeps*	'what investigate the land'
Hant Iiha Quimx itaajc ootp	*Theodoxus luteofasciatus*	'saliva that *Hant Iiha Quimx* spit out'
hant iiha quimxoj oeenec	*Cypraecassis coarctata*	'what the *hant iiha quimxoj* carried'
hant itcoj	*Cantharus macrospira*	'bases of the land'
hant yapol ipapl	*Knefastia dalli*	'necklace of the dark place'
hasajitij	*Crucibulum scutellatum*	'basket…'
hastisol	*Mitrella ocellata*	'yellow rock'
hatx cöcazoj	*Aplysia californica*	'rump/buttocks' and 'one'
hax cöoaal	*Hipponix antiquatus*	'water servant'
haxölinaail isoj	an unidentified gastropod	'real shell'
hitcz ihizcoofizlc	*Vokesimurex elenensis, V. tricoronis*	'the firewood my younger sister gathered'
iicj	*Conus princeps*	'fine sand'
isliicot itaail	*Arca pacifica*	'wide-based upper back'
matcz ital coofizlc	*Crepidula onyx*	'who accompanies your younger sister to gather firewood'

najmís	an unidentified shelled mollusk	perhaps from ***naj*** and *cmis* 'what resembles ***naj***'
pahotmazexe	*Anomia peruviana*	'if I had seen one of them I would have cut him'
pohaapx camoz	an unidentified gastropod	'what thinks it is an ***haapx***'
queeex	*Chionopsis gnidia*	'who grunts'
quiimosim xepenoyaafc	*Anadara multicostata, Cardites laticostatus*	'what the beggar pounds in the sea'
seeex	*Glycymeris gigantea*	'will grunt'
stac ano coom	*Leiosolenus spatiosus*	'what lies in coral'
xahícosa	*Vokesimurex elenensis, V. tricoronis*	possibly related to *cahícosa*, a variety of mule deer
xepe an hapafc	*Chione undatella*	'what is pounded in the sea'
xepe quih iti cöiixi iiqui izooxolcam	*Jenneria pustulata*	'eighth tide line'
xepe quis	*Theodoxus luteofasciatus*	possibly a variant of ***xepe quilp***
xepe taacoj iixi	an unidentified shelled mollusk	'where the high tide finished'
xlehaxöl	*Tagelus peruvianus*	the second syllable is '***haxöl***' (*Leukoma grata*)
xnaa caaa ano haxölinaail	*Melampus mousleyi*	'shell in the *Salicornia*'
xnahaail	*Anadara multicostata, A. tuberculosa*	meaning unclear
xnahooil	*Arca pacifica*	meaning unclear
xojmás	terrestrial pulmonate and freshwater snails	from *hax* 'water', *ano* 'in', ***mas***, *Olivella dama*
xomcahaail	*Arca pacifica, Barbatia reeveana*	meaning unclear
xpacaao iiha	*Vokesimurex elenensis, V. tricoronis*	'the mermaid's possession'
xpajísotoj	*Pholas chiloensis, Tagelus peruvianus*	*ísotoj* is 'narrow (pl.)'
xpanooil	*Glycymeris gigantea*	meaning unclear
zojiime	*Costoanachis coronata*	snails found at *Zoj Iime*

APPENDIX 3
Species, including Seri names

Standard taxonomic notation is used in this appendix. As Coan and Valentich-Scott (2012:11) explain: "In formal scientific writing, the name of the author who first published the taxon and the date of its publication are generally given on first citation. . . . If the original generic assignment has been changed, the original author's name is shown in parentheses." The lack of a Seri name indicates that none was provided for that species.

Acanthochitona exquisita (Pilsbry, 1893)	***haxz heeque, tamax cosi***
Agaronia testacea (Lamarck, 1811)	***iplmateemij, mayoocö ixpaleemelc, oot ixpaleemelc***
Amiantis callosa (Conrad, 1837)	——
Anachis scalarina (G. B. Sowerby I, 1832)	***haxölinaail coopl***
Anadara formosa (G. B. Sowerby I, 1833)	——
Anadara grandis (Broderip & G. B. Sowerby I, 1829)	——
Anadara multicostata (G. B. Sowerby I, 1833)	***quiimosim xepenoyaafc, xnahaail***
Anadara tuberculosa (G. B. Sowerby I, 1833)	***xnahaail***
Anomia peruviana d'Orbigny, 1846	***haxölinaail cmasl, haxölinaail cxatlc, haxölinaail hamisj istj cmis, pahotmazexe***
Aplysia californica Cooper, 1863	***hatx cöcazoj***
Arca pacifica (G. B. Sowerby I, 1833)	***isliicot itaail, xnahooil, xomcahaail***
Architectonica nobilis Röding, 1798	***itx imonaaaij***
Arcopsis solida (G. B. Sowerby I, 1833)	——
Argopecten ventricosus (G. B. Sowerby II, 1842)	***hasla***
Astraea unguis (Wood, 1828)	***Hast Án ano haapx, xepenyaaan***
Atrina maura (G. B. Sowerby I, 1835)	***oot iseeten, seeten cmaam ctam, seeten iizax coocp, seeten inaail cquihjö***
Atrina oldroydii Dall, 1901	***oot iseeten, seeten comihj, seeten cooil***
Atrina tuberculosa (G. B. Sowerby I, 1835)	***seeten cmaam***
Barbatia reeveana (d'Orbigny, 1846)	***xatoscö ihaxöl, xomcahaail***
Berthellina ilisima (Marcus & Marcus, 1967)	***xepenozaah***

Bulla gouldiana Pilsbry, 1895	***cacaapxom***
Cantharus elegans (Griffith & Pidgeon, 1834)	——
Cantharus macrospira (Berry, 1957)	***hant itcoj, haxölinaail caaoloj***
Carditamera affinis (G. B. Sowerby I, 1833)	***quiit***
Carditamera affinis (an odd growth form)	***oot iquiit***
Cardites laticostatus (G. B. Sowerby I, 1833)	***haxölísotoj, mooxon iixz, quiimosim xepenoyaafc***
Cerithidea mazatlanica Carpenter, 1857	***haxölinaail quítajij, xtapacaj coopol, xtapacaj heeque***
Cerithium maculosum Kiener, 1841	***napxeec***
Cerithium stercusmuscarum Valenciennes, 1833	***napxeec***
Chama buddiana C. B. Adams, 1852	***hapos cahaaxat, zeel***
Chicoreus erythrostomus (Swainson, 1831)	***xiica iteencoj cheel, xnoocat an cheel***
Chione californiensis (Broderip, 1835)	***spiitquim***
Chione tumens Verrill, 1870	***haxöl aapa, oot ihaxöl***
Chione undatella (G. B. Sowerby I, 1835)	***spiitquim coospoj, xepe an hapafc***
Chionista fluctifraga (G. B. Sowerby II, 1853)	***haan***
Chionopsis gnidia (Broderip and G. B. Sowerby I, 1829)	***queeex, spiitquim cpaxz, spiitquim ctamoo, spiitquim quictamo***
Chiton virgulatus G. B. Sowerby II, 1840	***capoclim, capoclimiil, tamax capoclim***
Codakia distinguenda (Tryon, 1872)	***xtamaxiictoj***
Columbella fuscata G. B. Sowerby I, 1832	***hast iti caatolam, iteel it hacaptxö, mosnaact***
Columbella strombiformis Lamarck, 1822	***hast iti caatolam, iteel it hacaptxö, mosniil itaajc, mosniil oozt***
Conus princeps Linnaeus, 1758	***hant iiha cooml, iicj, mosniil yaaspoj, xpaleemelc***
Conus regularis G. B. Sowerby I, 1833	***xpaleemelc***
Conus ximenes Gray, 1839	***mosnipol oopl, xpaleemelc***
Costoanachis coronata (G. B. Sowerby I, 1832)	***haxölinaail cooscl, haxölinaail quimaaxat, zojiime***
Crassispira incrassata (G. B. Sowerby I, 1834)	***oot ixpaleemelc, ziic coopol ixpaleemelc***
Crassostrea columbiensis (Hanley, 1846)	***haxt, stacj caacöl***
Crassostrea corteziensis (Hertlein, 1951)	***haxt, xat***
Crassostrea gigas (Thunberg, 1793)	——
Crepidula onyx G. B. Sowerby I, 1824	***casquim quih iti ihiij, matcz, ital coofizlc, miizj hant quixquim quih iti ihiij, oot iscám***
Crepidula striolata Menke, 1851	***casquim quih iti ihiij, miizj hant quixquim quih iti ihiij, oot iscám***
Crucibulum scutellatum (Wood, 1828)	***cahiixona, casooca, catamax, hasajitij, tásona***
Crucibulum spinosum (G. B. Sowerby I, 1824)	***cahiixona, catamax, tamax cosi***
Ctenocardia biangulata (Broderip and G. B. Sowerby I, 1829)	——
Cymatium keenae (Beu, 1970)	***hacx xah cöscaap, xepe yamaasij ctam***
Cyphoma emarginatum (G. B. Sowerby I, 1830)	——
Cypraecassis coarctata (G. B. Sowerby I, 1825)	***hant iiha quimxoj oeenec, xepe yamaasij***

Dentalium spp. Linnaeus, 1758	***cozaaij***
Diodora inaequalis (G. B. Sowerby I, 1835)	***hast iti queemij, mosniil inoosj, mosniscl inoosj***
Donax punctatostriatus Hanley, 1843	——
Dosinia ponderosa (J. E. Gray, 1838)	***halít cahooxp***
Echinolittorina aspera (Philippi, 1846)	***haxölinaail cooscl***
Elysia diomedea (Bergh, 1894)	——
Enaeta cumingii (Broderip, 1832)	——
Entodesma pictum (G. B. Sowerby I, 1834)	***oot isatoj***
Eucrassatella antillarum (Reeve, 1842)	——
Eupleura muriciformis (Broderip, 1833)	***hant iti yeemej coospoj***
Euvola vogdesi (R. Arnold, 1906)	***cascanoohcö*** (convex valve); ***cascanoohcö ooix, cascapeetij*** (concave, nearly flat valve)
Ficus ventricosa (G. B. Sowerby I, 1825)	***xasecö***
Fusinus ambustus (Gould, 1853)	——
Fusinus dupetitthouarsi (Kiener, 1840)	***hacx xah cöscaap, siip isoj, sipisot, xasecö ctam, xpal̲eemelc cacösxaj***
Gari helenae Olsson, 1961	——
Glycymeris gigantea (Reeve, 1843)	***naapxa yaat, satim, saxáp, seeex, xpanooil***
Heliacus bicanaliculatus (Valenciennes, 1832)	——
Heterodonax pacificus (Conrad, 1837)	***taijitiquiixaz***
Hexaplex nigritus (Philippi, 1845)	***xnoocat, xnoocat cmaacoj, xnoocat coopol, xnoocat coospoj, xnoocat cooxp, xnocatiil***
Hipponix antiquatus (Linnaeus, 1767)	***cmatjij oot quih yaa, hax cöoaal, mox***
Hysteroconcha lupanaria (Lesson, 1831)	——
Isognomon janus P. P. Carpenter, 1857	***copas hapapl, oot icopas***
Isognomon recognitus (Mabille, 1895)	***copas hapapl, oot icopas***
Janthina prolongata Blainville, 1822	***hant haaco iima ano haxölinaail, xepe ano xojmás***
Jenneria pustulata (Lightfoot in Solander, 1786)	***hatajeen camós cahtxima, hatajeen camós coopol, xepe quih iti cöiixi iiqui izooxolcam***
Knefastia dalli Bartsch, 1944	***hant yapol ipapl***
Laevicardium elatum (G. B. Sowerby I, 1833)	***xtiip***
Laevicardium substriatum (Conrad, 1837)	***xtiip heeque***
Lamelliconcha concinnus (G. B. Sowerby I, 1835)	——
Leiosolenus spatiosus P. P. Carpenter, 1857	***stac ano coom***
Leukoma grata (Say, 1831)	***haxöl, haxöl áa***
Limaria pacifica (d'Orbigny, 1846)	***comcaai him oocatx, xepenozaah***
Liocerithium judithae Keen, 1971	***haxölinaail quítajij, xtapacaj heeque***
Lithophaga aristata (Dillwyn, 1817)	——
Lithophaga attenuata (Deshayes, 1836)	——
Lottia dalliana (Pilsbry, 1891)	***cmacjij, hast iti cöcooyam, haxölinaail cooil, haxölinaail cooil caacoj, haxölinaail cooil ctopis***

Lottia stanfordiana (Berry, 1957)	***hast iti cöcooyam, haxölinaail cooil, haxölinaail cooil ctopis, naj***
Lottia strigatella (Carpenter, 1864)	***haxölinaail cooil cooxi***
Macrocypraea cervinetta (Kiener, 1843)	***camós caacoj, haxölinaail cheel caacoj***
Mactrotoma nasuta (A. A. Gould 1851)	——
Malea ringens (Swainson, 1822)	***xasecö cmaam, xepe yamaasij***
Mancinella speciosa (Valenciennes, 1832)	——
Mancinella tuberculata (G. B. Sowerby I, 1835)	***haxölinaail ancoj cmasl, honcaacoj imazaj, napxeec***
Megapitaria aurantiaca (G. B. Sowerby I, 1831)	***camazjij caacoj, zaah***
Megapitaria squalida (G. B. Sowerby I, 1835)	***camazjij***
Melampus mousleyi Berry, 1964	***haxölinaail comihj, xepe yamaasij heeque, xnaa caaa ano haxölinaail***
Melongena patula (Broderip and G. B. Sowerby I, 1829)	***napxeec***
Mexacanthina lugubris angelica (Oldroyd, 1918)	***hast imocl cooxalca, iquihimz iic cöihiipe***
Mitra belcheri Hinds, 1843	***haxölinaail cpotyoj caacoj***
Mitra tristis Broderip, 1836	——
Mitrella ocellata (Gmelin, 1791)	***hastisol, haxölinaail coopl***
Modiolus capax Conrad, 1837	***satoj***
Modiolus capax (large)	***saacj***
Modiolus tumbezensis Pilsbry & Olsson, 1935	***oot isatoj***
Morum tuberculosum (Reeve, 1842)	***haxölinaailc imahcaptxö, haxölinaailc impafc, oot yácmolca***
Myrakeena angelica (Rochebrune, 1895)	***stacj caacöl***
Mytella charruana (d'Orbigny, 1842)	***satoj aapa***
Mytella guyanensis (Lamarck, 1819)	——
Mytilus californianus Conrad, 1837	***oot isatoj, satoj aapa***
Nassarius brunneostomus (Stearns, 1893)	***haxölinaail anxö icaticpan, haxölinaail cnoosc, haxölinaail quicös, xpa̱leemelc quicös***
Nassarius tiarula (Kiener, 1841)	***haxölinaail cnoosc***
Natica chemnitzii Pfeiffer, 1840	***xepeiinoj mayoocö quih yaa***
Navanax inermis (Cooper, 1863)	***hant iti queemij***
Nerita funiculata Menke, 1851	***comcaii inoosj heeque***
Nerita scabricosta Lamarck, 1822	***comcaii inoosj***
Neverita reclusiana (Deshayes, 1839)	***xepeiinoj***
Niso splendidula (G. B. Sowerby I, 1834)	——
Nodipecten subnodosus (G. B. Sowerby I, 1835)	***hanl***
Octopus alecto Berry, 1953	***hapaj inloj caact, hapaj inloj cacöla, hapaj inloj cozla***
Octopus bimaculatus Verrill, 1883	***hapaj áa, hapaj isoj, hapaj quimaaxat***
Octopus fitchi Berry, 1953	***hapaj cosni***
Octopus hubbsorum Berry, 1953	***hapaj cheel***

Octopus sp.	***hapaj caacoj***
Octopus sp.	***hapaj cooil***
Octopus sp.	***hapaj cmaaitj***
Oliva incrassata (Lightfoot in Solander, 1786)	***xpaleemelc caacoj, xpaleemelc comihj caacoj, xpaleemelc coospoj, xpaleemelc quipcö***
Oliva porphyria (Linnaeus, 1758)	***xpaleemelc caacoj***
Oliva spicata (Röding, 1798)	***haxölinaail comihj, xpaleemelc comihj***
Olivella alba (Marrat, 1871)	***xiica cooxp heecto***
Olivella dama (Wood, 1828)	***mas, sayana, xiica cooxp, xiica hant cosítyalca***
Olivella steveni Burch & Campbell, 1963	***mas, xiica cooxp, xiica coxpísotoj, xiica hant cosítyalca***
Onchidella binneyi Stearns, 1894	***tamax***
Papyridea aspersa (G. B. Sowerby I, 1833)	***snoomtxö***
Parametaria dupontii (Kiener, 1849–50)	——
Paroctopus digueti (Perrier & Rochebrune, 1894)	——
Pascula ferruginosa (Reeve, 1846)	——
Periglypta multicostata (G. B. Sowerby I, 1835)	***spiitquim ctamoo***
Petricolaria cognata (C. B. Adams, 1852)	——
Pholas chiloensis Molina, 1782	***haxöl icaai ctam, heexoj, xpajísotoj, ziic ccap cooxp ihaxöl***
Pinctada mazatlanica (Hanley, 1856)	***caaitmoj, casopaj, copas, xiica quiictoj, ziix iiquet inaail***
Pinna rugosa G. B. Sowerby I, 1835	***seeten ctam***
Placunanomia cumingii Broderip, 1832	***Hant Ihiini stacj, haxölinaail quihócaloj***
Pleuroploca princeps (G. B. Sowerby I, 1825)	***xasecö caacoj***
Plicopurpura pansa (Gould, 1853)	***iquihimz iic cöihiipe***
Polinices bifasciatus (Griffith & Pidgeon, 1834)	***xepeiinoj heecto***
Polinices uber (Valenciennes, 1832)	***xepe quilp cooxp, xepeiinoj cooxp***
Psammotreta cognata (Pilsbry and Vanatta, 1902)	***haxöl icaai sleecoj iti yazám, sleecoj cooxp ihaxöl***
Psammotreta pura (A. A. Gould, 1853)	***hatj ihaxöl cooxp***
Pteria sterna (A. A. Gould, 1851)	***soocajam, xiica quiictoj, ziix iiquet inaail***
Pteropurpura centrifuga (Hinds, 1844)	***xepe án icös***
Pteropurpura erinaceoides (Valenciennes, 1832)	——
Raeta undulata (A. A. Gould, 1851)	***Juan Tomás ihaxöl***
Saccostrea palmula (P. P. Carpenter, 1857)	***stacj, stacj hapahit***
Saccostrea palmula (an odd growth form)	***stacj coiicöt, stacj quítajij***
Sanguinolaria tellinoides A. Adams, 1850	***hatj ihaxöl, haxöl icaai***
Semicassis centiquadrata (Valenciennes, 1832)	***haxölinaail ctocnij, haxölinaail quihimca, xepe yamaasij***
Simomactra dolabriformis (Conrad, 1867)	***haxöl icaai***
Sinum debile (Gould, 1853)	——
Sonorella sp. Pilsbry 1900	***xojmás***
Spondylus leucacanthus Broderip, 1833	***oot iteexoj, teexoj ctam***

Spondylus limbatus G. B. Sowerby II, 1847	***iimox, teexoj***
Stramonita biserialis (Blainville, 1832)	***haxölinaail ancoj cmasl***
Strombus galeatus Swainson, 1823	***haxölinaail an icoopxoj, napxeec, xepe yamaasij, xpaḻeemelc caacoj***
Strombus gracilior G. B. Sowerby I, 1825	***haxölinaail cotázita, haxölinaail cotitla cmaam, haxölinaail cotítzilca***
Strombus granulatus Swainson, 1822	***cotitla oot quih yaa, haxölinaail cotitla ctam, haxölinaail cotítzilca***
Tagelus peruvianus Pilsbry & Olsson, 1941	***ctam yazexe, haxöl icaai ctam, haxöl icaai sleecoj iti yazám, ipac casa ihaxöl, xlehaxöl, xpajísotoj, xtipisot***
Tegula rubroflammulata (Koch, 1843)	***cotopis quislitx***
Tegula rugosa (A. Adams, 1853)	***haapx***
Tellina cumingii Hanley, 1844	——
Tellina regia Hanley, 1844	***hatj ihaxöl, slenapzj ihaxöl***
Terebra ornata Gray, 1834	***xtapacaj coospoj***
Terebra variegata Gray, 1834	***xtapacaj heeque***
Theodoxus luteofasciatus (Miller, 1879)	***Hant Iiha Quimx itaajc ootp, xepe quilp, xepe quis***
Tivela byronensis (J. E. Gray, 1838)	***oot icamazjij***
Trachycardium consors (G. B. Sowerby I, 1833)	***oot izacz, siml ihataxoj, zacz, zacz ctam, zcaczj quictamo***
Trachycardium procerum (G. B. Sowerby I, 1833)	***siml ihataxoj, zacz***
Tripsycha tripsycha (Pilsbry & Lowe, 1932)	***xtoozajö***
Trivia solandri (G. B. Sowerby I, 1832)	***camonlc, hasit camós, haxölinaail cheel***
Turbo fluctuosus Wood, 1828	***cotopis***
Turritella banksi Reeve, 1849	***xtapacaj heeque***
Turritella gonostoma Valenciennes, 1832	***xtapacaj***
Turritella leucostoma Valenciennes, 1832	***xtapacaj***
Vitularia salebrosa (King & Broderip, 1832)	——
Vokesimurex elenensis (Dall, 1909)	***hant iti yeemej coospoj, hitcz ihizcoofizlc, motcz, quitxísotaj, xahícosa, xpacaao iiha***
Vokesimurex tricoronis (Berry, 1960)	***hant iti yeemej coospoj, hitcz ihizcoofizlc, motcz, quitxísotaj, xahícosa, xpacaao iiha***
Zonaria annettae (Dall, 1909)	***hatajeen camós***

APPENDIX 4
Species listed by family, including Seri names

Bivalvia

Mytilidae

Leiosolenus spatiosus	***stac ano coom***
Lithophaga aristata	——
Lithophaga attenuata	——
Modiolus capax	***satoj***
Modiolus capax (large)	***saacj***
Modiolus tumbezensis	***oot isatoj***
Mytella charruana	***satoj aapa***
Mytella guyanensis	——
Mytilus californianus	***oot isatoj***, ***satoj aapa***

Arcidae

Anadara formosa	——
Anadara grandis	——
Anadara multicostata	***xnahaail***
Anadara tuberculosa	***xnahaail***
Arca pacifica	***isliicot itaail***, ***xnahooil***, ***xomcahaail***
Barbatia reeveana	***xatoscö ihaxöl***, ***xomcahaail***

Glycymerididae

Glycymeris gigantea	***naapxa yaat***, ***satim***, ***saxáp***, ***seeex***, ***xpanooil***

Noetiidae

Arcopsis solida	——

Pteriidae	
Pinctada mazatlanica	***caaitmoj, casopaj, copas, xiica quiictoj, ziix iiquet inaail***
Pteria sterna	***soocajam, xiica quiictoj, ziix iiquet inaail***
Isognomonidae	
Isognomon janus	***copas hapapl, oot icopas***
Isognomon recognitus	***copas hapapl, oot icopas***
Pinnidae	
Atrina maura	***oot iseeten, seeten cmaam ctam, seeten iizax coocp, seeten inaail cquihjö***
Atrina oldroydii	***oot iseeten, seeten comihj, seeten cooil***
Atrina tuberculosa	***seeten cmaam***
Pinna rugosa	***seeten ctam***
Ostreidae	
Crassostrea columbiensis	***haxt, stacj caacöl***
Crassostrea corteziensis	***haxt, xat***
Crassostrea gigas	——
Myrakeena angelica	***stacj caacöl***
Saccostrea palmula	***stacj, stacj hapahit***
Saccostrea palmula (an odd growth form)	***stacj coiicöt, stacj quítajij***
Anomiidae	
Anomia peruviana	***haxölinaail cmasl, haxölinaail cxatlc, haxölinaail hamisj istj cmis, pahotmazexe***
Placunanomia cumingii	***Hant Ihiini stacj, haxölinaail quihócaloj***
Pectinidae	
Argopecten ventricosus	***hasla***
Euvola vogdesi	***cascanoohcö*** (convex valve); ***cascanoohcö ooix, cascapeetij*** (concave, almost flat valve)
Nodipecten subnodosus	***hanl***
Spondylidae	
Spondylus leucacanthus	***oot iteexoj, teexoj ctam***
Spondylus limbatus	***iimox, teexoj***

Limidae	
Limaria pacifica	***comcaai him oocatx, xepenozaah***
Lucinidae	
Codakia distinguenda	***xtamaxiictoj***
Carditidae	
Carditamera affinis	***quiit***
Carditamera affinis (an odd growth form)	***oot iquiit***
Cardites laticostatus	***haxölísotoj, mooxon iixz, quiimosim xepenoyaafc***
Crassatellidae	
Eucrassatella antillarum	——
Cardiidae	
Ctenocardia biangulata	——
Laevicardium elatum	***xtiip***
Laevicardium substriatum	***xtiip heeque***
Papyridea aspersa	***snoomtxö***
Trachycardium consors	***oot izacz, siml ihataxoj, zacz, zacz ctam, zcaczj quictamo***
Trachycardium procerum	***siml ihataxoj, zacz***
Chamidae	
Chama buddiana	***hapos cahaaxat, zeel***
Mactridae	
Mactrotoma nasuta	——
Raeta undulata	***Juan Tomás ihaxöl***
Simomactra dolabriformis	***haxöl icaai***
Tellinidae	
Psammotreta cognata	***haxöl icaai sleecoj iti yazám, sleecoj cooxp ihaxöl***
Psammotreta pura	***hatj ihaxöl cooxp***
Tellina cumingii	——
Tellina regia	***hatj ihaxöl, slenapzj ihaxöl***
Donacidae	
Donax punctatostriatus	——

Psammobiidae

Gari helenae	——
Heterodonax pacificus	***taijitiquiixaz***
Sanguinolaria tellinoides	***hatj ihaxöl, haxöl icaai***

Solecurtidae

Tagelus peruvianus	***ctam yazexe, haxöl icaai ctam, haxöl icaai sleecoj iti yazám, ipac casa ihaxöl, xlehaxöl, xpajísotoj, xtipisot***

Veneridae

Amiantis callosa	——
Chione californiensis	***spiitquim***
Chione tumens	***haxöl aapa, oot ihaxöl***
Chione undatella	***spiitquim coospoj, xepe an hapafc***
Chionista fluctifraga	***haan***
Chionopsis gnidia	***queeex, spiitquim cpaxz, spiitquim ctamoo, spiitquim quictamo***
Dosinia ponderosa	***halít cahooxp***
Hysteroconcha lupanaria	——
Lamelliconcha concinnus	——
Leukoma grata	***haxöl, haxöl áa***
Megapitaria aurantiaca	***camazjij caacoj, zaah***
Megapitaria squalida	***camazjij***
Periglypta multicostata	***spiitquim ctamoo***
Tivela byronensis	***oot icamazjij***

Petricolidae

Petricolaria cognata	——

Pholadidae

Pholas chiloensis	***haxöl icaai ctam, heexoj, xpajísotoj, ziic ccap cooxp ihaxöl***

Lyonsiidae

Entodesma pictum	***oot isatoj***

Gastropoda

Fissurellidae

Diodora inaequalis	***hast iti queemij, mosniil inoosj, mosniscl inoosj***

Lottiidae

Lottia dalliana	***cmacjij hast iti cöcooyam, haxölinaail cooil, haxölinaail cooil caacoj, haxölinaail cooil ctopis***
Lottia stanfordiana	***hast iti cöcooyam, haxölinaail cooil, haxölinaail cooil ctopis, naj***
Lottia strigatella	***haxölinaail cooil cooxi***

Trochidae

Tegula rubroflammulata	***cotopis quislitx***
Tegula rugosa	***haapx***

Turbinidae

Astraea unguis	***Hast Án ano haapx, xepenyaaan***
Turbo fluctuosus	***cotopis***

Neritidae

Nerita funiculata	***comcaii inoosj heeque***
Nerita scabricosta	***comcaii inoosj***
Theodoxus luteofasciatus	***Hant Iiha Quimx itaajc ootp, xepe quilp, xepe quis***

Littorinidae

Echinolittorina aspera	***haxölinaail cooscl***

Turritellidae

Turritella banksi	***xtapacaj heeque***
Turritella gonostoma	***xtapacaj***
Turritella leucostoma	***xtapacaj***

Vermetidae

Tripsycha tripsycha	***xtoozajö***

Cerithiidae

Cerithium maculosum	***napxeec***
Cerithium stercusmuscarum	***napxeec***
Liocerithium judithae	***haxöl inaail quítajij, xtapacaj heeque***

Potamididae

Cerithidea mazatlanica	***haxölinaail quítajij, xtapacaj coopol, xtapacaj heeque***

Strombidae

Strombus galeatus	***haxölinaail an icoopxoj, napxeec, xepe yamaasij, xpal̲eemelc caacoj***
Strombus gracilior	***haxölinaail cotázita, haxölinaail cotitla cmaam, haxölinaail cotítzilca***
Strombus granulatus	***cotitla oot quih yaa, haxölinaail cotitla ctam, haxölinaail cotítzilca,***

Janthinidae

Janthina prolongata	***hant haaco iima ano haxölinaail, xepe ano xojmás***

Eulimidae

Niso splendidula	——

Hipponicidae

Hipponix antiquatus	***cmatjij oot quih yaa, hax cöoaal, mox***

Calyptraeidae

Crepidula onyx	***casquim quih iti ihiij, matcz ital coofizlc, miizj hant quixquim quih iti ihiij, oot iscám***
Crepidula striolata	***casquim quih iti ihiij, miizj hant quixquim quih iti ihiij, oot iscám***
Crucibulum scutellatum	***cahiixona, casooca, catamax, hasajitij, tásona***
Crucibulum spinosum	***cahiixona, catamax, tamax cosi***

Naticidae

Natica chemnitzii	***xepeiinoj mayoocö quih yaa***
Neverita reclusiana	***xepeiinoj***
Polinices bifasciatus	***xepeiinoj heecto***
Polinices uber	***xepe quilp cooxp, xepeiinoj cooxp***
Sinum debile	——

Triviidae

Trivia solandri	***camonlc, hasit camós, haxölinaail cheel***

Cypraeidae

Macrocypraea cervinetta	***camós caacoj, haxölinaail cheel caacoj***
Zonaria annettae	***hatajeen camós***

Ovulidae

Cyphoma emarginatum	——

Jenneria pustulata	***hatajeen camós cahtxima, hatajeen camós coopol, xepe quih iti cöiixi iiqui izooxolcam***

Tonnidae

Malea ringens	***xasecö cmaam, xepe yamaasij***

Cassidae

Cypraecassis coarctata	***hant iiha quimxoj oeenec, xepe yamaasij***
Semicassis centiquadrata	***haxölinaail ctocnij, haxölinaail quihimca, xepe yamaasij***

Ficidae

Ficus ventricosa	***xasecö***

Ranellidae

Cymatium keenae	***hacx xah cöscaap, xepe yamaasij ctam***

Muricidae

Chicoreus erythrostomus	***xiica iteencoj cheel, xnoocat an cheel***
Eupleura muriciformis	***hant iti yeemej coospoj***
Hexaplex nigritus	***xnoocat, xnoocat cmaacoj, xnoocat coopol, xnoocat coospoj, xnoocat cooxp, xnocatiil***
Mancinella speciosa	——
Mancinella tuberculata	***haxölinaail ancoj cmasl, honcaacoj imazaj, napxeec***
Mexacanthina lugubris angelica	***iquihimz iic cöihiipe, hast imocl cooxalca***
Pascula ferruginosa	——
Plicopurpura pansa	***iquihimz iic cöihiipe***
Pteropurpura centrifuga	***xepe án icös***
Pteropurpura erinaceoides	——
Stramonita biserialis	***haxölinaail ancoj cmasl***
Vitularia salebrosa	——
Vokesimurex elenensis	***hant iti yeemej coospoj, hitcz ihizcoofizlc, motcz, quitxísotaj, xahícosa, xpacaao iiha***
Vokesimurex tricoronis	***hant iti yeemej coospoj, hitcz ihizcoofizlc, motcz quitxísotaj, xahícosa, xpacaao iiha***

Buccinidae

Cantharus elegans	——
Cantharus macrospira	***hant itcoj, haxölinaail caaoloj***

Columbellidae

Anachis scalarina	***haxölinaail coopl***
Columbella fuscata	***hast iti caatolam, iteel it hacaptxö, mosnaact***

Columbella strombiformis	***hast iti caatolam, iteel it hacaptxö, mosniil itaajc, mosniil oozt***
Costoanachis coronata	***haxölinaail cooscl, haxölinaail quimaaxat, zojiime***
Mitrella ocellata	***hastisol, haxölinaail coopl***
Parametaria dupontii	——

Melongenidae

Melongena patula	***napxeec***

Nassariidae

Nassarius brunneostomus	***haxölinaail anxö icaticpan, haxölinaail cnoosc, haxölinaail quicös, xpal̲eemelc quicös***
Nassarius tiarula	***haxölinaail cnoosc***

Fasciolariidae

Fusinus ambustus	——
Fusinus dupetitthouarsi	***hacx xah cöscaap, siip isoj, sipisot, xasecö ctam, xpal̲eemelc cacösxaj***
Pleuroploca princeps	***xasecö caacoj***

Volutidae

Enaeta cumingii	——

Harpidae

Morum tuberculosum	***haxölinaailc imahcaptxö, haxölinaailc impafc, oot yácmolca***

Olividae

Agaronia testacea	***iplmateemij, mayoocö ixpal̲eemelc, oot ixpal̲eemelc***
Oliva incrassata	***xpal̲eemelc caacoj, xpal̲eemelc comihj caacoj, xpal̲eemelc coospoj, xpal̲eemelc quipcö***
Oliva porphyria	***xpal̲eemelc caacoj***
Oliva spicata	***haxölinaail comihj, xpal̲eemelc comihj***

Olivellidae

Olivella alba	***xiica cooxp heecto***
Olivella dama	***mas, sayana, xiica cooxp, xiica hant cosítyalca***
Olivella steveni	***mas, xiica cooxp, xiica coxpísotoj, xiica hant cosítyalca***

Mitridae

Mitra belcheri	***haxölinaail cpotyoj caacoj***
Mitra tristis	——

Conidae	
Conus princeps	***hant iiha cooml, iicj, mosniil yaaspoj, xpal̲eemelc***
Conus regularis	***xpal̲eemelc***
Conus ximenes	***mosnipol oopl, xpal̲eemelc***
Terebridae	
Terebra ornata	***xtapacaj coospoj***
Terebra variegata	***xtapacaj heeque***
Turridae	
Crassispira incrassata	***oot ixpal̲eemelc, ziic coopol ixpal̲eemelc***
Knefastia dalli	***hant yapol ipapl***
Architectonicidae	
Architectonica nobilis	***itx imonaaaij***
Heliacus bicanaliculatus	——
Bullidae	
Bulla gouldiana	***cacaapxom***
Aglajidae	
Navanax inermis	***hant iti queemij***
Aplysiidae	
Aplysia californica	***hatx cöcazoj***
Pleurobranchidae	
Berthellina ilisima	***xepenozaah***
Plakobranchidae	
Elysia diomedea	——
Onchidiidae	
Onchidella binneyi	***tamax***
Hydrobiidae	
a mud snail	***xojmás***
Ellobiidae	
Melampus mousleyi	***haxölinaail comihj, xepe yamaasij heeque, xnaa caaa ano haxölinaail***

Succineidae	
a terrestrial pulmonate snail	***xojmás***
Bulimulidae	
a terrestrial pulmonate snail	***xojmás***
Helminthoglyptidae	
Sonorella sp.	***xojmás***

Polyplacophora

Chitonidae	
Chiton virgulatus	***capoclim, capoclimiil, tamax capoclim***
Acanthochitonidae	
Acanthochitona exquisita	***haxz heeque, tamax cosi***

Scaphopoda

Dentaliidae	
Dentalium sp.	***cozaaij***

Cephalopoda

Teuthida (squid)	***pajaas***
Octopodidae (octopus)	***hapaj, ziix cotopis***
Octopus alecto	***hapaj inloj caact, hapaj inloj cacöla, hapaj inloj cozla***
Octopus bimaculatus	***hapaj áa, hapaj isoj, hapaj quimaaxat***
Octopus fitchi	***hapaj cosni***
Octopus hubbsorum	***hapaj cheel***
Octopus sp.	***hapaj caacoj***
Octopus sp.	***hapaj cmaaitj***
Octopus sp.	***hapaj cooil***
Paroctopus digueti	——

APPENDIX 5
Mollusks and their uses

The uses of mollusks given here are based on the data available. The absence of a particular use for a species does not necessarily mean that it was not used in that way (see Species Accounts for details). For example, occasionally a previously unused shell is seen in jewelry and might not be noted here.

Species	Food	Utilitarian	Adornment	Medicine and magic	Recreation	Sale or trade	Bait
Acanthochitona exquisita	eaten in past						
Agaronia testacea			strung in shell jewelry		feather throwing toy		
Amiantis callosa							
Anachis scalarina			strung in shell jewelry				
Anadara formosa							
Anadara grandis	eaten in past	scraper			throwing, flipping game		
Anadara multicostata	eaten in past	scraper			throwing, flipping game		
Anadara tuberculosa							
Anomia peruviana	eaten in past		strung in shell jewelry				
Aplysia californica							incidental

Species	Food	Utilitarian	Adornment	Medicine and magic	Recreation	Sale or trade	Bait
Arca pacifica	eaten in past	hygienic implement, scraper					
Architectonica nobilis				strung as amulet			
Arcopsis solida							
Argopecten ventricosus	eaten in past		strung in shell jewelry			shells traded	
Astraea unguis							
Atrina maura						meat sold	
Atrina oldroydii							
Atrina tuberculosa	eaten in past	food platter, paddles for balsa (probably incidental)			musical rasp	meat sold	
Barbatia reeveana							
Berthellina ilisima				medicinal tea			
Bulla gouldiana			strung in shell jewelry		throwing game		
Cantharus elegans							
Cantharus macrospira							
Carditamera affinis	eaten in past		strung in shell jewelry (rare)				
Cardites laticostatus	possibly eaten in past						
Cerithidea mazatlanica			strung in shell jewelry				
Cerithium maculosum			strung in shell jewelry				
Cerithium stercusmuscarum			strung in shell jewelry				
Chama buddiana	eaten in past						

Species	Food	Utilitarian	Adornment	Medicine and magic	Recreation	Sale or trade	Bait
Chicoreus erythrostomus	eaten		opercula strung in shell jewelry			meat sold	
Chione californiensis	eaten			good luck cache, vision circle	throwing games	meat sold	
Chione tumens							
Chione undatella	eaten						
Chionista fluctifraga	eaten					sold live	
Chionopsis gnidia	eaten						
Chitonidae	eaten in past		strung in shell jewelry				
Chiton virgulatus	eaten in past						
Codakia distinguenda	eaten in past	digging tool, pigment container	strung as earrings in past	fetish			
Columbella fuscata			strung in shell jewelry				
Columbella strombiformis			strung in shell jewelry				
Conus princeps			strung in shell jewelry (pendant)		feather throwing toy		
Conus regularis			strung in shell jewelry				
Conus ximenes			strung in shell jewelry				
Costoanachis coronata			strung in shell jewelry				
Crassispira incrassata							
Crassostrea columbiensis	eaten					sold live	
Crassostrea corteziensis	eaten					sold live	
Crassostrea gigas	eaten						
Crepidula onyx	possibly eaten in past						

Species	Food	Utilitarian	Adornment	Medicine and magic	Recreation	Sale or trade	Bait
Crepidula striolata	possibly eaten in past						
Crucibulum scutellatum	eaten in past	pigment container					
Crucibulum spinosum							
Ctenocardia biangulata							
Cymatium keenae							
Cyphoma emarginatum							
Cypraecassis coarctata					child's rattle or toy		
Dentalium spp.							
Diodora inaequalis							
Donax punctatostriatus	possibly eaten in past						
Dosinia ponderosa	eaten	scraper, digging tool					
Echinolittorina aspera			strung in shell jewelry				
Elysia diomedea							
Enaeta cumingii							
Entodesma pictum							
Eucrassatella antillarum							
Eupleura muriciformis							
Euvola vogdesi (concave, flat valve)	eaten in past		strung in shell jewelry (pendant)		child's toy or rattle		
Euvola vogdesi (convex valve)	eaten in past	scraper, pigment container					

Species	Food	Utilitarian	Adornment	Medicine and magic	Recreation	Sale or trade	Bait
Ficus ventricosa		dipper or funnel, drinking vessel, container					
Fusinus ambustus							
Fusinus dupetitthouarsi					doll	shell traded	
Gari helenae							
Glycymeris gigantea	eaten in past	eating utensil, scraper					
Heliacus bicanaliculatus			strung in shell jewelry				
Heterodonax pacificus	eaten by children at play		strung in shell jewelry, clothes decoration				
Hexaplex nigritus	eaten					meat sold	
Hipponix antiquatus							
Hysteroconcha lupanaria							
Isognomon janus			strung in shell jewelry				
Isognomon recognitus			strung in shell jewelry				
Janthina prolongata							
Jenneria pustulata			strung in shell jewelry				
Knefastia dalli			strung in shell jewelry				
Laevicardium elatum	eaten	eating utensil, cup, food and pigment container, dipper, digging tool, scraper			broken pieces in rattles		

Species	Food	Utilitarian	Adornment	Medicine and magic	Recreation	Sale or trade	Bait
Laevicardium substriatum							
Lamelliconcha concinnus	possibly eaten in past						
Leiosolenus spatiosus							
Leukoma grata	eaten	tweezers, clam markings sources for basket designs	strung in shell jewelry (rare)		game pieces	meat sold	
Limaria pacifica							
Liocerithium judithae			strung in shell jewelry				
Lithophaga aristata							
Lithophaga attenuata							
Lottia dalliana	eaten in past		strung in shell jewelry				
Lottia stanfordiana	eaten in past		strung in shell jewelry				
Lottia strigatella							
Macrocypraea cervinetta				strung as amulet	child's toy or rattle		
Mactrotoma nasuta							
Malea ringens							
Mancinella speciosa							
Mancinella tuberculata	eaten in past					meat sold	
Megapitaria aurantiaca	eaten						
Megapitaria squalida	eaten						
Melampus mousleyi			strung in shell jewelry				
Melongena patula							

Species	Food	Utilitarian	Adornment	Medicine and magic	Recreation	Sale or trade	Bait
Mexacanthina lugubris angelica	possibly eaten in past			topical skin ointment			
Mitra belcheri							
Mitra tristis							
Mitrella ocellata			strung in shell jewelry				
Modiolus capax	eaten	knife, scraper, potter's tool, eating utensil, pigment container		medicinal powder, byssal fibers as amulet or charm		meat sold	
Modiolus capax (large)	eaten	knife, scraper, pigment container					
Modiolus tumbezensis	eaten						
Morum tuberculosum			strung in shell jewelry (pendant)		feather throwing toy		
Myrakeena angelica	eaten					sold live	
Mytella charruana							
Mytella guyanensis							
Mytilus californianus		knife					
Nassarius brunneostomus			strung in shell jewelry				
Nassarius tiarula			strung in shell jewelry				
Natica chemnitzii			strung in shell jewelry (pendant)				
Navanax inermis							
Nerita funiculata			strung in shell jewelry	good luck charm in boats			
Nerita scabricosta	eaten in past			good luck charm in boats			on hook or chumming

Species	Food	Utilitarian	Adornment	Medicine and magic	Recreation	Sale or trade	Bait
Neverita reclusiana		to listen to the ocean when inland					
Niso splendidula							
Nodipecten subnodosus	eaten in past	digging tool, container for oil or pigments		fetish			
Octopodidae	eaten		suction cups strung as a necklace				on hook
Octopus alecto							
Octopus bimaculatus	eaten						
Octopus fitchi				medicinal tea, stop the wind			
Octopus hubbsorum	eaten						
Oliva incrassata			strung in shell jewelry (pendant)		feather throwing toy		
Oliva porphyria							
Oliva spicata			strung in shell jewelry		feather throwing toy		
Olivella alba			strung in shell jewelry				
Olivella dama			strung in shell jewelry				
Olivella steveni			strung in shell jewelry				
Onchidella binneyi	eaten in past						
Papyridea aspersa							
Parametaria dupontii							
Paroctopus digueti							
Pascula ferruginosa			strung in shell jewelry				

Species	Food	Utilitarian	Adornment	Medicine and magic	Recreation	Sale or trade	Bait
Periglypta multicostata	eaten						
Petricolaria cognata							
Pholas chiloensis		knife, potter's tool					
Pinctada mazatlanica	eaten in past		inlay	medicinal powder		pearls sold	
Pinna rugosa	eaten	food platter, musical rasp, paddles for balsa (probably incidental)		byssal fibers in charm or amulet	byssal fibers used as doll hair	pearls sold	
Placunanomia cumingii							
Pleuroploca princeps							
Plicopurpura pansa				topical medicine			
Polinices bifasciatus			strung in shell jewelry (pendant)				
Polinices uber			strung in shell jewelry				
Psammotreta cognata							
Psammotreta pura			strung in shell jewelry				
Pteria sterna	eaten in past	pigment containers		medicinal powder		pearls sold	
Pteropurpura centrifuga							
Pteropurpura erinaceoides							
Raeta undulata		eating utensil					
Saccostrea palmula	eaten						
Saccostrea palmula (odd growth form)							

Species	Food	Utilitarian	Adornment	Medicine and magic	Recreation	Sale or trade	Bait
Sanguinolaria tellinoides			strung in shell jewelry				
Semicassis centiquadrata							
Simomactra dolabriformis	possibly eaten	potter's tool, eating utensil					
Sinum debile							
Sonorella spp.			strung in shell jewelry (rare, in past)				
Spondylus leucacanthus							
Spondylus limbatus	eaten in past	dish for food				meat sold	
Stramonita biserialis	eaten in past					meat sold	
Strombus galeatus	eaten in past	shell trumpet			doll	meat sold	
Strombus gracilior	eaten in past		opercula and small shells strung in shell jewelry			meat sold	
Strombus granulatus	possibly eaten in past						
Tagelus peruvianus		eating utensil, potter's tool, knife					
Tegula rubroflammulata							
Tegula rugosa	eaten in past						on hook or chumming
Tellina cumingii							
Tellina regia			strung in shell jewelry				
Terebra ornata							
Terebra variegata			strung in shell jewelry				
Teuthida (squids)	eaten						

Species	Food	Utilitarian	Adornment	Medicine and magic	Recreation	Sale or trade	Bait
Theodoxus luteofasciatus			strung in shell jewelry, clothes decoration				
Tivela byronensis	possibly eaten in past						
Trachycardium consors	eaten in past	scraper					
Trachycardium procerum	eaten in past	scraper					
Tripsycha tripsycha		smoking pipe, drinking straw					
Trivia solandri			strung in shell jewelry, earrings on dolls				
Turbo fluctuosus	eaten			medicinal powder	opercula used as game pieces or gambling counters		on hook or chumming, lobster traps
Turritella banksi			strung in shell jewelry				
Turritella gonostoma		punch or auger			gambling and games, used in play	shells traded	
Turritella leucostoma		punch or auger			gambling and games, used in play	shells traded	
Vitularia salebrosa							
Vokesimurex elenensis							
Vokesimurex tricoronis							
Zonaria annettae			strung in shell jewelry	strung as amulet	child's toy or rattle		

APPENDIX 6
Names from previous records

In previous years outsiders interacted with the Seri people, some visiting for short periods, others staying for years. Short or extended lists of Seri words were recorded, a few of which are mollusks names. Such transcriptions reveal continuity of names and give clues to past names or uses, but are inherently difficult to decipher because of the different sounds, the absence of an internationally used linguistic alphabet in the past, and the fact that most of these people had no linguistic training. The table in this appendix attempts to relate the transcriptions of mollusk names with the species referred to, without the actual mollusk at hand. Where the species name is indecipherable, none is supplied.

Plausible identification	**Pinart 1879**	**Sheldon 1921–1922**	**Davis 1936 (see appendix 7 for his notes accompanying the names)**	**G. Harrington 1941**	**Malkin 1953, 1955**	**E. and M. Moser 1951–1976**
Olivella dama		muss 'necklace'				mas
Chione californiensis		spáetkim 'clam (one variety)'	Pat ke		spít ́kɛmenáł [inạail, shell], pīt ́ham	spíitkim
Chionopsis gnidia						spíitkiŋ ktamóo
Leukoma grata	aX'X'uuill 'almeja, muscle [*sic*] shell fish'				aẋł	ʔáXWł
Saccostrea palmula	stakX 'osteón [*sic*] (ostra), oysters'					stakx

Plausible identification	Pinart 1879	Sheldon 1921–1922	Davis 1936 (see appendix 7 for his notes accompanying the names)	G. Harrington 1941	Malkin 1953, 1955	E. and M. Moser 1951–1976
various gastropods	Xualem 'caracolitos, a small snail, perriwinkle [*sic*]'		Palemˊex Pa̤lam egh		x̣paˏlémmeł, x̣paˏlé^{t}mmɛłk, xpaˏlemēł, x̣pāˊlemmeł	x̣paˈlæmæłk
Crucibulum scutellatum				t'aasonah		
Euvola vogdesi (concave, almost flat, valve)			Ca̤sca̤ Ba̱ˊtish	casca-peh-tiht		kaskapéttiix
Euvola vogdesi (convex valve)			Caca-No̱ˊ			kaskanóoʔk^{w}
Hexaplex nigritus			No̱ˊcaht		no^{k}kāt	nókkaat
Chicoreus erythrostomus						nókkaat an kʔéeł
Ficus ventricosa			Ha̤ sa̱ˊo̱k		xaseú^{k}	
Dosinia ponderosa			A̤ se̱t ka̤ o̱ˊ-p,ˏ A̤ sit ka̤ oapkˊ		asct kaōxp	ʔałít kaʔóox̣p
——			Sta̱ˊa̱p			
Trivia solandri			ˏA̤ sa̱ ka̤ mo̱the̱ ˊah [*iha* 'it is']	ah-set-ah-mos	asyt kamós	ˈʔássit kamós
Trachycardium sp.			Cahk s			
Turbo fluctuosus			Ca̤to̱k pia	ko-tō-peh sen-ais [*inaail*]	kato^{p} pɛł	kotóppis
Simomactra dolabriformis			Ous-le ki, O̱kle ki̱			
——			La̤ to̱k te se̱ i̱ mo̱s			
Laevicardium elatum			Sta̱ˊip		x̣tɛ^{p}	x̣tíip
Crassostrea corteziensis			O̱kt			ˈʔax̣t
Megapitaria squalida			Co̱ mas kee			kaˈmaṣ̌xix
Turritella sp.			To̱ ba̤ˊkag͜h		x̣taba^{k}kāx, xtabák ˊkāx, x̣tapa^{k}kāḥ,	x̣tapákkaax

Plausible identification	**Pinart 1879**	**Sheldon 1921–1922**	**Davis 1936 (see appendix 7 for his notes accompanying the names)**	**G. Harrington 1941**	**Malkin 1953, 1955**	**E. and M. Moser 1951–1976**
Argopecten ventricosus			Ka̲st logh			
Crucibulum sp.			Co̲ ɛgh o̲n -nine - nine [*inaail*]		kaéx̣ ′x̣ōna, kaɛ ′óna,	kaʔíXXoona
——			Sto̲mpt ′			
Spondylus limbatus			Tass′ho̲			té·x̣ox, téXXoox
Pinnidae			Sɛt ten		seᵗ′tɛn	′sættæn
Dentalium sp., Vermetidae			Co̲ shi ′			košáaix
——			Sho̲′ck			
Trachycardium sp.			Suck se̲ ni̲c ′		šak ′šynał [*inaail*]	ṣakṣ̌
Dosinia ponderosa			Ut ka̲ oupt ′			
Vermetidae			Tou′ sau			tóoṣ̌aox̣W, ′Xt̲ōṣ̌àòW
Spondylus limbatus			A̲′ mo̲h			ímox̣
Carditamera affinis			Ka̲′ it		kīᵗ	kíit
Turbo fluctuosus operculum			Ca took′ pais At te̲ i̲′mos, Ca̤-to̲-pe̲-hiss-e̲-te̲-yah é̲-mos			
Modiolus capax			Sah′ toh		šɛᵗtox	′sattox
Pinctada mazatlanica				co'k-paz		
various gastropods				nap-hehck		naapX éek
Octopus sp.					ap′pax	ʔáppax
Octopus fitchi						ʔapax kósni
Onchidella binneyi					tām ′māx̣	támmaax̣
Acanthochitona exquisita, Crucibulum spinosum						tamax̣ kóssi

Plausible identification	Pinart 1879	Sheldon 1921–1922	Davis 1936 (see appendix 7 for his notes accompanying the names)	G. Harrington 1941	Malkin 1953, 1955	E. and M. Moser 1951–1976
Anomia peruviana					šopao^{t}mašex̣ˊxε̄	
Olivella dama					ˀε^{k}kaan gose^{t}·tyaxłkáł [pl.]	ṣ̌ìx̣ ˈʔān̪t̪ˀ kòˈsīt̪t̪òòx̣ [sg.]
Tegula rugosa					aʔpx̣	ˈʔaapx̣
Neverita reclusiana					xe^{p}ˊpenox	x̣æppæˈinox
Bulla gouldiana					xakápxom	kakáapx̣om
Nodipecten subnodosus					anˊsynáł [*inaail*]	ʔánł
——					x̣asák^{x}	
Theodoxus luteofasciatus					x̄ε^{p}pεkíł^{p}	x̣æpæ kíłp
Tellina regia					aˊtxeox̣ł	ʔátx iʔáx̣Wł
Architectonica nobilis						ítx̣ imonáaix
Chitonidae						kapókłiŋ
Nerita sp.						koŋkái inósx
Nassarius sp.						ʔax̣Wł ináił knóosk
Anomia peruviana						ʔaXWł ináił kXátłk
Tagelus peruvianus						íppaak kássa iʔáx̣Wł
Mexacanthina lugubris angelica, Plicopurpura pansa						ikiʔímš ik k^{w}iʔíippe
Columbella sp.						itéeł it ʔakáptx̣W
Agaronia testacea						mayóok^{w} ix̣paléemmeełk
Hipponix antiquatus						mox̣
Morum tuberculosum						oot yákmołka
Modiolus capax						ˈsaakx̣

Plausible identification	Pinart 1879	Sheldon 1921–1922	Davis 1936 (see appendix 7 for his notes accompanying the names)	G. Harrington 1941	Malkin 1953, 1955	E. and M. Moser 1951–1976
terrestrial and freshwater snails						Xoxmás
Glycymeris gigantea						x̣panóiɬ
various gastropods						x̣éppee yamáassiix
Anadara multicostata, Cardites laticostatus						kímmoosiŋ Xéppe ano yáafk
Zonaria annettae						ʔata'xæn ka'mos

APPENDIX 7

Names recorded by Edward H. Davis

The first serious attempt at transcribing Seri mollusk names was made by Edward H. Davis in March 1936 on Isla Tiburón, based on shells he collected on the island's shores.[1] Davis's list of names and notes are reproduced below as he wrote them. My interpretation of his transcription, when clear, is given in the third column. When I am only guessing, I enclose the name in square brackets. For the convenience of the reader, the scientific name is given as well.

1. Palem´ex	food	***xpaḻeemelc***, various gastropods
2. Ca̤sca̤ Ba̱´tish		***cascapeetij***, *Euvola vogdesi,* concave, almost flat valve
3. Caca-No̱´		***cascanoohcö***, *Euvola vogdesi,* convex valve
4. No̱´caht		***noocat***, *Hexaplex nigritus*
5. Ha̤ sa̱´o̱k	cup to drink	***xasecö***, *Ficus ventricosa*
6. A̤ se̱t ka̤ o̱´-p		[***halít cahooxp***, *Dosinia ponderosa*]
7. Sta̱´a̱p		——
8. A̤ sa̱ ka̤ mo̱the̱´ah		***hasit camós*** *iha*, 'that is an ***hasit camós***', *Trivia solandri*
9. Cahk´s		***xcacz***, *Trachycardium* sp.
10. Ca̤ to̱k´pia	food shell	***cotopis***, *Turbo fluctuosus*
11. Ous-le̱ ki´		***haxöl icaai***, *Simomactra dolabriformis*
12. La̤ to̱k´ te se̱ i̱´mo̱s		——
13. Sta̱´ip	used drinking cup	***xtiip***, *Laevicardium elatum*
14. A̤ sit´ ka̤ o̱apk´	white hair	***halít cahooxp***, *Dosinia ponderosa*
15. O̱kt´	oyster food	***haxt***, *Crassostrea corteziensis*
16. Co̱ mas´ kee	food	***camazjij***, *Megapitaria squalida*
17. Pa̱t´ ke	food	***spiitquim***, *Chione californiensis*
18. To̱ ba̤´ kag͜h	used gambling with carrizo	***xtapacaj***, *Turritella* sp.
19. Pa̤la̱m egh		***xpaḻeemelc***, various gastropods

1. Davis (1910–1944), Notebook 20, 21.

20. Ka̱st ´ logh	ears	***quisloj***, *Argopecten ventricosus*
21. O̱kle̱ ki̱	rub ollas smooth	***haxöl icaai***, *Simomactra dolabriformis*
22. Co̱ ɛgh ´ o̱n -nine - nine		***cahiixona*** *inaail*, *Crucibulum* sp. shell
23. Sto̱mpt ´	food	——
24. —	no good, cuts the feet	[***cozaaij***, Vermetidae, *Dentalium* sp.]
25. Tass´ho̱	food	***teexoj***, *Spondylus limbatus*
26. Sɛt ´ ten	knife	***seeten***, Pinnidae
27. Co̱ shi ´	knife, to cut fruit of pitahalla [*sic*] used as pocket knife today	[*icoozix*, 'what one saws with']
28. Sho̬´ck	to hear ocean-ear	[possibly another name for ***xepeiinoj***, *Neverita reclusiana*]
29. Nop hapt´	necklace	*hapapl* '(what is) strung'
30. same		
31. Suck se̱ ni̱c ´	scrape the bisnaga [*sic*] for water	***zacz*** *inaail*, *Trachycardium* spp. shell
32. Ut ka̱ oupt ´		[***halít cahooxp***, *Dosinia ponderosa*]
33. Tou´ sau	long tube used to suck through	***xtoozajö***, Vermetidae
34. A̱´ moh	food	***iimoj***, *Spondylus limbatus*
35. Ka̱´ it	food	***quiit***, *Carditamera affinis*
36. —	mussel, knife to cut babies [*sic*] navels	[probably *Modiolus capax*]
37. (?) [Davis's note]		
38. Operculum	Seri money	[probably *Turbo fluctuosus* operculum]

From other lists Davis made

Ca took´ pais At te̱ i̱´mos	Old Seri money	***cotopis*** *iti yahimoz*, *Turbo fluctuosus* operculum
Ca̤-to̱-pe̱-hiss-e̱-te̱-yah é̱-mos	Operculum shell money of Seris—old time	***cotopis*** *iti yahimoz*, *Turbo fluctuosus* operculum
Sah´ toh	mussel shell knife old time	***satoj***, *Modiolus capax*

APPENDIX 8
Seri consultants

Many Seris contributed to this study. A few were born while their families still lived as hunter-gatherers, when contact with the outside world was sporadic. The knowledge of their own world did not come through books or electronic media; it was a daily experience, reinforced through listening to those who carried knowledge and passed it on to others. Indeed, some were recognized as being more knowledgeable and were known as *hant iiha cöhacomxoj* 'those who have been informed about the ancient things'. Below, I briefly introduce some who shared their lives and stories with me.

María Luisa Astorga Flores (b. 1945). Naturally inquisitive and greatly knowledgeable about the world around her, María Luisa was trained as a nurse in Hermosillo. For years she has worked with doctors sent into the Seri villages by the Mexican government, and it is not unusual that she is called on to handle medical emergencies in Desemboque in their absence. The daughter of José Astorga and Rosa Flores (a primary consultant for the Seri ethnobotany), María Luisa is a special friend who never hesitates in answering my many odd queries with honesty and candor. (She is referred to in the text as María Luisa Astorga.)

María Victoria Astorga Encinas (b. ca. 1922, d. 2013). Victoria was married to Miguel Barnett, a Seri well known to outsiders who was featured in early photographs by Edward H. Davis. Much of what Victoria knew about the Seri world she learned while walking the desert and shores with her grandmother Angelita Díaz and grandfather Manuel Encinas, both of whom are listed in the 1890 census.[1] Victoria's stories recounting a nomadic childhood are amazing, and her recent death severs for us a personal link with that era. She will be greatly missed by the whole community. (Referred to in the text as Victoria Astorga.)

Efraín Estrella Romero (b. 1945, d. 2010). Efraín was married to María Luisa Astorga. He was the grandson of Chico Romero, from whom he learned much about the Seri world; as an acknowledged "bearer of knowldege," Efraín loved sharing songs, stories, and oral historical accounts with his family and others. His death left an enormous gap among the remaining few whose lives bridged the traditional hunter-gatherer life with the modern era. (Referred to in the text as Efraín Estrella.)

Juana Herrera Casanova (b. ca. 1938). Sister of Roberto Herrera Marcos, Juanita patiently and kindly explains the old ways of life she knew as a child. She is known as a midwife and herbalist well acquainted with the desert's medicinal plants. Through long years of chronic illness in her family, and with very few resources, she has been a steadfast provider and caregiver. Juanita is one of my heroes. (Referred to in the text as Juanita Herrera.)

1. This was the first known census of the Seris, taken in 1890 by order of Sonora governor Ramón Corral (Bowen 2000, citing Monteverde 1891:567–73). The census, though incomplete due to the people's nomadic lifestyle, was compiled at Rancho Costa Rica, and included data such as age, sex, place of birth, marital status, residence, proficiency in Spanish, relationship to others in the census, and sometimes a physical characteristic.

Lorenzo Herrera Casanova (b. 1947). Son of Roberto Herrera Marcos, Lorenzo has a keen mind and loves pondering the old ways, often bringing up information new to me. He is an excellent writer and enjoys recording on paper, in Seri, his knowledge of the Seri world.[2] (Referred to in the text as Lorenzo Herrera.)

Roberto Herrera Marcos (b. 1916, d. 1988). A well-known and colorful character, Roberto was the son of Antonio Herrera. Antonio was a friend and companion of Roberto Thomson Encinas, a non-Seri Mexican who did much to promote Seri interests during the early decades of the twentieth century, acting as a liaison between the Sonoran government and the Seri people. In a common practice of that period, a Seri child was sometimes "baptized" by a non-Seri, who as the child's godfather would then give the child his own name. In this way, Roberto took on Roberto Thomson Encinas's name and was subsequently known as Roberto Thomson or *Cmiique* Roberto 'Scri Roberto'. *Cmiique* Roberto's father was an early Seri constable, a role Roberto himself later assumed in Desemboque. Roberto was a primary consultant to my father, working with him on documentation and analysis of the language and the translation of the New Testament. (Referred to in the text as Roberto Herrera.)

Ramón López Barnett (b. 1947, d. 2009). Only a few months after I recorded the two songs sung by Ramón that are included here, he died from an aggressive cancer. Ramón proudly sang his songs for me, saying that he wanted to be known as their composer. (Referred to in the text as Ramón López.)

Alfredo López Blanco (b. 1943). The brother of Evangelina López Blanco, Alfredo is an expert fisherman and guide; because of his deep knowledge of the natural world, he has been a wonderful source of fascinating information. (Referred to in the text as Alfredo López.)

Evangelina López Blanco (b. ca. 1925, d. 1997). Gregarious and entertaining, and with a quick sense of humor, Eva delighted in sharing her knowledge of shells, often including humorous stories and quips about them. She explained that because she acquired her knowledge from her mother, who really knew shells, her own information was reliable. (Referred to in the text as Eva López.)

René Montaño Herrera (b. 1963). René is the son of José Ángel Montaño and Juana Herrera Casanova. A gifted writer and thinker who loves his first language, he has worked on researching, writing down, and archiving his knowledge of the Seri world in that language, even expanding this interest into teaching Seri young people how to write it.[3] (Referred to in the text as René Montaño.)

José Ángel Montaño M. (b. ca. 1932, d. 2012). Husband of Juana Herrera Casanova, José Ángel, along with his parents and siblings, was part of the last group of Seris living on Isla Tiburón in the 1950s. He was a gentle, quiet, and good-humored man who enjoyed talking about the old way of life. On a visit to his home, as we talked of traditional foods, Juanita indicated a jar containing an amber liquid high on a shelf—it was some sea turtle oil that José Ángel still enjoyed for flavoring his food. (Referred to in the text as José Ángel Montaño.)

Cleotilde Morales Colosio (b. 1942). Cleotilde is the daughter of Jesús Morales, a well-known storyteller and singer, and María Antonia Colosio, *Comcaii Mayoor*, a central figure in the Seri ethnobotany. Cleotilde is married to Alfredo López and frequently accompanies him on boat excursions in the area. A special memory I have is of Cleotilde sitting low in a boat, with her eyes closed, softly singing songs of the sea as we navigated offshore Isla Tiburón. (Referred to in the text as Cleotilde Morales.)

Ernestina Morales Colosio (b. ca. 1949). Daughter of Nacho Morales, who was considered by many to be the last traditional Seri shaman, Ernestina was one

2. Lorenzo wrote an essay about a group of Seris from Baja California (Herrera Casanova 2012b).
3. René wrote an essay about the Seri relationship with the leatherback sea turtle (Montaño Herrera 2012c).

of my childhood playmates. Special were the times we spent together on the dunes near Desemboque, gathering very fragrant sand verbena blossoms to string together like daisy chains. (Referred to in the text as Ernestina Morales.)

Francisco Morales Herrera (b. 1950, d. 2013). Gentle, soft-spoken, and knowledgeable, Francisco often kept special shells to show me and always enjoyed sharing details about them. (Referred to in the text as Francisco Morales.)

Claudia Raquel Moreno Díaz (b. 1961, d. 2013). Raquel was the granddaughter of Porfirio Díaz, a central figure in the book *Unknown Island* (Bowen 2000) and one of the last Seris to have lived on Isla San Esteban. Raquel was extremely knowledgeable about Seri oral history and the natural world. (Referred to in the text as Raquel Moreno.)

Francisco Xavier Moreno (b. 1964). Xavier recalls peering through a window as his grandfather Roberto Herrera worked with my father, listening as they discussed Seri words, and being especially fascinated seeing my father type without looking at the typewriter's keys. According to Xavier, that is when his interest in knowing more about his language began, an interest which led to his key participation in helping to compile the Seri dictionary. A fisherman by trade, Xavier is an excellent writer of Seri and often shares his enjoyment of pondering words and their meanings.[4] (Referred to in the text as Xavier Moreno.)

José Juan Moreno (b. 1925, d. 2007). According to his daughter, Raquel, despite José Juan's birth certificate giving his birthplace as Isla Tiburón, he was actually born at a ranch near Bahía de Kino owned by the non-Seri Moreno family. Because the Seri family was cared for by the Morenos, José Juan was given their name. José Juan recalled being present when Edward H. Davis visited the Seris on one of his later trips. A good reader, José Juan was delighted to read anything written in Seri. (Referred to in the text as José Juan Moreno.)

Ángela Torres Cubillas (b. ca. 1941). Angelita spent much of her early life on Isla Tiburón and in camps along the Infiernillo Channel. An excellent basket maker, she has won awards for her work, traveled outside of the Seri region, and counts many non-Seris as friends. Angelita easily recalls the ways of the traditional nomadic life, including special times of travel through the desert with her family and playing with dolls made from bones of sea turtle and deer. (Referred to in the text as Angelita Torres.)

Armando Torres Cubillas (b. ca. 1944, d. 2011). Unable to walk since becoming disabled as a child, Armando was internationally known for his beautiful ironwood carvings of sea turtles and ocean birds. He enjoyed singing traditional Seri songs and talking about the things he learned as a child on Isla Tiburón. Armando's absence on the seaside dune at Desemboque's edge where his house and workshop stood is keenly felt by many non-Seri friends who visited him there. (Referred to in the text as Armando Torres.)

Chavela Torres Molina (b. ca. 1924, d. 2013). Chavela was married to the expert hunter Cayetano Montaño. Although blind since young adulthood, Chavela was a midwife and was said to be one of the best. She was recognized as a rich repository of knowledge and songs and loved sharing her memories with anyone who showed an interest. She is greatly missed. (Referred to in the text as Chavela Torres.)

Sara Villalobos (b. ca. 1900, d. 1982). Sara was born at the dune camp of *Masaanaj* near Estero Sargento, where her family was part of the *Xnaamotat* people who lived around Cerro Tepopa. Sara's more distant

4. Xavier wrote an essay about the highest mountain on Isla Tiburón and its place in the Seri world (Moreno Herrera 2012b).

ancestors were from the *Xiica xnaai iicp coii*, a group that came from the southern part of the former Seri territory. Sara's trademark cloth dolls on cradleboards no doubt reside in many collections, and they are special reminders of a gregarious woman who touched the lives of many visitors who interacted with the people during much of the twentieth century. (Referred to in the text as Sara Villalobos.)

Photography credits:
María Victoria Astorga Encinas (seated) with her daughter Rosa Amalia Barnett. Photograph courtesy of Lynn Johnson.
Lorenzo Herrera Casanova. Photograph courtesy of Lynn Johnson.

NOTES

Introduction

1. See Sheridan (1999:9) for the early population estimates. Smith (1954) reports that a series of measles and influenza epidemics in 1925 and 1926 took many lives, reducing the Seri tribe from "perhaps 350 to about 160 population." According to M. Moser (n.d.) there were 241 Seris in 1956.
2. The Instituto Nacional de Estadística y Geografía Informática reports 595 speakers of the language aged five and over, in 2005 (INEGI 2009:35). The number given by the Seris differs, reflecting such parameters as language use and parentage. In 2011, the Seri government recorded around 600 adults of voting age (Miguel Estrella, via Carolyn O'Meara, personal communication).
3. Bahre (1980:200–202); Sheridan (1999:9–12).
4. See S. Marlett (2008a) and M. Moser and S. Marlett (2010) for Seri place names and naming practices. A project to record place names on Isla Tiburón and adjacent mainland area was completed in 2006 by Diana Luque-Agraz and Antonio Robles-Torres (Luque-Agraz and Robles-Torres 2006).
5. Bowen (2009) explores human contact with the Gulf's islands based on historical information and the archaeological record, with particular attention to the Seris.
6. See Spicer (1962), Sheridan (1999), and Bowen (2000) for postcontact Seri history.
7. The coordinates for El Desemboque del Río San Ignacio (*Haxöl Iihom*) are (29° 30′ 13″ N, 112° 23′ 43″ W); those of Punta Chueca (*Socaaix*) are (29° 00′ 51″ N, 112° 09′ 41″ W).
8. The ISO 639–3 code for the language is [sei]. The origin of the name Seri (and its spelling variants), an exonym that has been in use since the seventeenth century, has never been demonstrated. No scholarly work has even discussed it since no evidence of substance has been discovered; proposed meanings appear to be complete fabrications. Nor is there evidence that the name is or was pejorative (S. Marlett 2011).
9. See Johnston (1968) for a description of Seri ironwood carving. Ryerson (1976) discusses ironwood carving as well as other Seri crafts, focusing on the economic factors of the industry.
10. See the trilingual Seri dictionary (M. Moser and S. Marlett 2010) and the Seri grammar (S. Marlett, in preparation) for a comprehensive description of the language.
11. Campbell (1997:160); S. Marlett (2008a).
12. M. Moser and S. Marlett (2010:18).
13. Hendrickx et al. (2007:639) give its length as 4,512 kilometers. Length estimates differ in the literature because of such variables as scale and endpoints and whether or not the islands are included in the calculation (Richard Brusca and Michel Hendrickx, personal communication).
14. Felger et al. (2007:617) describe an estero as "Spanish for 'estuary', 'tidal marsh', 'marine wetland'. Esteros are crucial habitat for young fish, shorebirds, and wetland plants." A hypersaline environment is created since a regular freshwater inflow does not exist; hence, these areas have been called "negative estuaries." There are ten esteros in the Seri region (this number includes a very small estero north of Punta Chueca), on both the eastern shore of Isla Tiburón and the mainland coast. The largest, Estero Sargento, is the northernmost mangrove ecosystem in North America, with the exception of a small outlier population of black mangrove at Puerto Lobos, Sonora (Felger et al. 2001:10; Felger et al. 2012:9). As part of the historical Seri territory, the extensive Estero Santa Cruz, south of Bahía de Kino, figures in this book as well. The estero's name varies in common use as well as in the literature; I use names given by the Instituto Nacional de Estadística y Geografía (INEGI).

    In 2009 the Infiernillo Channel and its esteros (29,700 hectares; 29° 10' N, 112°14' W) were included in the international Ramsar list by the Mexican government, with Seri tribal agreement. See http://www.ramsar.org/cda/es/ramsar-news-archives-2009-mexico-names-marine-site/main/ramsar/1-26-45-84%5E24252_4000_2__.
15. Brusca and Findley (2005:14); Brusca et al. (2004:xv).
16. The name El Desemboque del Río San Ignacio (or, less formally, El Desemboque de los Seris) distinguishes it from another village to the north, also called El Desemboque. I use the commonly used name El Desemboque (or just Desemboque) for the Seri village. The people live today on communal land, called an *ejido*, with legal rights granted by the Mexican government in the 1970s. This coastal strip is a portion of their historic territory, lying between an area north of Bahía de Kino and the coastal mountains just north of El Desemboque.
17. Brusca and Findley (2005:16); Brusca and Hendrickx (2010:73–81); Hendrickx et al. (2007:641, 644).
18. Brusca (2007:423).
19. Brusca (2007:418).
20. Other tide lines are not commonly numbered. No explanation has been given for the reference here to the number eight; however, in the Seri beliefs, the numbers four and eight are significant.
21. The term *haxoj* implies the shore area as seen from the seaward side, as when arriving at shore in a boat.
22. Juan Ortiz Zapata, from 1678 (Sheridan1999:28); Adamo Gilg, from 1692 (Di Peso and Matson 1965:51–54); and Tomás Miranda, from 1749 (Sheridan 1999:148).

23. Di Peso and Matson (1965:unnumbered plate).

24. Pfefferkorn (1989:202). Pfefferkorn includes various indigenous groups in his description of the "Sonoran," likely including the Seris. Although he does not specify which he is describing in this detail, the Seris would have had the easiest access to shells.

25. Hardy (1829:281, 286, 298).

26. McGee (1898:101).

27. Pinart (1879).

28. McGee (1898). The shells he described are in the National Museum of Natural History collections, Smithsonian Institution.

29. Sheldon's notebooks (1921–1922) about the Seris are archived in the Special Collections of the University of Arizona Libraries, Tucson, Arizona. The word lists were not published.

30. Kroeber (1931).

31. Some of Davis's notebooks and sketchbooks are archived in the Cornell University Library (Division of Rare and Manuscript Collections), Ithaca, New York; others are at the San Diego History Center, San Diego, California. The shells he described are in the National Museum of the American Indian collections, Washington, DC, and the Southwest Museum in Los Angeles, California.

32. Harrington's collection and field notes are on file at the Arizona State Museum, Tucson, Arizona.

33. Smith's field notes, unpublished manuscript, photographs, and films are archived in the Special Collections of the University of Arizona Libraries, Tucson, Arizona.

34. Whiting's (1951) unpublished, typewritten copy of his diary is on file at the Museum of Northern Arizona, Flagstaff.

35. Malkin received his MA at the University of Minnesota in 1960. I am not aware if the shells he identified are in any collection.

36. Malkin (1960, 1962).

37. Griffen (1959). His Seri notes and photographs are archived in the Special Collections of the University of Arizona Libraries, Tucson, Arizona.

38. Bowen and E. Moser (1968).

39. Felger and M. Moser (1985).

40. M. Moser and S. Marlett (2010).

Preliminaries

1. Formerly known as the Summer Institute of Linguistics.

2. Their primary goal was to translate the New Testament and related scripture portions, and they based their work on a solid foundation of language learning, linguistic analysis, and anthropological documentation. Their ethnographic publications, sometimes coauthored with other researchers, describe such key aspects of the Seri culture as basketry, music, pottery, childbirth, and botanical knowledge. Other publications detail accounts from oral history or document features of the language. The practical application of language development for community use involved publications in the Seri language, such as a small dictionary, reading primers, and books on folklore and history. The trilingual dictionary (completed after Ed's death) is a product of over half a century of meticulous and extensive recording of the language.

3. For those interested, I describe some of my childhood experiences with Ike and others in C. Marlett (2000, 2002).

4. M. Moser and S. Marlett (2010).

5. See www.lengamer.org/admin/language_folders/seri/user_uploaded_files/links/File/Enciclopedia/EncicloSeri.htm.

6. Francisco "Chico" Romero (b. ca. 1888, d. 1974). Chico was present when early ethnographers visited the Seris and is pictured in photographs taken by both Edward H. Davis and Alfred Kroeber. Considered to be one of the "bearers of knowledge," Chico was a rich source of information. Affable and tall in stature, he was a true gentleman, and his handsome white beard fit well with the dark suit jacket he liked to wear. After losing his eyesight in old age, he was often led around Desemboque by his walking staff, his large white enamel cup hung by a cord around his neck and readily available for a hot cup of coffee. As he sat and sipped the coffee he would share his wonderful stories and knowledge of his desert world.

7. The people today are becoming much more aware of current issues involving their communities and how they fit into the world at large. There are several ongoing initiatives in which young people are being mentored in ecology and conservation. These include ornithology, botany, and sea turtle and bighorn sheep ecology.

8. Scott Ryerson provided a short interview on cassette tape with Roberto Hererra, made in 1969, in which Roberto named and described eleven shells.

9. The films, located in the Arizona State Museum and the University of Arizona Libraries Special Collections, are being digitized and repatriated to the *Comcaac* by Rodrigo T. Rentería-Valencia and Carolyn O'Meara. This is part of a long-term project to produce a new "annotated" version of the films in collaboration with the community, involving comments from elders and young people about what these films show and mean to them. At the time of this writing, digitization is not complete, so only short sections are available.

10. The collection was purchased from a private collector by Michael Higgins.

11. The shells were identified by Paul Valentich-Scott and the tortoise shell was identified by Richard S. White.

12. Beach drift in some areas can be heavy with ocean floor dwellers such as gorgonians (sea fans) or unfamiliar mollusks dredged by commercial shrimp boats. Brusca and Hendrickx (2010:86–92) discuss the huge impact of this industry and other forces on the biodiversity of the Gulf.

13. For an introduction to Seri songs, see Astorga et al. (1998). Vennum (2000) discusses Seri music and its relationship to other indigenous music of North America.

14. See Bowen (1976:103–107) and E. Moser and White (1968:146–153) for a discussion of the Giants.

15. See S. Marlett (2006) for history and details of the alphabet and writing system.

16. See S. Marlett (2007).

17. In a photo essay of the Seri people, Burckhalter (1976) includes photographs of some of my earliest consultants.

18. S. Marlett (2008b).

Topical Ethnography

1. Felger and M. Moser (1985:43–44).

2. Ibid., 59–78, 334–342.

3. Ibid., 59.

4. I thank Xavier Moreno for pointing this out to me.

5. The phrase implies something being from a common base, instead of simply indicating a similarity of some attribute where the verb *cmis* 'resemble' would be far more commonly used. The expression *hant cöquiih* is used with two noun phrases: the subject—the item to be classified—and another noun phrase—the item that is the point of comparison and the prototype for that category.

6. Felger and M. Moser (1985:370). The word *xoját* is *joját* for some people (M. Moser and S. Marlett 2010).

7. Two different terms for 'living or alive'—*quiisax* and *ccam*—reflect the difference between human and animal life, respectively. The words *quiistox* and *ccámotam* are the plural forms.

8. While the phrase covers all marine life, the in-depth classification of that other than mollusks is beyond the scope of this work.

9. Acorn and goose barnacles are included as *haxölinaailc* by most Seris. As one person explained, the barnacles are not made of bone or soft skin, but do contain an animal. However, no one I asked included either barnacles found on sea turtles (*Chelonibia testudinaria*) or whales (*Cryptolepas rhachianecti*).

10. Some such names incorporating the singular term *haxölinaail* are actually plural, since the modifiers are clearly plural. In this instance, *cotítzilca* is the plural form of *cotitla* 'to pinch'. One might think that to be grammatically correct, such names should use the plural form *haxölinaailc*. However, names in standard use (as recorded by different people in past years from various speakers and also verified by direct questioning in recent years) clearly incorporate the singular form *haxölinaail* in otherwise plural names.

11. I have not been able to elicit a term corresponding to a single valve of a bivalve. The whole shell is considered the *inaail*, since it covers only one animal—there is no term for "half shell." One person gave me the phrases *haxöl ihaxöl quih coocj* 'clam with two clam[shell]s' (for bivalves), and *haxöl ihaxöl quih tazo* 'shell with one clam[shell]' (for univalves), but these were obviously improvised in an attempt to oblige me.

12. The word for 'clam' in the Yaqui (Yoeme) language is *auli* (Estrada Fernández 2004:213). Most likely the word was borrowed from Seri ***haxöl*** (David Shaul, personal communication). As late as the nineteenth century the Seris had regular contact with the Yaqui people, something reflected in a few shared vocabulary items.

13. In considering the gastropod shell shapes given in Keen (1971:912), René Montaño identified the obvate, turriform, and obconic shapes as being those most closely corresponding to his concept of a ***xpa̲leemelc***. He described these shapes as *cooij* 'tubular'.

14. Eugene Hunn (2008:92–93) discusses this issue in his work with Zapotec biological data.

15. Griffen (1959) and E. Moser (1963) discuss the bands in detail. Bahre (1980) interprets such accounts given by the Seris in light of other factors.

16. Hunn (2008:93).

17. Felger and M. Moser (1985:62).

18. Ibid.

19. For example, the diamond stingray (*Dasyatis dipterura*) is called *hacat cmaam* 'female shark' because of its round shape, in contrast to the long streamlined shape of a shark. See Felger and M. Moser (1985:66) for a brief discussion of the sexual reference as relating to plants.

20. Loreta (occasionally spelled Loreto) Marcos (b. ca. 1890, d. 1968) is sometimes referred to as Loreta Margos in unpublished notes. She was married to Santo Blanco and was a gifted potter.

21. Felger and M. Moser (1985: 244).

22. Plant names reflecting this custom are discussed by Felger and M. Moser (1985:108); see also S. Marlett (2008a). Today the practice of name taboo continues somewhat, evidenced as a reluctance to directly name a recently deceased person, and instead using a system of reference terminology. The practice, however, seems to be less rigidly followed, and the deceased's name is occasionally used.

23. Three forms of a similar name recorded from Seri oral history are *Mas Quiho* '(one who) Sees ***Mas***'; *Xojmás Quiho* '(one who) Sees ***Xojmás***'; and *Masquillo* (M. Moser n.d.). The change of the snail's name was probably recent, as many older people today recall the former name. The word *muss* was transcribed by Charles Sheldon (1921–1922) as meaning 'necklace' (see Olivellidae). Probably when asking for the word for necklace, he was given the name of the shell with which it was strung. The word used today for necklace is *hapapjc* 'what is strung'.

24. Anatomical terms are presented in the third person possessed form (usually beginning with the prefix *i-*) and are translated with a third person possessor ('its' in these cases, as in 'its foot'). While English and Spanish anatomical terms occur without indicating the possessor in the word itself, such terms without possessors are rarely used in Seri.

25. J. Harrington (1945:297).

26. The antennules of the lobster are called the *itxaloms*, a term also referring to stray locks of hair on a person's head. It is unclear which usage is primary. The larger antennae of the lobster, as well as those of other invertebrates, are called *itaamalca* 'its horns or antlers'.

27. Mammals, birds, fish, reptiles, and amphibians are said to contain *ipxasi*; the word implies something edible. Also included are marine arthropods such as the lobster, large shrimp, and swimming crabs. However, most other marine invertebrates, such as the sea cucumbers, sea urchins, sea anemones, small arthropods, and worms, are not considered to have *ipxasi*, but instead are described as having a less firm internal consistency or being more "mushy."

28. Exploration with René Montaño Herrera about the terms used to describe various shapes of gastropods, looking at the illustrations of the shapes in Keen (1971:912), produced the following correlations: *csaamij* 'spiral' for involute; *cpeetij* 'circular' for patellate; *cneepni* 'blunt, stooped' for turbinate and trochoid; *ctocnij* 'round, spherical' for globose and bulloid; *cooipj* 'oval' for lenticular and cylindrical; and *coiij* 'tubular' for obovate, obconic, and turriform. Terms that relate to the shell sculpture include the following: *comihj* 'smooth, polished'; *cnoosc* 'rough'; *ihaaoloj cooiplam* 'whose grooves are transverse' for concentric ridges; *ihaaoloj cacöla* 'whose grooves are long' for radial sculpture; and *cpotyoj* 'what has small ridges' for ridged sculpture.

29. Felger and M. Moser (1985:100–102) present in much greater detail the traditional accounts of the origin story, as well as Seri beliefs about the supernatural.

30. In the translations of such accounts I refer to this figure as the Name Giver, even though the name does not have that literal meaning.

31. "Before diluvio [flood] came—a man (unknown name)—came and asked people what animal or bird it wanted to become—fish—whale—bird—horned toad—coyote—deer, etc. etc. Ocean flooded land—people died and became their chosen animal—or fowl or fish" (E. Moser n.d.).

32. Clay figurines found in the area are called *ziix coosyat yacaam*. The term *yacaam* is a special variant (in this context) of the obsolete from of *acaam* 'her husband'. See E. Moser and White (1968:146–153) for a discussion of figurines as relating to the Giants.

33. E. Moser (n.d.). Perhaps the way the bivalve grows attached to a rock evokes the way a small child is carried on one's hip.

34. See Felger and M. Moser (1985:100–108) for a detailed description of the Seri supernatural.

35. The spirit was said to also attack adults. Anyone who was weak, thin, and listless, with "no *iisax* (spirit)" was known as *Coen ooicöt* 'what the *Coen* has killed'. The term *cacoene* 'cause to be cursed (by the *Coen*)' means to harm someone through witchcraft.

36. Felger and M. Moser (1985:386).

37. Felger and M. Moser (1985:230–231).

38. Fontana and Fontana (2000:145–146, endnote 84).

39. McGee (1898:287–293).

40. For a discussion of burial practices, see Griffen (1959:28–30; 44–46) and Bowen and E. Moser (1968:123).

41. In describing the custom, María Luisa Astorga recalled seeing some of Sara Villalobos's possessions buried with her in 1982. She added that this was the last burial at which she had seen this done.

42. These *Arca pacifica* shells, now empty, were pictured in his report and are in the National Museum of Natural History collections (McGee 1898:213, fig. 25).

43. In relating such a custom with food use, McGee (1898:209–212) mentions the Seri practice of collecting cactus seeds contained in dry feces for food use in a second harvest. Davis and Dawson (1945:201) note the use of large seashells in collecting such feces. It is unclear whether Davis witnessed this or learned of it secondhand. The process is described in Felger and M. Moser (1985:253).

44. Nolasco (1967:187) describes the wreath ("corona") as a grave decoration. She writes that graves were covered with spiny branches or seashells to keep the coyotes from digging into the graves.

45. Griffen (1959:49–51) and Felger and M. Moser (1985:103–104) describe the process in detail.

46. Bowen (1976:40–42) and Bowen et al. (2004:26–35) describe such circles. Although Bowen quotes E. Moser as saying the circles were only made with stones, a one-line note of Moser's reports Roberto Herrera as saying the circles could be made with shells as well. Felger and M. Moser (1985:103) briefly mention the circles, but note that present day Seris do not know much about them.

47. Other methods used to influence the weather are detailed in Felger and M. Moser (1985:107).

48. Felger and M. Moser (1985:106–107, 254). More recently Juanita Herrera commented that this was sometimes done to "mark something—to say that you had been there."

49. One older woman in her sixties humorously quipped that pork was *hapáx* to her—she just didn't like the meat. She added that she did not consider any shellfish or other foods to be *hapáx*.

50. Angelita Torres suggested that eating such meat would cause the woman to go blind later in life, and the man who supplied the meat would have poor hunting results thereafter.

51. *Carditamera affinis* contains a red-colored blood. As a child, I can recall seeing older women harvest this clam, and I somehow had the idea that this was "an old lady's food."

52. Griffen (1959:4) records having been told that pearls were sold in Hermosillo, along with animal skins, baskets, and other items.

53. M. Moser (n.d.).

54. Smith (1954) records that between 1926 and 1928 Thomson transported small quantities of fish purchased from the Seris in Bahía de Kino to sell in Hermosillo.

55. Smith (1954) discusses the impact of the new economy, noting the advantages as well as the problems that newly acquired money brought, one of these being the effects of easily accessed alcohol.

56. Smith (1954) describes the early years of the cooperative.

57. María Luisa Astorga provided the year 1964.

58. In writing about their trip to Bahía de Kino in 1932, Dane and Mary Coolidge describe how the Seri men cracked the oyster shells, and the women removed the meat for sale. The Coolidges include details of how much money was paid to the people and how the oysters were packed for transport (Coolidge and Coolidge 1939:19, 21, 42).

59. Malkin (1962:39) notes being told that murex shells and those of *Turbo* sp. were sold in bulk in Hermosillo, and that the Seris were not aware of any use for the shells.

60. Basurto (2002) records that the harvesting of pen shells by hookah diving began in the late 1960s in nearby Bahía de Kino. Basurto (2006) describes the history, harvesting process, and current status of the fishery in the region.

61. Although this is certainly the case, Angelita Torres chuckled and said matter-of-factly that the people "just liked to move!"

62. E. Moser (1963) presents information based on Seri oral history; others discuss such evidence in light of historical references and other factors (see Bahre 1980).

63. E. Moser (n.d.) records that long ago an inland group of Seris known as the *Caail iti Comcaac* 'the dry lake *Comcaac*' lived near the Playa de San Bartolo, a large dry lake north of Bahía de Kino. The people did not fish, only collecting clams and oysters. Little was known of them, and it is said they disappeared without a trace.

64. Although most shellfish were available all year long, the best time for eating them was in January and into March, when the clams and snails were fat. According to Raquel Moreno, the adductor muscles from the pen shells are fat in the cooler months; in the summer they are thin and "like plastic," and not good to eat. Octopuses were best eaten in the cool months as well.

65. Felger et al. (1976) documented this previously unrecorded phenomenon of winter dormancy in sea turtles that was well known to the people.

66. Felger and M. Moser (1985:36–37) discuss marine invertebrates in the Seri diet.

67. Kroeber (1931:20): "My impression is that clams, other mollusks, and crabs constitute more than a tenth of the total consumption. Clams especially can usually be obtained at low tide when wind and water prevent other food taking."

68. According to René Montaño, an unwritten code is understood by the Seri people today: use the food in the sea wisely; take care of it. When other food is gone, the sea will be there to provide for you.

69. Large piles of shells can be found near the two Seri villages. Most are of the pink-mouthed murex (*Chicoreus erythrostomus*) and are left from the cooked meat sold to non-Seris in recent years. Older piles of shells are much smaller and are probably the food remains of smaller Seri groups in the past.

70. According to Smith (1954) a red tide event occurred for six days in 1940 near Desemboque: "Killed great number of fish, large and small. Once only occurred. Fish were hauled in by boat-load and sold [commercially]. No sickness resulted to Seris, only to fish. Totuaba [*sic*], Bahia [*sic*], Pinta, etc. died. (Also happened in Nov.–Dec. 1950 near Cabo Tepopa and Seris gained as in 1940.)" Unfortunately there are no further notes about the effects of eating such fish.

71. This seems a rather sweeping statement, as the people are very much aware of such things as the highly toxic organs of the pufferfish (*Sphoeroides annulatus*).

72. *Yaxi imxizi* was a derogatory term used behind the back of a person who was "not good," who would not be mourned at death. The conversation about this subject seemed somewhat hypothetical, rather than describing something actually practiced in the culture.

73. According to Basurto (personal communication), while harvesting pen shells commercially, shellfish such as *Megapitaria aurantiaca* and other large clams seen by divers are collected as well, to be eaten raw in the boat or taken home. Although plentiful in the same area, mussels are apparently not collected in this way.

74. The term *hapaznij* is used for something cooked that retains its form, such as clams. The term *hapaztoj* is used for a gruel or solution, such as coffee.

75. Felger and M. Moser (1985:48). Del Barco ([ca. 1770s] 1980:265) briefly notes the use of such a cooking method by earlier Gulf peoples.

76. This method was also used for roasting *haamjö* (*Agave subsimplex*) although hot rocks were used to cover the food, and the process took much longer (Felger and M. Moser 1985:225).

77. In 2007, however, while on a botanical collecting trip on Isla Tiburón with several Seri men, Richard Felger helped one man collect large numbers of tiny whelks to take back to the man's wife to make necklaces (Richard Felger, personal communication).

78. Some places are said to be cursed because they belonged to a Giant. Eating a mullet (*Mugil* sp.) from a certain estero is said to produce dangerous dreams and hallucinations, so eating that fish is strongly avoided. Such a place and the things found there are called *caxápotol* 'what produce hallucinations'.

79. I am aware of two instances in which a woman was involved in such a role. Long ago, when reed balsas were still in use, a woman named *Cmaam Quiscama* 'woman who has a balsa' dressed as a man and liked to accompany the men fishing. Another woman in the mid-twentieth century enjoyed fishing with men as well (E. Moser n.d.).

80. René Montaño categorized the information given in the table.

81. Seri baskets are sewn from splints of limberbush (*Jatropha cuneata*). For detailed descriptions see E. Moser (1973) and Felger and M. Moser (1985).

82. According to Juanita Herrera, the trip by foot from the coast near Bahía de Kino to Hermosillo (about sixty-five miles) took six days. She described, "as the sun went down on the sixth day we would arrive outside the city." No information is available about live mollusks taken as far as Hermosillo. The distance from Punta Chueca to Desemboque (about forty miles) was normally walked in three days; one man was said to have made the trip in one day "because he was angry." The walk from Punta Chueca to Bahía de Kino (about eighteen miles) could be made in one day.

83. See Felger and M. Moser (1985:233) for this use of the plant.

84. McGee (1898:233–234).

85. In the early 1950s my mother noted the manner in which older Seris drank from a cup—not by holding the handle, but by picking the cup up by its opposite rim area and drinking from the near side. María Luisa Astorga demonstrated how the *Laevicardium elatum* shell was held in this way when used for drinking.

86. I know of no cases in which shells were worked or formed into eating utensils. McGee (1898:186) notes that he did not observe any "wrought" shells.

87. McGee (1898:230) recorded "then hatcheled [the fibers] with the *hupf* [*xap*, (archaic) grinding stone] or the edge of a shell."

88. In January 2009, Juanita Herrera gave me one valve of a very large mussel (***saacj***, *Modiolus capax*), saying that she was going to give the other one to a friend to contain pigments with which to paint her face.

89. Hardy (1829:286) and McGee (1898:165–66, plate 27) both mention shells used to hold pigments.

90. See Felger and M. Moser (1985:152–156) for a description of pigments and face painting, and M. Moser (1964) for details of the process for making the natural blue pigment used by the people.

91. According to Raquel Moreno, larger mussel shells were sharpened against a stone.

92. Davis recorded that a mussel shell (Davis 1910–1944; also see appendix 7) and the shell of the *Pholas chiloensis* clam were used for severing an umbilical cord (M. R. Harrington 1938:123).

93. The exterior surfaces of many prehistoric sherds display characteristic striations from this shell (Bowen 1976:60). Bowen discusses in depth the construction of clay vessels and the use of shells in the process.

94. Felger and M. Moser (1985:91).

95. The use of the barrel cactus for emergency liquid is described in Felger and M. Moser (1985:262–264.)

96. Bowen (1976:73–74, 79) describes smoking pipes from the archaeological record in the Seri region; Felger and M. Moser (1985:165) discuss the smoking of plant products in the Seri culture.

97. Tom Bowen, personal communication.

98. Father Ortiz Zapata, 1678 (in Sheridan 1999:28).

99. Di Peso and Matson (1965:51–54).

100. Ibid.

101. Sheridan (1999:148).

102. Bowen (1976:87–88) describes such artifacts.

103. Many disc beads appear to be made from the shells of cone snails, however. The people today have no recollection of the manufacture of shell or stone beads, and such prehistoric shell ornaments are said to have belonged to the Giants.

104. Sheldon (1979:figure 6-1) includes a photograph of a shell necklace he collected that appears to be made of such beads.

105. Etched shells have been recovered archaeologically in the area by tourists; none have been preserved in museum collections nor formally described (Richard White, personal communication).

106. Felger and M. Moser (1985:148–152) describe such uses in detail.

107. Ibid., p.148.

108. Hardy (1829:281) describes a Seri man wearing "a straw hat, ornamented with shells, and an eagle's feather." Probably hatbands were an unusual adornment, and those in museum collections may have been made for sale to tourists. Shells used on Seri hats are uncommon in photographs.

109. A cord worn as a man's belt was called *oaaf*. I am unaware of any photograph showing a belt of shells in actual use; as with hatbands, they were more commonly made for sale to tourists.

110. Malkin (1962:39) reports that he was told that the Seris made buttons from *Pteria sterna* shells, something that I have been unable to confirm.

111. Ignaz Pfefferkorn (1989:190), writing in the eighteenth century, includes a tantalizing brief passage describing the dress of the indigenous Sonoran people (a group that apparently included Seris): the women wore long deerskin skirts ornamented with "little sea shells, snail shells" and other attached items that produced a rattling sound that delighted the women when they walked or ran.

112. See Ryerson (1976:122–123) for a discussion of contemporary Seri crafts, including the manufacture of necklaces.

113. Objects that are not shells have also recently been used. Since 2008, fish scales from the *hatoj* (barred pargo, *Hoplopagrus guentherii*) fish have been strung. The scales are usually brightly colored with commercial dye and strung with glass beads or *xneeejam is hayaa* (*Viscainoa geniculata*) seeds. These necklaces are very popular with tourists.

114. Felger and M. Moser (1985:109–112) present a revised and expanded list of medicinal plants, first published in Felger and M. Moser (1974).

115. Ibid., p.113.

116. Ibid.

117. Ibid.

118. G. Harrington (1941:14).

119. Kroeber (1931:15–16) described some such games.

120. M. Moser (n.d.).

121. Felger and M. Moser (1985:160–161).

122. Davis (1910–1944). Unpublished Notebook 16, p. 94, 1929.

123. The use of leg rattles was borrowed from the Yaqui culture. For a description of Seri musical instruments, see Bowen and E. Moser (1970).

124. Edward H. Davis, as mentioned in M. R. Harrington (1938:123).

125. Quinn and Quinn (1965:161).

126. Malkin (1962:38, 41) mentions such use.

127. Strips of the cortex of the *Echinocereus grandis* cactus were used as bait by people living on Isla San Esteban (Felger and M. Moser 1985:130, 261).

128. See the discussion of place names in S. Marlett and M. Moser (1995) and the revised and expanded set presented in M. Moser and S. Marlett (2010).

Bivalvia

1. Herrera Casanova (2012a).

2. Solórzano also bought fish and sea turtles from the Seris during 1937–1948; it was he who helped establish a fishing camp at Desemboque in the winter of 1940 (Smith 1954).

3. Similar uses of *coiicöt* (or some related form based on the transitive verb *quicö* 'to kill') are found elsewhere: *Coen oiicö* 'whom the *Coen* spirit killed' (referring to a sickly, weak person); ***stacj coiicöt*** '*stacj* that kill', an undesirable form of the rock oyster *Saccostrea palmula*; *siml oiicö* 'whom the *siml* barrel cactus (*Ferocactus emoryi*) killed'; and *psaac oiicö* (meaning unclear), a menstruating woman. In such cases a sickness or condition is implied rather than physical death.

4. Felger and M. Moser (1985:113).

5. According to María Luisa Astorga, no other mollusks were seeded in this or any other way.

6. Because of the mussel's size, color, and the observation that it does not grow with other mussels, the people consider it to be very different. This warrants further investigation (Paul Valentich-Scott, personal communication).

7. There is a lack of consensus as to what the term ***aapa*** describes. The term generally means huge; however, it also connotes very or true. Although some say that a ***satoj aapa*** is just a normal-sized mussel, most say that it is larger. One woman said that it refers to an uncommon mussel having an extremely curved foot, adding that someone she knew had a nose of this shape, and it could be said that he had a nose "like a ***satoj*** *aapa*." Finally, a few people have said that ***satoj aapa*** refers to a mussel differing in both size and color from the most common mussel.

8. E. Moser (n.d.).

9. Such use of the name ***saacj*** today no doubt indicates the loss through time of the taboo associated with the word.

10. This illustration was made from a photograph of that shell (catalog no. 211087.001). It is interesting to note that this one very unusual mussel shell was collected by Davis especially since the ubiquitous *Modiolus capax* shells were in much more common use, possibly indicating that this shell was considered a special possession. There are sporadic records of this mussel from the Gulf, but not of an established population (Paul Valentich-Scott, personal communication).

11. M. R. Harrington (1938:123).

12. Malkin (1960:170) recorded the name ***xasaacj*** for this clam, but no Seri today has confirmed it. According to Raquel Moreno, ***xasaacj*** was described by her father as a clam that was "reddish, pretty, and occurred on the west side of Isla Tiburón."

13. The non-Seri camp name comes from Alberto "Dollar [*sic*]" Durrazo, who along with others first made a road to the camp in 1935 (Smith 1954).

14. According to Seri folklore, animals were once people; many received their names as they became animals.

15. McGee (1898:213). There is no corroboration of this claim today; in fact, the idea is met with derision and declared nonsense. One consultant who has seen such areas described by McGee on Isla Tiburón suggested that the shells had a hygienic use, a common use in the past for this and other similar shells, raising the probability that McGee had simply stumbled on a toilet area and then supplied a rather exotic explanation for a purely natural situation.

16. Keen (1971:55).

17. Bowen (1976:87) notes that "although *Glycymeris* bracelets were produced on the central coast . . . it is obvious that they were made by the Trincheras people, not the indigenous inhabitants."

18. Bowen (2000) gives a detailed background of the pearling industry in the Gulf.

19. "The Seri traded skins—baskets—pearls—with barco [ship] run by a Yaqui—he approached them with a cross raised on his barco to show who he was. Traded contraband—rifles—bullets—clothes—merchandise. In years of plundering of [by?] Seri. Traded mainly with Band III, seems to be after destruction of Band II. Perhaps one of those Yaquis was Juan Comichi" (E. Moser n.d.).

20. Miguel del Barco recorded in the 1770s that people dove for pearls in the Gulf (del Barco 1980). Although he mentions other indigenous people, the name Seri does not occur in his work, and most of his description centers on the pearling industry on the Baja California peninsula. There is general agreement among the Seris today that their ancestors never participated in diving for pearls, or for anything else.

21. I use the term mermaid when referring to the *xpacaao*, as many Seris today associate it with that concept. The *xpacaao* is described as a dangerous sea creature, the size of a person, and according to some, possessing female physical characteristics. It keens, announcing a coming storm, and it is said that if one sees it, he will lose a member of his family. One Seri man was said to have seen one while riding in a boat; it grabbed him around the waist and tried to pull him into the sea. When he escaped and arrived on shore he reportedly had slime around his waist. In 2007 María Luisa Astorga showed me some earrings she had been given by a non-Seri friend, made with pearls purchased in Toledo, Spain. She said that when her husband saw them, he told her to get rid of them because they belonged to the *xpacaao* and would bring her bad luck. She added that she kept them since they came from somewhere else and wouldn't hesitate to wear them.

22. Sheridan (1999:28).

23. The three Topete brothers, Antonio, Alfredo, and Oscar, were non-Seri Mexicans who figure in early commercial ventures

in the Seri fishing camps. Antonio and Alfredo bought fish, sea turtles, and other seafood, trucking them on ice to Hermosillo. Each of the brothers owned small stores in Desemboque or Punta Chueca during the latter half of the twentieth century.

24. E. Moser (n.d.); M. Moser and Marlett (2010).

25. Although the people know that other bivalves contain pearls, the name is only given to the pearl oysters.

26. Stratford [1746] (1958:62); Sheridan (1999:122–124).

27. Malkin (1960:169) recorded the name ***zopahotmazexe*** for this shell. Variations of that name are more commonly given to *Anomia peruviana* today.

28. According to people presently involved with harvesting, the sex cannot be determined until the shells are opened (Xavier Basurto, personal communication).

29. In 2007, we were given a bag of frozen adductor muscles ("scallops") by a Seri friend, who said that they had been a gift to her, and "aren't something we eat."

30. For a detailed description of current harvesting practices, see Basurto (2006).

31. The foot and byssal gland of *Pinna rugosa* are large enough to sell, while those of the other pen shells are not (Xavier Basurto, personal communication).

32. Malkin (1962:39) mentions the sale of pearls from pearl oysters and pen shells, saying that those of the pen shell had "immense value, for the local market" compared with those of the *Pteria* sp. He was probably mistaken.

33. McGee (1898:219). Mashém (identified as Francisco Astorga [M. Moser n.d.]) is pictured working with McGee at Rancho Costa Rica (McGee 1898: plates II, XVII). Mashém (*Maziim*) means 'it/he/she is handsome, pretty'; the basis for his name is unclear. In one note by Moser his name is recorded as *Imaziim* 'who is not handsome'.

34. Davis and Dawson (1945:199). Since Davis reported seeing only one balsa in use, and by a Seri man, he was probably reporting something that he was told.

35. M. R. Harrington (1938:123).

36. This was probably a very early note; the spelling is not consistent, and the orthography has since changed. The word *setoj* or *settoj* is unclear—it could refer to the byssal fibers of either ***seeten*** (pen shells) or ***satoj*** (mussels), although it most likely describes those of the pen shell, as they are longer and more striking.

37. In January 2009, while observing pen shell harvesting in the northern end of the Infiernillo Channel, Xavier Basurto recorded those taken by one boat during two consecutive days. Each day approximately one hundred *Atrina tuberculosa* were harvested by the diver, compared with only two or three *Atrina maura*. Basurto suggests that the *A. tuberculosa* is well established in the area, while *A. maura* occurs in greater numbers to the south and elsewhere (Xavier Basurto, personal communication).

38. Moreno Herrera (2012a:71–72).

39. Xavier Basurto, personal communication.

40. Basurto (2002:147) notes that "according to the [pen shell] fishermen the two most abundant species [in the area] are the kidney and the round [shape of adductor muscle] pen shell. Fishermen use the scallop meat, shell shape, and mantle's color as the main identifying characteristics. They claim that the half moon and pink pen shell are new to the Infiernillo Channel . . . Some fishers said that half moon pen shell became the pink pen shell once it grew older. While half moon pen shells were clearly identified as *Atrina maura* . . . it was not possible to clearly identify the pink pen shells due to the lack of enough samples in the fishermen's catch."

41. Felger and M. Moser (1985:287).

42. I am including the name ***xat*** with this oyster. Victoria's description also somewhat fits *Crassostrea gigas*, but as a recent arrival to the Gulf, that species would not have been known to Victoria's aunt, who died early in the twentieth century.

43. José Carlos Estrada Ortega, personal communication. Estrada Ortega is a biologist working in the commercial oyster fishery operating in Estero Santa Cruz, where large numbers of *Myrakeena angelica* and *Crassostrea* spp. are harvested.

44. Lindsey Groves at the Natural History Museum of Los Angeles County identified the shell.

45. M. R. Harrington (1938:123).

46. Sheridan (1999:12–13).

47. Lionett (1986:115); Pennington (1979:24). Mission records from Sonora tell of Seris wearing shell pendants, specifically mentioning a beautifully polished round shell. Drilled shells of pectens and pearl oysters have been found archaeologically.

48. M. R. Harrington (1938:123).

49. Three of these shells containing pigment made from the sap of the elephant tree are in the National Museum of the American Indian collections.

50. Herrera Casanova (2012b).

51. Herrera Casanova, personal communication. Del Barco (1980:265) in the 1770s described early Gulf peoples drying and transporting such shellfish.

52. The root of the Seri word for ten is *hanl*.

53. M. R. Harrington (1938:123).

54. A burial sponsor is a person responsible for handling the details of a person's burial.

55. Among the shells in the National Museum of Natural History collections is a fire-blackened *Nodipecten subnodosus* shell that McGee reported collecting from a woman's grave.

56. Other species of this genus in the area are given the same names as the one illustrated.

57. Richard Brusca, personal communication.

58. Steinbeck and Ricketts (1941:505) note that this was eaten by "natives" in Puerto Escondido, brought up by harpoons or fish spears at low tide. This method is somewhat vague, as the shell is firmly cemented to rocky substrates.

59. This is characteristic of Sara, a gregarious woman. Photographs from the 1940s and 1950s show her wearing different

necklaces and earrings. And I well remember the rings she wore, made from brass, copper, or the shell of a hawksbill sea turtle.

60. E. Moser (n.d.) suggests this relationship.

61. José Ángel's son, René Montaño, recorded this information in an essay (Montaño Herrera 2012a).

62. One consultant said that all four designs were painted on the shell, rather than just two.

63. For a discussion of the term *xpacaao* see Pteriidae.

64. Edward H. Davis (1910–1944). Notebook 10, p. 76, March 1924.

65. Bowen (1976:48–49); Bowen and E. Moser (1968:123); Felger and M. Moser (1985:7).

66. McGee (1898:186, 191, 195).

67. M. Moser (1970:206). Such a small ***xtiip*** is sometimes called a ***xtiip*** *zaac* 'small ***xtiip***'. This is one of only a few expressions incorporating the term *zaac*; more commonly used are *heeque* or *quisil*.

68. Bowen (1976:43) describes such a cache.

69. Several Seris recall their families burying sealed glass jars or clay vessels containing sugar for future use upon returning to the camp. According to E. Moser (in Bowen 1976:92) such items might be buried for later use in times of danger.

70. Davis (1910-1944). Notebook [Memorandum] 9.

71. To entertain the people, Davis would throw flash powder into a fire, and leap around the display, singing, hooting, and barking. Needless to say, this was highly effective and memorable.

72. See Felger and M. Moser (1985:241). Some early descriptions of the people note that their faces were painted with "blood"—most certainly this was *Bursera microphylla* sap.

73. See M. Moser (1964:27–32) for details of the pigment-making process.

74. The text was recorded and transcribed by Edward Moser, probably with the assistance of Roberto Herrera, shortly before 1968. The text was included in a concordance project directed by Joseph E. Grimes (run at the University of Oklahoma Computer Laboratory under the Project for Computer Support for Linguistic Research, partially supported by the National Science Foundation Grant GS-1605).

75. Davis (1910-1944). Notebook [Memorandum] S-9.

76. Cocoon rattles are described and named in Kroeber (1931:14). See also Bowen and E. Moser (1970:188–192).

77. Malkin (1960:169) recorded the name ***zacz*** for this cockle.

78. Evidence suggests that ***zacz*** and ***zacz ctam*** were the original names for the *Trachycardium procerum* and *T. consors,* respectively, and that the new name given to *T. procerum* is ***siml ihataxoj***.

79. See Felger and M. Moser (1985:233) for *Batis maritima* use and (1985:262–264) for a complete description of extracting liquid from the barrel cactus.

80. A regional Spanish name for the clam is *catarro* 'head cold'. The similarity between the two names is probably coincidental.

81. Bowen and E. Moser (1968:125–129) discuss this type of pottery. They also describe all Seri pottery and its manufacture.

82. After their use was discontinued because of easily obtained metal containers and cooking pots, clay vessels were made for sale to tourists, but were of poorer quality than those made previously.

83. McGee mentions "a modern shell with a cotton string attached to a bush nearby" collected during his second expedition in 1895 (Fontana and Fontana 2000:75). The shell is not identified.

84. McGee (1898:291).

85. Both the clam and crab occur in the sand in the swash zone, where sea foam is deposited by the waves. The name *iicj ano moosni* 'sea turtle in fine sand' aptly describes this curious little crab with its turtle-like carapace, that, when uncovered, in a flash reburies itself in the sand.

86. Although it is clear that *Simomactra dolabriformis* is the true ***haxöl icaai***, the use of the shell by a potter is a very rare sight today, and the distinction may have been blurred through time.

87. Edward H. Davis (1910–1944, Notebook 20) recorded the name Co shi' and wrote that the shell was "knife to cut fruit of pitahalla [*sic*] used as pocket knife today" (appendix 7). Davis was probably referring to this shell and the word he wrote down may have been *icoozix* 'what one cuts with [using a sawing action]'.

88. M. Moser and S. Marlett (2010). It is perhaps interesting to note here the connection between the connotation of the word *queeex* and another of the shell's names, ***spiitquim ctamoo***, since a shaman could curse people. No one has suggested it, and the connection remains speculative.

89. No explanation is given for the shell's name. A possible source might be seen in one of its other names, ***spiitquim quictamo*** 'fierce ***spiitquim***'.

90. Edward H. Davis collected the shell and recorded that its meat was eaten (M. R. Harrington 1938:123).

91. Bowen (2009:54–57) notes finding seventeen *Dosinia ponderosa* shell tools and flakes at the Seri site (SL-6) on Isla San Lorenzo. The worked shells are found at sites on other Gulf islands as well (Tom Bowen, personal communication).

92. M. R. Harrington (1938:123).

93. Ugarte [1722] (1958:46) in Bowen (2000:67).

94. M. R. Harrington (1938:123). Accompanying the note and listed together as food clams are: *Pitar concinna* [=*Lamelliconcha concinnus*]; *Tivela byronensis*; *Paphia grata* [=*Leukoma grata*]; *Pecten circularis* [=*Argopecten ventricosus*]; and *Pecten excavatus* [=*Euvola vogdesi*].

95. According to a note by E. Moser (n.d.), all but a few men plucked their whiskers. In the past, two shells held together were used; since the mid-twentieth century the lid of a metal can, bent in half, has served the purpose. This, and tweezers, are called *hateems ipapt* 'what whiskers are pulled out with'.

96. W. N. Smith (1954) mentions this use.

97. Cattle were probably first seen by the people sometime in the late seventeenth or early eighteenth century, by which time herds had been developed in Sonora.

98. Beach deposits rich with fossil shells at that location are another possible source for the name, though this was not mentioned when the area was described to me.

99. M. R. Harrington (1938:123).

100. Ibid.

Gastropoda

1. Richard White, personal communication.

2. The word *cooil* spans the blue and green spectrum. To specify a certain blue or green a modifier can be used: *sleecoj iipx yaail cöcooil* 'blue/green like the blue of (a) heron's egg' or *xpanaams yaail cöcooil* 'blue/green like the green of seaweed'.

3. This type of name occurs in other animal and plant names as well, for example, *pohaas camoz* 'what thinks it is an *haas* (mesquite)' refers to both *Desmanthus covillei* and *Acacia farnesiana* (Felger and M. Moser 1985:326). Such similarity is humorously noted as a desire to emulate the real thing.

4. Kroeber (1931:15) records: "Counters are little natural shell disks." M. R. Harrington (1938:124) notes that the Seris "formerly used [the opercula] as coins, and Mr. Davis was told their value was twenty-five cents Mexican in old times." This has not been corroborated by anyone today.

5. One of the species that Steinbeck and Ricketts (1941:234, 532) recorded finding in great numbers on the south side of Isla Tiburón was *Callopoma fluctuosum* [= *Turbo fluctuosus*]. A large midden in the area contains many of these shells (Richard Felger, personal communication).

6. The word *inoosj* is ambiguous, and could mean fingernail, toenail, or claw. However, my consultants have clarified that the snail's name references the toenail.

7. This snail was a good candidate for surviving in the boats, as *Nerita* spp. snails are found in the upper intertidal area or the splash zone.

8. Lorenzo Herrera provided the name ***haxölinaail isoj*** 'real shell' for an unidentified snail. He described the snail as having been collected by his mother on Isla Tiburón, at camps located near large seaside rocks where the snails were found. Though purely speculative, it is tempting to posit that these snails, also found on Isla Tiburón, might be those referenced by the name.

9. Felger and M. Moser (1985:248).

10. Davis (1910-1944). Notebook 16, 1929.

11. Bowen (1976:73–74) discusses such pipes.

12. It was an informative day when we happened upon this beach on a boat trip to Isla Tiburón in 2009; I had been told of such a place name, but the description was vague and the meaning puzzling. The startling and rather obvious explanation became clear when I saw the location, and the episode impressed on me the difference between an oral description and seeing something firsthand.

13. A live worm snail is a curious creature to watch; the animal retreats deep into its shell tube and disappears in a flash—perhaps Sara was referring to this when she made the comment.

14. A note by E. Moser (n.d.) recorded, "'Turkey Vulture's daughter'. Not eaten. Girls used them for dolls (note—squat, round shape, female shape) 'this is a female shell'."

15. Bowen and E. Moser (1970:186).

16. More commonly, *xapij an iquipl* 'cane in which one puts one's tongue', whistles made of a section of cane (*Phragmites australis*), were easily carried and used in the desert as a means of communication.

17. Coolidge and Coolidge (1939:210).

18. The term *inoosj* might be translated as either nail or claw. Here it clearly means claw, as a rooster's claw was referenced when the name was given.

19. The traditional reed balsas did not have seats. The people began using wooden boats in the early 1900s, which corresponds nicely with Loreta's lifetime (ca. 1890–1968) and her naming the shell.

20. The hard wood left from the knothole in the *Bursera microphylla* tree sometimes resembles a tiny boat, known as *oot icanoaa* 'the coyote's boat'. See Felger and M. Moser (1985:244).

21. Malkin (1960:169) recorded the name ***cahiixona*** for the Lottiidae (limpets without an internal septum) and Calyptraeidae (the slipper snails and the cup-and-saucer snails). The term might have earlier been a general name for the limpets, although I have chosen to use it for the cup-and-saucer snails, as today it is most clearly applied to them.

22. The interview was recorded on cassette tape by Scott Ryerson in 1969.

23. The name ***casooca*** was given to me in 2003 by Celsa Félix, and was confirmed by several women, one of whom commented, "Celsa is from the *Xnaamotat* people (a southern group of Seris). They know a lot about such things—more than we do."

24. G. Harrington (1941).

25. A note by E. Moser (n.d.) records "the meat causes a rough voice (choked-up throat) which affects speaking, not eaten."

26. Sara Villalobos and others showed some hesitation when I first asked the name of this shell. It appeared they were talking of another creature with similar characteristics, of which I was not then aware.

27. The museum record accompanying this object gives both cowries the name ah-set-ah-mos [***hasit camós***], a name most often given to the *Trivia solandri* snail. It is not implausible that the similarly shaped shells previously shared the name.

28. The term *hatajeen* is used for either the male or female genital area.

29. This is not an unreasonable suggestion. Bird heads with necks attached were stuffed, wrapped with cloth, and used as dolls. Sea turtle or deer bone dolls were common toys. Sometimes larger bones were used: as a child Rosa Amalia Barnett had a large doll made from a piece of whale bone carved into a simple human shape, for which she sewed clothes.

30. Richard Felger, personal communication.

31. Malkin (1960:38) recorded that this snail was eaten by the people. No one confirms this today; in fact, the shells are usually found only as solitary specimens in beach drift.

32. Also mentioned in E. Moser and White (1968:145–146).

33. G. Harrington (1941).

34. The words *cquihjö* and *cheel* both mean 'red' and are often used interchangeably.

35. In an early recording, Eva López identified *Chicoreus erythrostomus* as ***xnoocat***, and *Hexaplex nigritus* as ***noocat***. No one else has made this distinction.

36. I have not seen this snail eaten by the people today, possibly because it is the less common of the two largest murexes. One day I brought a very large live specimen and showed it to Sara Villalobos, who grimaced and dropped it—then laughed and said that it was a *ziix coosyat oohit* 'Giant's food', commenting on its large size. She then said that the broken shells are found on dune middens because the Giants broke open the snails with stones to remove the meat, adding that the meat is sweet tasting. Victoria Astorga confirmed that the meat was eaten in the past.

37. The word *hacaacp* is also used to refer to a large bone, such as that of a deer, that is chipped open to remove the marrow.

38. The *Turbo fluctuosus* snail was cooked and eaten in this way as well.

39. E. Moser (n.d.) noted that only the outer end of the meat (the muscle) was eaten. The note added that these shells occur on Isla Tiburón, at *Haanc* and to the north.

40. The trail I followed to identify this snail by its Seri name was long. The name was on an early file slip of my father's, with no drawing, and in a short note I made much later, but with no specimen. No one I asked had any idea what it was. Through clues from some spontaneous conversations on different trips to the area, and finally with a potential candidate shell in hand, I was able to get a clear and unprompted identification from two older women.

41. Since the snail is rare in the region, details of its use are sketchy. The information was given by only a few older Seris and seems to overlap with that of *Mexacanthina lugubris angelica*, which shares the name. One might posit that *Plicopurpura pansa* is the original bearer of the name because of its distinctive colored secretion, a characteristic that is not obvious in *Mexacanthina lugubris angelica*. The staining capability (and strong color) mirror the well-used gentian violet, a topical substance that has long been used in Mexico (and is well known to the Seris) to treat thrush and other skin conditions. Any connection, though intriguing, remains speculative.

42. Malkin (1960:170) recorded the name ***xpa<u>l</u>eemelc*** for one member of this genus.

43. Male dolls were made from such things as the scapula of a deer—a flat shape that contrasted with the more bulbous sea turtle flipper bones used to make female dolls.

44. I thank Gary Nabhan for his help in getting a final confirmation of this species with its Seri name, first given in a note by E. Moser (n.d.).

45. Sheldon (1921–1922) made a list of Seri words. The name he transcribed for necklace is *muss*, a probable indication that the name ***mas*** was still in use at that time. The Seri word for necklace is *hapapjc* 'what is strung'.

46. This was in a very early note of E. Moser's (n.d.) and was later confirmed by several older people. It is purely coincidental that the name occurs in the name of the *Oliva sayana* snail, found in the Atlantic.

47. Malkin (1960:169) recorded the name ***xiica hant cosítyalca*** for this shell.

48. The shell is also common in the archaeological record in northern Mexico and the Southwestern United States, a result of the extensive Gulf shell trade.

49. When María Luisa Astorga saw a drawing of a cone snail's anatomy, including the radular dart, she gave the name *yacotni* 'what it stabs with' to the dart. This term is given to the barb of the stingray as well.

50. *Hant yapol* is also the name for the whale shark (*Rhincodon typus*) which occurs in the Gulf. Alfredo made it clear that the shell's name does not refer to the animal.

51. The slug was found and photographed by Xavier Basurto and identified by Hans Bertsch.

52. Richard Felger, personal communication.

53. When talking about the octopus ink sac, Raquel Moreno suggested that perhaps the sea hare had one, since it releases a similar fluid. She said she had never seen inside the animal, as handling it was said to cause hemorrhaging during childbirth. She laughed as I suggested she just might look now, since she isn't bearing any more children, and said she would let me know if she found the ink sac.

54. Bertsch and C. Marlett (2011) present the information given here.

55. Nemer Narchi (personal communication) first told me about such a use, and I have since confirmed it.

56. M. Moser (n.d.).

57. E. Moser (n.d.).

58. The snail was identified from examination of the actual shell as belonging to the family Hydrobiidae by Robert Hershler, Smithsonian Institution.

59. Since the internal animal was not available, and the shell size is only approximate, more certain identification is not possible (Lance Gilbertson, personal communication).

Polyplacophora

1. The suffix *-iil* may refer to color (as in this case), covering blue and green, or a large size, as in the name *mosniil*, a large black sea turtle.

2. The use of the sequence *tamax* in these names suggests an archaic meaning for *tamax*. However, since the word is hardly recognized today, any proposal would be highly speculative. Malkin (1960:169) recorded the name ***tamax*** for two species of chitons.

3. The name ***haxz heeque*** is also given to the hairy caterpillar (Arctiidae). Clearly, there is a similarity in the "furry" appearance of both the caterpillar and the chiton, resulting in the shared descriptive name.

Scaphopoda

1. As the scaphopod apparently does not occur on Desemboque shores, I was not familiar with it. Not until finding a note and drawing recorded by my mother did I begin to look for it and first collect several of the shells in beach drift at Bahía Sargento.

Cephalopoda

1. According to René Montaño, in November 2007 great numbers of large squids were being caught by the Seris and sold to a non-Seri fish buyer. The squids were caught at night with hook and line, each boat bringing in a ton or more per trip. The squids were transported by truck to Tijuana, Baja California.
2. Although other bivalves have the name ***oot isatoj***, *Modiolus tumbezensis* is the only one noted as having this effect.
3. For a discussion and analysis of the source of the off-flavor in the meat of *Dosidicus gigas* (the jumbo squid, common in the Gulf of California) see Sánchez-Brambila et al. (2004).
4. J. Harrington (1945:297) mentions that he was not aware of any North American Indians making any use of the octopus ink.
5. The term *hapát* has two meanings: a wooden rack or shrub on which meat or other things were placed for drying or keeping off the ground, and the method of cooking, in which meat was baked under hot sand and coals.
6. Griffen (1959:8) recorded such use.
7. E. Moser (n.d.). This group was known as the *Hast ano Comcaac* and was said to have strange customs and speak a different dialect, a few words of which E. Moser transcribed from the last person who recalled them, Porfirio Díaz, in the 1950s. Tom Bowen (2000) gives details of the people passed down through oral accounts.
8. Malkin (1962:41).
9. Some people say that *xpanaamalc* (or a variant pronunciation *xpanoomalc*) is a general term referring to any small (and hence "worthless") sea creatures, often found beneath intertidal rocks. The term includes creatures that are not eaten, such as sea urchins, fireworms, sea anemones, and sea cucumbers, or small fishes, crabs, and tiny octopuses.
10. Xavier described the octopus as a good-sized ***hapaj cheel*** (*Octopus hubbsorum*).
11. Among the many songs Dane and Mary Coolidge include in their book is one that takes up this theme even more explicitly, sung by Santo Blanco. The song includes two stanzas about a coyote's encounter with an octopus. The coyote runs through the desert with the octopus on its head; when the octopus finally drops off, it is eaten by the coyote (Coolidge and Coolidge 1939:189–190).
12. When I verified this with Victoria Astorga, she was *very* surprised that I would even have to ask about this, saying it was common knowledge.
13. Felger and M. Moser (1985:113).
14. In churning seas, octopus lairs in the intertidal zone would probably be filled with sand. Possibly the most easily collected octopus would be *Octopus fitchi*, which is found beneath rocks.

Other Marine Invertebrates

1. The term is probably a variant of *xpanoomalc* (see Octopodidae).
2. Malkin (1962:43–46) briefly discusses crustaceans and a few other marine invertebrates. Felger and M. Moser (1985:36–39) give some names and information; most of the names in table 9.1 can be found in the Seri dictionary (S. Marlett and M. Moser 2010).

REFERENCES

Astorga de Estrella, María Luisa, Stephen A Marlett, Mary B. Moser, and E. Fernando Nava L. 1998. "Las Canciones Seris: Una Visión General." In *Cuarto Encuentro Internacional de Lingüística en el Noroeste*, edited by Zarina Estrada Fernández, Max Figueroa Esteva, Gerardo López Cruz, and Andrés Acosta, tome 1, vol. 2, pp. 499–526. Hermosillo, Sonora: Editorial Unison.

Bahre, Conrad J. 1980. "Historic Seri Residence, Range, and Sociopolitical Structure." *The Kiva* 45:197–209.

Basurto, Xavier. 2002. "Community-Based Conservation of the Callo de Hacha Fishery by the Comcáac Indians, Sonora, Mexico." Master's thesis, University of Arizona.

———. 2006. "Commercial Diving and the Callo de Hacha Fishery in Seri Territory." *Journal of the Southwest* 48:189–209.

Berlin, Brent. 1973. "Folk Systematics in Relation to Biological Classification and Nomenclature." *Annual Review of Ecology and Systematics* 4:259–271.

Bertsch, Hans and Cathy M. Marlett. 2011. "The Seris, the Sun and Slugs: Cultural and Natural History of *Berthellina ilisima* and Other Opisthobranchia in the Central Sea of Cortez." *Thalassas* 27:9–21.

Bowen, Thomas. 1976. *Seri Prehistory: The Archaeology of the Central Coast of Sonora, Mexico.* Anthropological Papers, no. 27. Tucson: University of Arizona Press.

———. 2000. *Unknown Island.* Albuquerque: University of New Mexico Press.

———. 2009. *The Record of Native People on Gulf of California Islands.* Arizona State Museum Archaeological Series, no. 201. Tucson: University of Arizona Press.

Bowen, Thomas, and Edward W. Moser. 1968. "Seri Pottery." *The Kiva* 33:89–132.

———. 1970. "Material and Functional Aspects of Seri Instrumental Music." *The Kiva* 35:178–200.

Bowen, Thomas, Richard S. Felger, and R. James Hills. 2004. "Chollas, Circles and Seris: Did Seri Indians Plant Cactus at Circle 6?" *Desert Plants* 20:26–35.

Broyles, Bill, Luke Evans, Richard S. Felger, and Gary P. Nabhan. 2007. "Our Grand Desert: A Gazetteer for Northwestern Sonora, Southwestern Arizona, and Northeastern Baja California." In *Dry Borders: Great Natural Reserves of the Sonoran Desert*, edited by Richard S. Felger and Bill Broyles, pp. 581–679. Salt Lake City: University of Utah Press.

Brusca, Richard C. 2007. "Invertebrate Biodiversity in the Northern Gulf of California." In *Dry Borders: Great Natural Reserves of the Sonoran Desert*, edited by Richard S. Felger and Bill Broyles, pp. 418–504. Salt Lake City: University of Utah Press.

Brusca, Richard C., and Lloyd T. Findley. 2005. "The Sea of Cortez." In *A Distributional Checklist of the Macrofauna of the Gulf of California, Mexico. Part I. Invertebrates*, edited by Michel E. Hendrickx, Richard C. Brusca, and Lloyd T. Findley, pp. 11–24. Tucson: Arizona-Sonora Desert Museum and Conservation International.

Brusca, Richard C., and Michel E. Hendrickx. 2010. "Invertebrate Biodiversity and Conservation in the Gulf of California." In *The Gulf of California: Biodiversity and Conservation*, edited by Richard C. Brusca, pp. 72–95. Tucson: University of Arizona Press.

Brusca, Richard C., Erin Kimrey, and Wendy Moore, eds. 2004. *A Seashore Guide to the Northern Gulf of California.* Tucson: Arizona-Sonora Desert Museum Press.

Burckhalter, David. 1976. *The Seris.* Tucson: University of Arizona Press.

Campbell, Lyle. 1997. *American Indian languages: The Historical Linguistics of Native America.* Oxford: Oxford University Press.

Coan, Eugene V., and Paul Valentich-Scott. 2012. *Bivalve Seashells of Tropical West America. Marine Bivalve Mollusks from Baja California to Northern Peru.* Santa Barbara Museum of Natural History Monographs 6, Studies in Biodiversity 4. Santa Barbara: Santa Barbara Museum of Natural History.

Coolidge, Dane, and Mary R. Coolidge. 1939. *The Last of the Seris.* New York: E. P. Dutton.

Davis, Edward H. 1910–1944. Papers #9166. Division of Rare and Manuscript Collections, Cornell University Library. Ithaca: New York.

Davis, Edward H., and E. Yale Dawson. 1945. "The Savage Seris of Sonora I and II." *Scientific Monthly* (March and April issues) 60:193–268.

del Barco, Miguel. [ca. 1770s] 1980. *The Natural History of Baja California*, edited by Miguel León-Portilla. Los Angeles: Dawson's Book Shop.

Di Peso, Charles C., and Daniel S. Matson. 1965. "The Seri Indians in 1692 as Described by Adamo Gilg, S. J." *Arizona and the West* 7:33–56.

Estrada Fernández, Zarina, ed. 2004. *Diccionario Yaqui-Español y Textos: Obra de Preservación Lingüística.* Hermosillo: Universidad de Sonora; Mexico City: Plaza y Valdés Editores.

Felger, Richard S., and Bill Broyles, eds. 2007. *Dry Borders: Great Natural Reserves of the Sonoran Desert.* Salt Lake City: University of Utah Press.

Felger, Richard S., Kim Clifton, and Philip J. Regal. 1976. "Winter Dormancy in Sea Turtles: Independent Discovery and Exploitation in the Gulf of California by Two Local Cultures." *Science* 191:283–285.

Felger, Richard S., Matthew B. Johnson, and Michael F. Wilson. 2001. *The Trees of Sonora, Mexico*. Oxford: Oxford University Press.

Felger, Richard S., and Mary Beck Moser. 1974. "Seri Indian Pharmacopoeia." *Economic Botany* 28:414–436.

———. 1985. *People of the Desert and Sea: Ethnobotany of the Seri Indians*. Tucson: University of Arizona Press.

Felger, Richard S., and Benjamin T. Wilder. 2012. *Plant Life of a Desert Archipelago: Flora of the Sonoran Islands in the Gulf of California*. Tucson, University of Arizona Press.

Fontana, Hazel M., and Bernard L. Fontana. 2000. *Trails to Tiburon: The 1894 and 1895 Field Diaries of W J McGee*. Tucson: University of Arizona Press.

García y Alva, Federico. 1905–1907. *México y Sus Progresos: Album-Directorio del Estado de Sonora*. Hermosillo: Antonio B. Monteverde.

Griffen, William B. 1959. *Notes on Seri Indian Culture, Sonora, Mexico*. The Latin American Monograph Series no.10, edited by A. Curtis Wilgus. Gainesville: School of Inter-American Studies, University of Florida Press.

Hardy, R. W. H. [1829] 1977. *Travels in the Interior of Mexico in 1825, 1826, 1827, and 1828*. Glorieta, New Mexico: Rio Grande Press.

Harrington, Gwyneth. 1941. Gwyneth Harrington Papers 1934–1962. Series 5: Seri Ethnographic Papers 1937–1944. Tucson: Arizona State Museum Archives.

Harrington, John P. 1945. "Mollusca Among the American Indians." *Acta Americana* 3:293–297.

Harrington, M. R. 1938. "The Use of Shells by the Seris." *Masterkey* 12:122–124.

Hendrickx, Michel, Richard C. Brusca, Mercedes Cordero, and Germán Ramírez R. 2007. "Marine and Brackish-Water Molluscan Biodiversity in the Gulf of California, Mexico." *Scientia Marina* 71:637–647.

Herrera Casanova, Lorenzo. 2012a. "Loomz coi iicp hac." In *Comcaac quih ziix quih ocoaaj hac*, 2nd edition, pp. 50–51. Socaaix and Haxöl Iihom, Sonora.

———. 2012b. "Those Who had Hast Quita as Their Birthplace." In *Inside Dazzling Mountains: Southwest Native Verbal Arts*, edited by David L. Kozak, pp. 44–53. Lincoln: University of Nebraska Press.

Hunn, Eugene S. 2008. *A Zapotec Natural History*. Tucson: University of Arizona Press.

INEGI. 2009. *Perfil Sociodemográfico de la Población que Habla Lengua Indígena*. Aguascalientes, Mexico: Instituto Nacional de Estadística y Geografía.

Johnston, Bernice. 1968. "Seri Ironwood Carving." *The Kiva* 33:155–166.

Keen, A. Myra. 1971. *Sea Shells of Tropical West America: Marine Mollusks from Baja California to Peru*. 2nd edition. Stanford: Stanford University Press.

Kozak, David, ed. 2012. *Inside Dazzling Mountains: Southwest Native Verbal Arts*. Lincoln: University of Nebraska Press.

Kroeber, Alfred L. 1931. *The Seri*. Southwest Museum Papers, no. 6. Los Angeles, California.

Lionnet, Andrés. 1986. *Un Idioma Extinto de Sonora: El Eudeve*. Mexico City: Universidad Nacional Autónoma de México.

Luque-Agraz, D., and A. Robles-Torres. 2006. *Naturalezas, Saberes y Territorios Comcáac (Seri)*. Instituto Nacional de Ecologia (INE-SEMARNAT), Mexico City.

Malkin, Borys. 1960. "Ethnozoology of the Seri Indians of Sonora, Mexico." Master's thesis, University of Minnesota.

———. 1962. *Seri Ethnozoology*. Occasional Papers, no. 7. Pocatello: Idaho State College Museum.

Marlett, Cathy M. 2000. "A Desemboque Childhood." *Journal of the Southwest* 42:411–426.

———. 2002. "A Fine Day for Playing Hooky." In *Backcountry Pilot: Flying Adventures with Ike Russell*, edited by Thomas Bowen. Tucson: University of Arizona Press.

Marlett, Stephen A. 2006. "La Evolución del Alfabeto Seri." *Octavo Encuentro Internacional de Lingüística en el Noroeste*, vol. 3, edited by María del Carmen Morúa, pp. 311–325. Hermosillo: Editorial UniSon.

———. 2007. "Bringing it Home: The Implications of Documentation for a Vibrant Endangered Language." *Proceedings of Conference on Language Documentation and Linguistic Theory*, edited by Peter K. Austin, Oliver Bond, and David Nathan, pp. 157–163. London: School of Oriental and African Studies, University of London.

———. 2008a. "The Form and Use of Names in Seri." *International Journal of American Linguistics* 74:47–82.

———. 2008b. "Stress, Extrametricality and the Minimal Word in Seri." *Linguistic Discovery* 6:1–14.

———. 2011. "The Seris and the Comcaac: Sifting Fact from Fiction about the Names and Relationships." *Work Papers of the Summer Institute of Linguistics, University of North Dakota Session* 51:1–20. www.und.nodak.edu/dept/linguistics/wp/2011Marlett.

———. In preparation. *A Grammar of Seri*. Manuscript available at: http://www.und.nodak.edu/instruct/smarlett/Stephen_Marlett/GrammarDraft.html.

Marlett, Stephen A., and Mary B. Moser. 1995. "Presentación y Análisis Preliminar de 550 Topónimos Seris." In *Coloquio de Toponimia: Los Nombres de los Pueblos del Noroeste*, edited by José Gaxiola López and Everardo Mendoza Guerrero, pp. 143–204. Culiacán, Mexico: El Colegio de Sinaloa.

McGee, W J. 1898. *The Seri Indians*. Seventeenth Annual Report of the Bureau of American Ethnology, Part 1. Washington, DC: Government Printing Office.

Montaño Herrera, René. 2012a. "Xtiip cop iicp hac." In *Comcaac quih ziix quih ocoaaj hac*, pp. 90–91. 2nd edition. Socaaix and Haxöl Iihom, Sonora.

———. 2012b. "Zamt hacaatol quij iicp hac." In *Comcaac quih ziix quih ocoaaj hac*, pp. 92–93. 2nd ed. Socaaix and Haxöl Iihom, Sonora.

———. 2012c. "Leatherback Sea Turtle—*Xiica Cmotómanoj*." In *Inside Dazzling Mountains: Southwest Native Verbal Arts*, edited by David L. Kozak, pp. 31–43. Lincoln: University of Nebraska Press.

Monteverde, Enrique. 1891. "Padrón de la Tribu Seris Sometidos a la Paz, con Expresión de sus Nombres, Sexo, Edad y Lugar de su Nacimiento." In *Memoria de la Administración Pública del Estado de Sonora, Presentada a la Legislatura del Mismo por el Gobernador Ramón Corral*, vol. 1, pp. 567–573. Guaymas, Sonora: E. Gaxiola.

Moreno Herrera, Francisco Xavier. 2012a. "Seeten ctamcö com iicp hac." In *Comcaac quih ziix quih ocoaaj hac*, pp. 71–72. 2nd edition. Socaaix and Haxöl Iihom, Sonora.

———. 2012b. "Twin Peaks—*Hast Cacöla*." In *Inside Dazzling Mountains: Southwest Native Verbal Arts*, edited by David L. Kozak, pp. 54–60. Lincoln: University of Nebraska Press.

Moser, Edward W. (n.d.). Unpublished, unarchived field notes in various formats, taken between 1951 and 1976.

Moser, Edward W. 1963. "Seri Bands." *The Kiva* 28:14–27.

———. 1973. "Seri Basketry." *The Kiva* 38:105–140.

Moser, Edward W., and Richard S. White. 1968. "Seri Clay Figurines." *The Kiva* 33:133–154.

Moser, Mary B. (n.d.). Unpublished, unarchived field notes in various formats, taken between 1951 and 2005.

Moser, Mary B. 1964. "Seri Blue." *The Kiva* 30:27–32.

———. 1982. "Seri: From Conception through Infancy." In *Anthropology of Human Birth*, edited by Margarita A. Kay, pp. 221–232. Phildelphia: F. A. Davis Company.

Moser, Mary B., and Stephen A. Marlett, compilers. 2010. *Comcaac quih yaza quih hant ihiip hac: Diccionario Seri-Español-Inglés*. 2nd edition. Hermosillo: Universidad de Sonora; Mexico City: Plaza y Valdés Editores.

Narchi, Nemer E. 2003. "Eficiencia del Muestreo Etnofarmacológico en la Detección de Compuestos Bioactivos a Partir de Organismos Marinos Utilizados en la Medicina Tradicional Comcáac." Ph.D. dissertation. Universidad Autónoma de Baja California.

Nava L., E. Fernando. 2012. *El Arte Verbal Musicalizado de los Comcaac: Canciones Seris con Palabras y sin Palabras.* Series "Categorías nativas musicales," number 5. Mexico City: Comisión Nacional para el Desarrollo de los Pueblos Indígenas.

Nolasco, Margarita. 1967. "Los Seris, Desierto y Mar." *Anales* 18:125–194.

Pennington, Campbell W., ed. 1979. *The Pima Bajo of Central Sonora, Mexico*, Vol. II: *Vocabulario en la Lengua Nevome.* Salt Lake City: University of Utah Press.

Pfefferkorn, Ignaz. [1794] 1989. *Sonora: A Description of the Province.* Translated and annotated by T. E. Treutlein. Tucson: University of Arizona Press.

Pinart, Alphonse. 1879. [Vocabulary of the Seri] Manuscript. Bureau of American Ethnology Collection, National Anthropological Archives, Smithsonian Institution, Washington, DC.

Quinn, Charles Russell, and Elena Quinn, eds. 1965. *Edward H. Davis and the Indians of the Southwest United States and Northwest Mexico.* Downey, California: Elena Quinn.

Ryerson, Scott. 1976. "Seri Ironwood Carving: An Economic View." In *Ethnic and Tourist Arts: Cultural Expressions from the Fourth World*, edited by Nelson H. H. Graburn, pp. 119–136. Berkeley: University of California Press.

Sánchez-Brambila, G. Y., G. Álvarez-Manilla, F. Soto-Cordova, B. G. Lyons, and R. Pacheco-Aguilar. 2004. "Identification and Characterization of the Off-Flavor in Mantle Muscle of Jumbo Squid (*Dosidicus gigas*) from the Gulf of California." *Journal of Aquatic Food Product Technology* 13:55–67.

Sheldon, Charles. 1921–1922. Charles Sheldon papers (AZ 408) filed at the University of Arizona Libraries Special Collections, Tucson.

———. 1979. *The Wilderness of Desert Bighorns and Seri Indians.* Phoenix: The Arizona Desert Bighorn Sheep Society.

Sheridan, Thomas E., ed. 1999. *Empire of Sand: The Seri Indians and the Struggle for Spanish Sonora, 1645–1803.* Tucson: University of Arizona Press.

Skoglund, Carol. 2001. "Panamic Province Molluscan Literature: Additions and Changes from 1971 through 2000. I–II: Bivalvia and Polyplacophora." Supplement, *The Festivus*, vol. 32. San Diego: San Diego Shell Club.

———. 2002. "Panamic Province Molluscan Literature: Additions and Changes from 1971 through 2000. III: Gastropoda." Supplement, *The Festivus*, vol. 33. San Diego: San Diego Shell Club.

Smith, William Neil. 1954. "The Ethno-History of the Seri Indians from 1890 to 1953: A Study in Native Conservatism and Adaptability." Unpublished manuscript filed at the University of Arizona Libraries Special Collections, Tucson.

Spicer, Edward H. 1962. *Cycles of Conquest: The Impact of Spain, Mexico and the United States on the Indians of the Southwest, 1533–1960.* Tucson: University of Arizona Press.

Steinbeck, John, and E. F. Ricketts. 1941. *Sea of Cortez: A Leisurely Journal of Travel and Research.* New York: Viking Press.

Stratford, Guillermo. [1746] 1958. "Descripción de las Californias desde el Cabo de San Lucas que está al Sur: Sus Misiones, Puertos, Bahías, Placeres, Naciones Reducidas y Gentiles de que se Tiene Noticia la Habitan. . . ." In *Documentos Para la Historia de Baja California*, edited by Roberto Ramos, vol. 1, pp. 52-65. Mexico City: Editorial Jus.

Torre, Jorge, and Lloyd Findley. 2007. "Seri Ethnoichthyology: Folk Fishers in a Desert Sea." Poster presented at the 39th Annual Meeting of the Desert Fishes Council (Consejo de los Peces del Desierto), Ventura, California.

Ugarte, Juan de. [1722] 1958. "Relación del Descubrimiento del Golfo de California o Mar Lauretano, por el Padre Juan de

Ugarte en el Año de 1722." In *Documentos para la Historia de Baja California*, edited by Roberto Ramos, vol. 1, pp. 15–50. Mexico City: Editorial Jus.

Vennum, Thomas. 2000. "Locating the Seri on the Musical Map of Indian North America." *Journal of the Southwest* 42:635–760.

Whiting, Alfred L. 1951. A Seri Diary. June 11 to 24, 1951. MS-3. Museum of Northern Arizona, Flagstaff.

INDEX

The letter *t* following a page number denotes a table, and the letter *f* following a page number denotes a figure.

ABOUT THE AUTHOR

Cathy Moser Marlett, the daughter of American linguist-missionaries, spent much of her childhood in the Seri fishing village of El Desemboque, speaking Seri and Spanish in addition to her native English. Her parents' interest in documenting the fascinating Seri culture was pivotal in giving her an appreciation of the world of the *Comcaac* and their home on the shores of the Gulf of California. From her friends there she learned early on such invaluable arts as catching octopuses and making Seri baskets (three that she completed remain her treasures).

The researchers who occasionally passed through the area and dropped in at the Mosers' for a meal or visit played a role in giving her a new appreciation of the amazing world that she took for granted, and there began her deeper interest in that world.

Cathy graduated from Wheaton College (Illinois) with a BA in biology and art, later studying linguistics. She contributed "A Desemboque Childhood" in a special issue of the *Journal of the Southwest* (2000) and "A Fine Day for Playing Hooky," in *Backcountry Pilot: Flying Adventures with Ike Russell* (2002). Her illustrations have appeared in numerous publications, including *People of the Desert and Sea: Ethnobotany of the Seri Indians* (Felger and M. Moser 1985), *Unknown Island* (Bowen 2000), and the most extensive, involving more than six hundred drawings, in the trilingual Seri dictionary (M. Moser and S. Marlett 2010).

Cathy, together with her husband Steve, a linguist, works with SIL International as an illustrator. She is never happier than when exploring isolated beaches in the Seri area, or visiting with Seri friends, enjoying their songs and stories.

Cathy lives in Tucson, Arizona, with her husband.

THE SOUTHWEST CENTER SERIES

Joseph C. Wilder, Editor

Ignaz Pfefferkorn, *Sonora: A Description of the Province*

Carl Lumholtz, *New Trails in Mexico*

Buford Pickens, *The Missions of Northern Sonora: A 1935 Field Documentation*

Gary Paul Nabhan, editor, *Counting Sheep: Twenty Ways of Seeing Desert Bighorn*

Eileen Oktavec, *Answered Prayers: Miracles and Milagros along the Border*

Curtis M. Hinsley and David R. Wilcox, editors, *Frank Hamilton Cushing and the Hemenway Southwestern Archaeological Expedition, 1886–1889,* volume 1: *The Southwest in the American Imagination: The Writings of Sylvester Baxter, 1881–1899*

Lawrence J. Taylor and Maeve Hickey, *The Road to Mexico*

Donna J. Guy and Thomas E. Sheridan, editors, *Contested Ground: Comparative Frontiers on the Northern and Southern Edges of the Spanish Empire*

Julian D. Hayden, *The Sierra Pinacate*

Paul S. Martin, David Yetman, Mark Fishbein, Phil Jenkins, Thomas R. Van Devender, and Rebecca K. Wilson, editors, *Gentry's Rio Mayo Plants: The Tropical Deciduous Forest and Environs of Northwest Mexico*

W J McGee, *Trails to Tiburón: The 1894 and 1895 Field Diaries of W J McGee,* transcribed by Hazel McFeely Fontana, annotated and with an introduction by Bernard L. Fontana

Richard Stephen Felger, *Flora of the Gran Desierto and Río Colorado of Northwestern Mexico*

Donald Bahr, editor, *O'odham Creation and Related Events: As Told to Ruth Benedict in 1927 in Prose, Oratory, and Song by the Pimas William Blackwater, Thomas Vanyiko, Clara Ahiel, William Stevens, Oliver Wellington, and Kisto*

Dan L. Fischer, *Early Southwest Ornithologists, 1528–1900*

Thomas Bowen, editor, *Backcountry Pilot: Flying Adventures with Ike Russell*

Federico José María Ronstadt, *Borderman: Memoirs of Federico José María Ronstadt,* edited by Edward F. Ronstadt

Curtis M. Hinsley and David R. Wilcox, editors, *Frank Hamilton Cushing and the Hemenway Southwestern Archaeological Expedition, 1886–1889,* volume 2: *The Lost Itinerary of Frank Hamilton Cushing*

Neil Goodwin, *Like a Brother: Grenville Goodwin's Apache Years, 1928–1939*

Katherine G. Morrissey and Kirsten Jensen, editors, *Picturing Arizona: The Photographic Record of the 1930s*

Bill Broyles and Michael Berman, *Sunshot: Peril and Wonder in the Gran Desierto*

David W. Lazaroff, Philip C. Rosen, and Charles H. Lowe Jr., *Amphibians, Reptiles, and Their Habitats at Sabino Canyon*

David Yetman, *The Organ Pipe Cactus*

Gloria Fraser Giffords, *Sanctuaries of Earth, Stone, and Light: The Churches of Northern New Spain, 1530–1821*

David Yetman, *The Great Cacti: Ethnobotany and Biogeography*

John Messina, *Álamos, Sonora: Architecture and Urbanism in the Dry Tropics*

Laura L. Cummings, *Pachucas and Pachucos in Tucson: Situated Border Lives*

Bernard L. Fontana and Edward McCain, *A Gift of Angels: The Art of Mission San Xavier del Bac*

David A. Yetman, *The Ópatas: In Search of a Sonoran People*

Julian D. Hayden, *Field Man: The Life of a Desert Archaeologist*, edited by Bill Broyles and Diane Boyer

Bill Broyles, Gayle Harrison Hartmann, Thomas E. Sheridan, Gary Paul Nabhan, and Mary Charlotte Thurtle, *Last Water on the Devil's Highway: A Cultural and Natural History of Tinajas Altas*

Thomas E. Sheridan, *Arizona: A History, Revised Edition*

Richard S. Felger and Benjamin Theodore Wilder, *Plant Life of a Desert Archipelago: Flora of the Sonoran Islands in the Gulf of California*

David Burckhalter, *Baja California Missions: In the Footsteps of the Padres*

Guillermo Núñez Noriega, *Just Between Us: An Ethnography of Male Identity and Intimacy in Rural Communities of Northern Mexico*

Cathy Moser Marlett, *Shells on a Desert Shore: Mollusks in the Seri World*